INFORMATION SCIENCE

INFORMATION SCIENCE

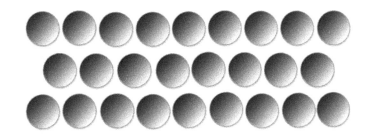

DAVID G. LUENBERGER

PRINCETON UNIVERSITY PRESS

Princeton and Oxford

Published by Princeton University Press, 41 William Street, Princeton, New Jersey 08540

In the United Kingdom: Princeton University Press,

3 Market Place, Woodstock, Oxfordshire OX20 1SY

Library of Congress Cataloging-in-Publication Data

Luenberger, David G., 1937–

Information science / David G. Luenberger.

p. cm

Includes bibliographical references and index.

ISBN-13: 978-0-691-12418-3 (alk. paper)

ISBN-10: 0-691-12418-3 (alk. paper)

1. Information science. 2. Information theory. I. Title.

Z665.L89 2006

004—dc22

2005052193

British Library Cataloging-in-Publication Data is available

This book has been composed in Times

Printed on acid-free paper. ∞

pup.princeton.edu

Printed in the United States of America

1 3 5 7 9 10 8 6 4 2

To Nancy

Information seems to be a characterizing theme of the modern age. It is mentioned everywhere. Yet information as a subject of study is so vast that it is impossible to fully define its various aspects in a simple succinct statement. It is, instead, perhaps more fruitful to assume a rough understanding of the term, and then seriously study some of the important and interesting facets of the subject it represents. That is the approach taken by this text, which is an outgrowth of an undergraduate course taught at Stanford for the past few years. The approach is based on exploring five general areas of information, termed the five E's. They are entropy, economics, encryption, extraction, and emission. In fact, the text is divided into five parts, corresponding to these five areas.

The text, of course, does not attempt to cover everything about information. It is limited to covering only these five aspects of the "science" of information. However, the text is not designed to be a survey or overview. It is packed with specific concepts, most of which can be cast into mathematical or computational form and used to derive important results or explain observed phenomena. These concepts are used in derivations, examples, and end-of-chapter exercises. Indeed, a major objective is to present concepts that can be used in a number of areas, even beyond those directly related to information. In that sense the text is as much about general methods of analysis and design as it is about the subject of information. Much of the "science" of information science is portable to other fields.

The text is organized in the standard way, by parts, chapters, sections, and subsections. The chapters are more or less independent. Chapter 2 is basic and should be covered by all. Chapter 3 is also useful background, and some other chapters refer to sections of chapters 3 and 5. Although largely independent, the chapters are tied together by frequent reference to the concept of entropy and by the use of several common methods of analysis.

Some sections or subsections are marked with an asterisk (*), indicating that the material may be more difficult or that it can be safely skipped without loss of continuity. Likewise, some end-of-chapter exercises are marked with an asterisk, indicating that they are more challenging than most.

The level at which the text can be used is variable. At Stanford, students ranging from sophomores to graduate students have taken the course. There is no specific prerequisite; however, students of this text should have some level of comfort with mathematical reasoning: both for modeling and for finding answers. In terms of a standard phrase, students should know how to solve "word problems." The actual mathematics used is of several types. Elementary calculus is employed in some sections. Other sections use algebraic theory. Still others use probability. However, the mathematics that is beyond elementary calculus or algebra is introduced and explained. In that sense, the text is essentially self-contained with respect to the mathematics required. And since the chapters are largely independent, it is possible to select topics at various mathematical levels.

The complete text includes far more material than can be treated in a single academic quarter or even a semester. At Stanford I have covered about fourteen or fifteen of the twenty-two chapters in one quarter, although the particular choice of chapters has varied. Even this somewhat reduced agenda includes an enormous amount of subject material; after all, there are entire texts devoted to some of the material in individual chapters. How can so much be covered in a single academic course without seriously compromising depth?

I believe that rapid progress hinges on genuine student interest and motivation. These are derived from five main sources. First, inherent interest is typically strong for this subject. Many students plan to seek careers in the information industry, and this motivates a desire to learn about the field. Second, students are naturally curious about things they work with. How do compression algorithms such as JPEG and ZIP work? How is it possible to have a secure digital signature that cannot simply be copied? How does the Internet route millions of messages to their proper destinations? Third, interest is enhanced when students witness or participate in illustrative examples and experiments. The text includes such examples, and many of them can be used as experiments, as explained in the instructor's manual. Fourth, subjects come alive when students learn something about the individuals who pioneered in the area of study, learning, for example, whether someone's particular contribution occurred by happenstance or as the result of intense struggle. These little stories add a human dimension to the subject. Fifth, if a student works with the material, adding to it, finding a new way to present it, or exploring a new application, he or she becomes an owner rather than simply an observer of the subject. In the Stanford class, students worked in teams of four to develop projects of their choice that were presented in written and oral form. Many students felt this was the highlight of the class.

One objective of the text is to create a desire for further study of information science and the methods used to explore it. I hope that students studying this material will see the relevance of the tools that are employed and the excitement of the areas presented.

Development of this text and the underlying course has been a rewarding experience. It was made all the more pleasant by the encouragement and help that I received from many colleagues and students. I wish especially to thank Martin Hellman, Fouad Tobagi, Ashish Goel, and Thomas Weber for detailed professional feedback on parts of the manuscript. I also wish to thank Kahn Mason, Christopher Messer, Geraldine Hraban, and Mareza Larizadeh for help with development of the class itself. I appreciate Charles Feinstein, Wayne Whinston, Robert Thomas, and Sharan Jagpal for providing helpful comments on the entire manuscript. I want to thank Leon Steinmetz for the special drawings in the text and the staff at Princeton University Press for their creativity and hard work. And certainly, I am grateful for the helpful suggestions of the many bright and enthusiastic students who greatly influenced the development of this text.

David G. Luenberger
Stanford
December 2005

INFORMATION SCIENCE

INTRODUCTION

This is the information age. Advances in information technology are transforming civilization more profoundly and more quickly than any other technical revolution in history. Yet if pressed to explain what defines the information age, we can't give a single answer: it is the technology of the Internet; it is the ability to access vast and diversified data with relative ease; it is greater freedom to communicate directly with others; and it is new forms of business, entertainment, and education.

Today many people wish to participate fully in the promise of the information age and contribute to its advance, and they seek knowledge and skills that will prepare them. Such preparation should emphasize fundamentals as well as details. But what are the fundamentals of information?

In this book there are five fundamentals of information, and the text is therefore divided into parts, each devoted to one of the five. The five essentials, the five E's, are the following:

1. **Entropy. The foundation of information.** It is the study of classical information and communication theory, based on bits, bandwidth, and codes, which underlies modern technology. The concept of entropy appears naturally in most aspects of information.

2. **Economics. Strategies for value.** Information is different from other commodities, such as apples or automobiles, since it is usually not consumed or worn out when it is used and often is easily duplicated. This has profound implications for how information is produced, priced, and distributed.

3. **Encryption. Security through mathematics.** Much of modern communication relies on secure transmission provided by encryption, enabling such advances as digital signatures and digital cash. This area has advanced profoundly in recent years.

4. **Extraction. Information from data.** Data is not the same as information. In practice, data must be organized, stored, searched, filtered, and

modeled in order that useful information can be obtained. The techniques for accomplishing this are expanding rapidly and are providing powerful ways to extract information.

5. **Emission. The mastery of frequency.** A large fraction of the information we obtain is transmitted electromagnetically, through radio, television, telephones, cell phones, or computer networks. These devices rely on electric currents, radio frequency waves, or light. Advances in our understanding and mastery of frequency have profoundly shaped the information age.

Certainly there are important aspects of information beyond these five, including, for example, the study of linguistics and the analysis of human information-processing capabilities. But as a general rule, study of the principles discussed in this text will prepare one for study of many important aspects of information science.

Abstract principles alone, however, do not provide a full understanding of information. It is important to study specific application issues, such as how coding theory has facilitated cell phones, compact disks, and pictures transmitted from satellites and surface modules we have sent to distant planets; how economic principles are combined with technology to invent new methods of pricing music and other information products; how encryption facilitates secure transmissions, electronic cash, and digital signatures; how large data banks are organized, searched, and analyzed; how advances in technology provide new services, cell phones, and access to vast stores of data. These examples give life to the principles and may inspire future creations.

It is helpful also, and sometimes entertaining, to study bits of history related to the development of information science. Of special interest are accounts of individual struggles, insights, successes, and disappointments of some of the people who have made contributions. These little stories add a special dimension to one's study, putting ideas in perspective and showing the excitement of the field.

The subject of information is a lacework of individual subtopics. This book hopes to convey appreciation for the overall patterns of this lacework. Such an appreciation will enrich the study of any one of the subtopics, and may provide a gateway to detailed study of related subjects. Engineering students may find that economics is fascinating and important. Scientists may find that there is great science in the study of information. Economists may find that understanding the technological principles of information helps in the formulation of efficient means for allocating information goods. Familiarity with the entire pattern enhances general understanding and highlights unifying principles.

Important principles and concepts are portable in that they can be applied in varied areas. This text emphasizes the portable concepts of information science.

1.1 Themes of Analysis

Several themes of analysis recur in the study of the five E's, and it is useful to watch for them.

1. **Average performance.** A basic concept used in the analysis of information is that of average performance. Often performance is measured, not with respect to a

particular instance, which may vary from other instances, but rather by averaging over many or all possible instances. This has two major consequences. First, and most obvious, is that an average characterizes, in a single number, performance over several instances. This is convenient and useful. The second consequence is more profound, and results from changing the definition of an instance. If several instances are grouped to form a single "big" instance, then by focusing on these new instances we may be able to improve performance. For example, suppose you mail packages of various sizes to the same person and you keep track of the average postage costs. This is done by averaging over the (small) instances of each package. Suppose instead that you redefine an instance as two consecutive packages that you repackage as a single large package. The cost of sending the two is often less than the sum of costs for the two individual packages. If you average over these big packages, you will likely find that the average cost is less than before. This simple idea is quite powerful. Indeed, it is one of the first concepts we encounter in Shannon's brilliant analysis of communication. Shannon showed that the average performance of a communication system can be improved by using big packages.

2. **Optimality.** The idea of finding an optimum arises frequently in information science. One may seek the lowest cost, the maximum profit, the minimum average time to completion, the maximum compression, the shortest search time, the maximum transmission rate, and so forth. The principle of optimization is now deeply rooted in analysis, and the basic analytical techniques of optimization are accessible to most people who will study this text; the main result being that the maximum or minimum of a function occurs at a place where the derivative of that function is zero. Constraints are treated with Lagrange multipliers. Sometimes, however, such as in combinatorial problems, special arguments must be used to characterize optimality.

3. **Complexity.** Much attention in information science is devoted to computational complexity. Surprisingly, complexity is regarded as bad in some cases but good in others. When it comes to finding a solution to a problem or computing something of interest, we desire simplicity. When protecting the security of messages, on the other hand, we want our code to be difficult for others to decipher, and hence we seek encryption methods that produce a high degree of complexity.

Standard measures of complexity are based on the average (or best or worst case) time required to carry out a task. This time is generally a function of the size of the components of the task. For example, the time that it takes to multiply two numbers together depends on how many digits they contain. Two main possibilities are distinguished for a function that measures time as a function of problem size: polynomial time and nonpolynomial time. A polynomial time algorithm completes its execution in a time proportional to a polynomial in the size variable. For example, multiplication of two numbers each with n digits can be attained on a sequentially operating computer in time proportional to n^2 (or less). Tasks that can be completed in polynomial time are generally considered fast and not terribly complex. As a converse example, the time required to factor an integer with n digits is roughly $e^{(\log n)^{1/3}}$ which is not polynomial. Thus factoring is considered complex (and hard).

Complexity is important in modern information science because of the huge mass of data that is processed by sorting, searching, coding, decoding, and so forth. Efficiency requires low complexity, which is generally taken to mean polynomial time.

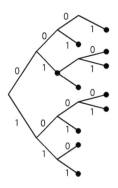

FIGURE 1.1 A code tree. Trees of various sorts are used frequently in information science.

4. **Structure.** Structure often reduces complexity. Without structure, there is chaos, and most likely attendant complexity. Of particular interest in information science are techniques for structuring relationships among data or among mathematical variables.

One of the most useful structures is that of a graphical tree, consisting of various nodes and branches that lead from node to node. An example of a tree is a code tree, like that shown in figure 1.1 and used in chapter 3. Trees are indeed used in many instances in the text. They facilitate study of the structure of codes (as in the figure), they represent possible stages of action or information, and they symbolize ways to arrange data.

Another common structure is a matrix or table—a two-dimensional array. A matrix may be used in the strict mathematical sense as relating vectors, or it may be simply a table of relationships, as for example a database matrix with students listed down the columns and classes taken listed across the rows. Sometimes trees are converted to matrices and vice versa.

Another structure is a system of equations. For example, the equations may describe the actions of several producers of information and how their profits depend on each other's actions. Or the equations may relate an output code to an input message.

One frequently attempts to tame a complex situation by seeking or imposing structure.

1.2 Information Lessons

Complementing methods of analysis are the results, general insights, and conclusions that analysis provides. Some of the general lessons obtained about information stand out.

1. **The measure of information.** Entropy (defined in chapter 2) proves itself repeatedly as a fundamental measure of information. Shannon originally introduced this measure in the context of efficient transmission of information, first in a noise-free environment and then in the presence of noise.

It is a short step from the basic definition of entropy to its application to data compression, which reduces the storage required to save data or increases the rate at which information can be sent. Standard compression methods for text, pictures, and music are based on entropy concepts.

Entropy provides a fundamental bound on the average time required to find a specific piece of data within a data file, and it suggests structures to best represent data to facilitate searches.

Entropy also provides the bound for the rate of information transmission through inherently noisy electromagnetic media, relating the transmission rate to the frequency bandwidth used.

Indeed, entropy is so basic that it arises in the study of all five E's of information.

2. **The value of information.** The value of a specific piece of information may be quite distinct from its information-theoretic content. You may find a certain long-winded book to have little personal value, even though it has a lot of words. Entropy cannot measure the value you place on the book's content. It cannot quantify

the "real" information in books, movies, and music. Yet the assignment of value can be approached systematically in many cases.

There are two main approaches. The first is based on economic supply and demand. Roughly, if lots of people want a certain piece of information and are willing to pay for it with time, money, or effort, it is deemed valuable. This viewpoint can be used to determine what information to produce and sell, and how to organize large masses of data so that what most people want is easily found.

Again the idea of averaging plays a role. A producer's profit may depend on the average value that people place on a product. But the producer can also use the big-package idea and increase average profit by selling products bundled together.

The second approach to assigning value is to measure its effect on decisions. For example, if knowledge of how much oil exists at a certain underground site would influence your decision to drill or not, then the value of that information can be related to the oil profit you might make. Again, this value of information is best defined as an average; in this case over all possible oil reports you are likely to receive.

The issue is more complex when one considers more than a single individual. Information can actually have negative average value in some group situations.

3. **Encryption.** Encryption can be achieved in basically two complementary ways. The first method is derived from the concept of entropy. The idea is to make the message appear completely random to an outsider (seeking to intercept and read the message). That is, the encrypted message should have no information content; it should have maximum entropy. The message must, of course, be decipherable by the intended recipient, and this generally requires that the sender and the recipient agree on a special key that is used to initially encrypt the message and later recover it.

The second method is based on complexity. A sender scrambles the message in a manner prescribed by the intended recipient, but it is extremely difficult to unscramble it for anyone who does not know the operation that is the inverse of the scrambling method. Some of these methods are called public key systems since the scrambling key can be made public, but the descrambling key is known only by the intended recipient. Some early mechanical coding devices, such as the Enigma machine used in World War II, employed both methods, making both the entropy and the complexity fairly high.

Future encryption methods may be based on quantum effects of light, and are likely to employ notions of either entropy or complexity.

4. **Storage, searching, and modeling.** These topics are essential in the information age and make up a huge segment of information science. Perhaps more than any of the other five E's, extraction uses to great advantage all the analytical principles discussed earlier: average performance, optimality, complexity, and structure. Huge masses of data in Internet files, in libraries, and in database systems for business and government are almost overwhelming in their sheer size and apparent complexity. The first attack on such masses of data is the imposition of structure. Structure renders the data manageable for storage, retrieval, intelligent searching, and ultimately for extraction of useful information.

Fortunately, the benefits of structure often can be quantified. We may deduce the average time required to search for an item in a database, the degree of security provided by a system of encryption, the probability that a classification scheme will correctly identify similar items, the closeness of fit of a simplified representation of data, or the amount of compression achieved by an algorithm.

5. **Physical and mathematical understanding.** Modern communication is largely carried out by electromagnetic media. Although the details are complex, many of the general information science principles are applicable to the understanding of this important aspect of information as well.

The continuing thread of achievement in information transmission has been what can be termed the "mastery of frequency." Continuous signals such as those of speech can be dissected into their frequency components. The understanding of this principle led to the conscious manipulation of frequency. Today frequency waves are altered by filters, modulated with speech, shifted higher or lower in the frequency spectrum, sampled and reconstructed, narrowed or spread in bandwidth, and purposely mixed with noise. These techniques led to radio, television, cell phones, beautiful pictures from Mars, and computer networks. This mastery is one of the highlights of information science.

ENTROPY

The Foundation of Information

2

INFORMATION DEFINITION

Cataclysmic events are rare in the development of applied mathematics, but the theory of information published by Claude E. Shannon deserves to be cataloged as such an occurrence. The theory was immediately recognized as so elegant, so surprising, so practical, and so universal in its application that it almost immediately changed the course of modern technology. Yet, unlike many other technological revolutions, the theory relied on no equipment, no detailed experiments, and no patents, for it was deduced from pure mathematical reasoning. Upon its publication, researchers and students applied it to all sorts of areas, and today it remains a central concept in information technology, providing a foundation for many information-processing procedures, a performance benchmark for information systems, and guidance for how to improve performance.

Shannon developed his theory in response to a vexing problem faced for years at the Bell Telephone Laboratories, where he was employed. Imagine that one wishes to send a long message consisting of zeros and ones by means of electrical pulses over a telephone line. Due to inevitable line disturbances, there is a chance that an intended zero will be received as a one, and likewise that a one will be received as a zero. There will be errors in communication. Engineers sought ways to reduce those errors to improve reliability.

A standard approach to this problem was to repeat the message. For example, if an intended zero is sent three times in succession and the disturbance level is not too great, it is likely that at least two out of the three zeros will be received correctly. Hence, the recipient will probably deduce the correct message by counting as zero a received pattern of either two or three zeros out of the three transmissions. The analogous majority-voting procedure would be applied to the interpretation of ones.

However, there is some chance with this repetition method that in a sequence of, say three zeros, two or more might be corrupted and received as ones. Thus, although repeating a digit three times reduces the chance of errors, it does not reduce that chance to zero.

Reliability can be further improved, of course, by repeating each message symbol several times. A hundred-fold repetition is likely to lead to an extremely small chance of error when a majority vote between zeros and ones is used by the receiver to decide on the likely symbol. But such repetition carries with it a huge cost in terms of transmission rate. As reliability is increased by greater repetition, the rate at which message symbols are sent decreases. Thus high reliability entails a low rate of transmission. In the limit of perfect reliability, the rate of transmission goes to zero, for it would take forever to send just a single message symbol, repeated an infinite number of times.

Shannon's brilliant theory showed that for a given level of disturbance, there is, in fact, an associated rate of transmission that can be achieved with arbitrarily good reliability.

Achievement of Shannon's promised rate requires coding that is much more sophisticated than simply repeating each symbol a number of times. Several symbols must be coded as a group and redundancy incorporated into that group.

The general idea can be understood by thinking of sending an English sentence. If one sends a single letter, say T, there is a chance that it will be corrupted in transmission and received as, say R. The T might have to be sent many times before it is received and interpreted as T with high reliability.

Instead, suppose that the T is part of the intended word message THIS. If that word is sent, there is again a chance that the T will be corrupted and received as R. However, if the other three letters are received correctly, the recipient would realize that RHIS is not a valid English word, and could deduce that the R should be a T. This explanation is not complete, but the rough idea is there. Namely, by sending blocks (or words) of symbols, new avenues for error correction become available.

2.1 A Measure of Information

Messages that are unusual and not easily predicted carry more information than those that are deemed likely even before they are received. That is the key idea of Shannon's measure of **information**.

For example, the message, "It is sunny in California today" normally embodies little information, because (as everyone knows) it is nearly always sunny in California. On the other hand, the message, "It is cloudy in California" represents significant information, since (as everyone knows) that is a rare occurrence.

As another example, if I verify that my watch is working, that is less information than if I find that it is not working.

Information is quantified by considering the probabilities of various possible messages. A message with low probability represents more information than one with high probability. For example, since cloudy weather in California has low probability, the message that it is cloudy represents a good deal of information.

Once the probability p of a message is specified, the associated information can be defined.

Information definition. The information associated with a message of probability p is

$$I = \log\left(1/p\right) \equiv -\log p, \tag{2.1}$$

where log stands for logarithm.

Any base can be used for the logarithm (such as the base e of the natural logarithms, base 10, or base 2). Different bases simply give different units to the information measure.

Notice that if p is small, $1/p$ is large and hence the information I will be large. This is in accord with the general notion that a report of an unlikely event provides more information than a report of a likely event.

Logarithms to the base 2 are used most often in information theory, and then the units of information are **bits**. If $p = 1/2$, then $I = -\log_2(1/2) = \log_2 2 = 1$ bit. As an example, if I flip a coin and tell you the outcome is heads, I have transmitted one bit of information because the probability of heads is one-half.

The measure of information in equation (2.1) was originally proposed by R.V.L. Hartley, who used base-10 logarithms, and when that base is used, it is customary to call the units of information **Hartleys**.

It is easy to transform from one base to another through the relation[1]

$$\log_b x = \log_a x / \log_a b. \tag{2.2}$$

In particular, when using base-2 logarithms, it is convenient to use $\log_2 x = \ln x / \ln 2$, where ln denotes logarithms to the base e. Since $\ln 2 = .693$, we can write $\log_2 x = \ln x / .693$.

Since base 2 is used most of the time in information theory, the explicit reference to the base is usually omitted and one simply writes $\log x$ for $\log_2 x$. (However, one must be careful, since most general purpose calculators and references use $\log x$ to mean $\log_{10} x$.)

Additivity of Information

Suppose I flip a coin twice, and the result is heads on the first flip and tails on the second. If I transmit this fact to you, how much information have I sent? There are, of course, four equally likely possibilities for the outcome of the two flips, namely HH, HT, TH, and TT. The particular outcome HT has probability 1/4, so the information content of that message (using base-2 logarithms) is $I = \log[1/(1/4)] = \log 4 = 2$ bits. This is also the sum of the information that would be transmitted by reporting the outcome of each flip separately—one bit each. Hence the information of the compound message is the sum of the information in the two individual messages.

The additive property is true in general for independent events.

Additive Property. If event A has probability p_A and event B has probability p_B and these are independent in the sense that one is not influenced by the other, then the probability of the joint event A and B is $p_A p_B$. The corresponding information is

$$I_{AB} = -\log p_A p_B = -\log p_A - \log p_B = I_A + I_B.$$

We might receive the message that it is sunny in California and John won the bowling tournament. The information content of this compound message is the sum

---

[1] To prove the relation, we write $b^{\log_b x} = a^{\log_a x}$. Taking the $\log_a$ of both sides yields $(\log_b x) \log_a b = \log_a x$. Hence $\log_b x = \log_a x / \log_a b$.

of the information that it is sunny and the information that John won the bowling tournament, assuming that the weather does not affect the tournament and vice versa.

Strong support for the definition of information as the logarithm of $1/p$ is given by the additive property. Indeed, the definition seems quite intuitive, but its importance will later become even more apparent.

2.2 The Definition of Entropy

We know how to measure the information of a particular message or event, such as the report that the weather is sunny, or that a coin flip was heads. Information is associated with knowledge of an event that has occurred. **Entropy** is a measure of information that we expect to receive in the future. It is the average information taken with respect to all possible outcomes.

Suppose, for example, that there are two possible events (such as sunny or cloudy weather). The first will occur with probability p and the second with probability $1 - p$. If the first event occurs, a message conveying that fact will have an amount of information equal to $I_1 = -\log p$. Likewise, if the second occurs, the corresponding information will be $I_2 = -\log (1 - p)$. On average, event 1 occurs with probability p, and event 2 occurs with probability $1 - p$. Hence, the average information is $pI_1 + (1 - p)I_2$. This average information is the entropy of the two event possibilities. This leads to the following definition.

Entropy definition. For two events with probabilities p and $1 - p$, respectively, the entropy is

$$H(p) = -p \log p - (1 - p) \log (1 - p). \tag{2.3}$$

Entropy has the same units as information, so when information is measured in bits (using base-2 logarithms), entropy is also measured in bits.

Example 2.1 (Weather). As a specific weather example, suppose that weather in California is either sunny or cloudy with probabilities 7/8 and 1/8, respectively. The entropy of this source of information is the average information of sunny and cloudy days. Hence

$$\begin{aligned}
H &= -(7/8) \log (7/8) - (1/8) \log (1/8) \\
&= -\frac{1}{8}[7 \log 7 - 7 \log 8 - \log 8] \\
&= -\frac{1}{8}[7 \times 2.81 - 7 \times 3 - 3] \\
&= -\frac{1}{8}[19.65 - 21 - 3] = \frac{1}{8}[4.349] = .54 \text{ bits.}
\end{aligned}$$

(In this calculation $\log 8$ is 3 because $2^3 = 8$. The log of 7 is found from $\log_2 7 = \ln 7/\ln 2 = 1.459/.693 = 2.81$.)

The entropy of two events is characterized by the probability p of one of the events since the other event must have probability $1 - p$. The function $H(p)$ given by equation (2.3) is plotted as a function of p in figure 2.1.

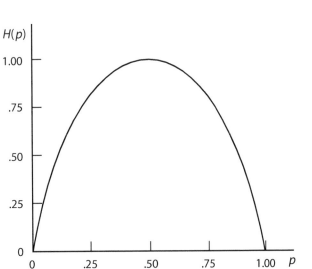

FIGURE 2.1 Entropy *H(p)* as a function of *p*. Entropy is symmetric about the point 1/2, where it attains a maximum of 1 bit. Entropy is 0 if *p* is zero or one.

If p is either zero or one, the event outcome is completely determined. The entropy is zero since if the outcome is certain, no information is conveyed by a report of what occurred.

Entropy is symmetric about the point $p = 1/2$ because p and $1 - p$ can be interchanged. That is, it makes no difference whether the labels of the two events are interchanged with event 1 being called event 2 and vice versa.

Finally, entropy is maximized at $p = 1/2$, where its value is 1 bit. This is the entropy of a single coin flip having a 50-50 chance of being heads or tails. A 50-50 chance represents the greatest uncertainty for two events, and hence the greatest entropy.

We may verify that $H(p)$ achieves a maximum at $p = 1/2$ by a simple application of calculus. A maximum occurs at the point where the derivative of $H(p)$ is zero. It is easiest to use logarithms to the base e and divide by $\ln 2$. Thus, in terms of bits, we may write

$$H(p) = -[p \ln p + (1 - p) \ln (1 - p)]/(\ln 2).$$

Then, since the derivative of $\ln p$ is $1/p$, setting the derivative of $H(p)$ to zero yields

$$0 = \frac{dH(p)}{dp} = -\left[\ln p + \frac{p}{p} - \ln (1 - p) - \frac{1 - p}{1 - p} \right] \Big/ \ln 2$$
$$= \left[-\ln p + \ln (1 - p) \right] / \ln 2.$$

This implies that $\ln p = \ln (1 - p)$, and this in turn implies $p = 1 - p$ or, finally, $p = 1/2$.

2.3 Information Sources

An **information source** or simply a **source** is defined to consist of the possible (mutually exclusive) events that might occur together with their probabilities; and the definition of entropy is easily extended to sources with several possible events. Suppose there are n possible events in a source, with the i-th event having probability p_i (for $i = 1, 2, \ldots, n$). None of the probabilities are negative and they must sum to 1. The information of the message that event i occurred is $I_i = \log(1/p_i)$. The entropy is the average information of these.

Entropy of an n-event source. The entropy of an n-event source with probabilities $p_1, p_2, \ldots, p_n$ is

$$H = p_1 \log(1/p_1) + p_2 \log(1/p_2) + \cdots + p_n \log(1/p_n) \tag{2.4}$$
$$= -[p_1 \log p_1 + p_2 \log p_2 + \cdots + p_n \log p_n].$$

This function is sometimes denoted $H(p_1, p_2, \ldots, p_n)$.

The following example illustrates the straightforward calculation of entropy.

Example 2.2 (Three-event source). Suppose there are three events with probabilities 1/2, 1/4, 1/4. The corresponding entropy is

$$H(1/2, 1/4, 1/4) = (1/2)\log 2 + (1/4)\log 4 + (1/4)\log 4$$
$$= 1/2 + (1/4) \times 2 + (1/4) \times 2$$
$$= 3/2.$$

The basic properties of entropy exhibited by figure 2.1 for the case of two events generalize to properties of entropy for n events.

Two properties of entropy.

1. (Nonnegativity) $H(p_1, p_2, \ldots, p_n) \geq 0$.
   Since $0 \leq p_i \leq 1$, each $\log p_i \leq 0$. Hence $-p_i \log p_i \geq 0$ for each i, which means $H \geq 0$.

2. $H(p_1, p_2, \ldots, p_n) \leq \log n$.
   As in the case with $n = 2$, the maximum of H occurs when all probabilities are equal, with $p_i = 1/n$ for each i. Hence $H \leq \sum_{i=1}^{n} (1/n)\log n = \log n$.

Example 2.3 (20 questions). The popular parlor game of 20 questions illustrates one facet of entropy. One person selects an object and tells another only whether the object is classified as animal, vegetable, or mineral. The other person may then ask up to 20 questions, which are answered either yes or no, to determine the object.

Clearly two possible objects, say A and B, can be distinguished with a single question, such as "Is it A?" (although if the answer is no, the question "Is it B?" must be asked to complete the game even though the answer is already known). One of four objects can be determined with two questions. In general one out of 2^n objects can be determined with n questions. The strategy for determining the one object from 2^n is of course to repeatedly divide in half the group of objects remaining under consideration.

If we suppose that the 2^n objects are equally likely (each with probability $1/2^n$), the entropy of this source is the sum of 2^n terms

$$\frac{1}{2^n}\log 2^n + \frac{1}{2^n}\log 2^n + \cdots + \frac{1}{2^n}\log 2^n = \log 2^n = n.$$

Thus the number of questions to determine the object is equal to the entropy of the source.

This is true only when the number of objects is a power of 2, in which case the entropy is an integer. For other cases, the entropy figure must be increased to the nearest integer to obtain the number of required questions to assure success.

As an interesting calculation, we note that $2^{20} = 1,048,576$, which is the number of objects that can be distinguished with 20 questions (although only 2^{19} can be definitely distinguished and stated as a final question).

2.4 Source Combinations

Entropy is additive in the same way that information itself is additive. Specifically, the entropy of two or more independent sources is equal to the sum of the entropies of the individual sources. For example, the entropy of two coin flips is twice the entropy of a single flip. The entropy of the California weather report and the report of John's performance in the bowling tournament is the sum of entropies of the two events separately. However, the entropy of the combination of weather conditions (sunny or cloudy) and outside temperature (warm or cool) is not the sum of the individual entropies because weather condition and temperature are not independent—sunny weather is likely to imply warm temperature, for example. Additivity of information depends on the two sources being independent.

Mathematically, two sources S and T are **independent** if the probability of each pair (s,t) with $s \in S$, $t \in T$ is $p_{st} = p_s p_t$, where p_s and p_t are the probabilities of s and t, respectively. Additivity follows from the property of logarithms; namely, $\log p_s p_t = \log p_s + \log p_t$.

Formally, the **product of two sources** S and T is denoted (S,T) and consists of all possible pairs (s,t) of events, one from S and one from T. We mentioned earlier the example of the source made up of California weather and John's bowling record, a product source of four events.

Additive property of entropy. If the sources S and T are independent, then the entropy $H(S,T)$ of the product source (S,T) satisfies

$$H(S,T) = H(S) + H(T).$$

The proof of the property is obtained by simplifying the expression for the combined entropy. Suppose that the probability of an event s in S is p_s and the probability of an event t in T is p_t. Then an event (s,t) in (S,T) has probability $p_s p_t$. The entropy of the product source is

$$H(S,T) = -\sum_{s\in S, t\in T} p_s p_t \log p_s p_t = -\sum_{s\in S, t\in T} p_s p_t \left[\log p_s + \log p_t\right]$$

$$= -\sum_{t\in T} p_t \left[\sum_{s\in S} p_s \log p_s\right] - \sum_{s\in S} p_s \left[\sum_{t\in T} p_t \log p_t\right] = H(S) + H(T).$$

It is always true, even if S and T are not independent, that $H(S, T) \leq H(S) + H(T)$. For example if two channels of TV both reported the California weather, the entropy would be equal to just one of them, not two. Proof of the general inequality is given in chapter 5, where **conditional entropy** is discussed.

An important special case where independence holds is when the product source is the result of independent repetitions of a single source—like two flips of a coin, or two unrelated days of weather reports. If the original source is denoted by S, the product source, consisting of independent pairs of events from S, is denoted S^2. Likewise we can consider a source that is the product of any number n of independent events from S and denote this source by S^n. For example, if S is derived from the heads and tails of a coin flip, then S^3 consists of three independent coin flips. We easily find the following result.

Entropy of S^n. When independent samples are taken from a source S with entropy $H(S)$, the entropy of the resulting source S^n is

$$H(S^n) = nH(S).$$

Mixture of Sources

Two or more sources can be mixed according to fixed probabilities. Let the independent sources S_1 and S_2 have entropies H_1 and H_2, respectively. They can be mixed with probabilities p and $1 - p$ by selecting a symbol from S_1 with probability p or a symbol from S_2 with probability $1 - p$. For example S_1 might be a coin, and S_2 a six-sided die. The mixed source would with probability p flip the coin to obtain Heads or Tails or otherwise (with probability $1 - p$) throw the die to obtain 1, 2, 3, 4, 5, or 6. The resulting source has possible symbols Heads, Tails, 1, 2, 3, 4, 5, 6. In general, if S_1 is chosen, then a specific item is selected from it according to the probabilities of items in S_1; likewise for S_2 if it is chosen.

Mixture entropy. The entropy of the source obtained by mixing the independent sources S_1 and S_2 according to probabilities p and $1 - p$, respectively, is

$$H = pH_1 + (1 - p)H_2 + H(p),$$

where H_1 is the entropy of S_1 and H_2 is the entropy of S_2.

For example, if each source has only a single element so that $H_1 = H_2 = 0$, the resulting entropy is not zero, but rather $H(p)$. (See exercise 5.) For the coin and die example, if $p = \frac{1}{2}$, then

$$H = \frac{1}{2}\left(1 + \log 6\right) + H\left(\frac{1}{2}\right) = 2 + \frac{1}{2}\log 3.$$

2.5 Bits as a Measure

The bit is a unit of measure frequently used in the information sciences. However, it has at least two slightly different meanings. In its most common use, a bit is a measure of the actual number of binary digits used in a representation. For example, the expression 010111 is six bits long. If information is represented another way, as for example, by decimal digits or by letters of the alphabet, these can be measured in bits by using the conversion factor of $\log_2 10 = 3.32$ and $\log_2 26 = 4.7$. Thus the string 457832 consists of $6 \times 3.32 = 19.92$ bits. In general anything that has n possibilities is commonly said to have $\log_2 n$ bits. Conversely, k bits can represent a total of 2^k things. This usage does not directly reflect information or entropy. For instance, the expression 010001 representing the California weather report for six specific days, with 0 for sunny and 1 for cloudy, contains far less than six bits of information. Likewise, the entropy of six days of weather (the average of the information over any six days) is less than six bits. In general, the direct measure of bits as they occur as symbols matches the entropy measure only if all symbols occur equally likely and are mutually independent.

Neither the raw combinatorial measure of bits nor the entropy measure says anything about the usefulness of the information being measured in bits. A string of 1,000 bits recording the weather at the South Pole may be of no value to me, and it may have low entropy, but it is still 1,000 bits from a combinatorial viewpoint.

A bit is a very small unit of measure relative to most information sources, and hence it is convenient to have larger-scale units as well. In many cases the **byte** is taken as a reference, where one byte equals eight bits. Common terms for large numbers of bits are shown in table 2.1.

Some of these are huge numbers representing enormous quantities of information. To provide a concrete comparison, two and a half kilobytes is roughly one page of text; a megabyte is about the equivalent of a 400-page book. A gigabyte is equivalent to a short movie at TV quality.

A popular unit is the LOC, representing 20 terabytes, which is roughly the contents of the U.S. Library of Congress when converted to digital form.

Information (at least in combinatorial bits) is being created at an enormous rate. It is estimated that during one year the information created and stored is on the order of one exabyte. Of this total, printed materials account for only about .003 percent.

Although human-generated and recorded information is vast, it is small compared to that in nature. The DNA of an amoeba contains about 10^9 bits of information. Human DNA potentially holds about one exabyte.

Our interest is primarily in human-generated information. This information is stored, manipulated, transported by various means, and absorbed by the human mind. Information theory helps us do this efficiently.

TABLE 2.1
Terms Defining Large Numbers of Bits.

byte = 8 bits
kilobyte = 10^3 bytes
megabyte = 10^6 bytes
gigabyte = 10^9 bytes
terabyte = 10^{12} bytes
petabyte = 10^{15} bytes
exabyte = 10^{18} bytes
zettabyte = 10^{21} bytes
yottabyte = 10^{24} bytes

2.6 About Claude E. Shannon

Claude Elwood Shannon was born in 1915 in Petoskey, Michigan. He attended the University of Michigan, where he obtained the degrees of both bachelor of science of electrical engineering and bachelor of science in mathematics. He then attended the Massachusetts Institute of Technology and obtained the S.M. degree in electrical engineering and the degree of doctor of philosophy in mathematics in 1940 (both at the same time).

His master's thesis was extremely innovative and important. Some have called it the most important master's thesis ever written in the area of digital circuit design.

Basically, he showed how to systematize the design of complex circuits by the use of Boolean algebra. For this work he was awarded the Alfred Noble Prize, an engineering award given to young authors.

His Ph.D. dissertation was no less novel. In it he showed how to depict genetic information with mathematical structures. His methods enabled calculations and conclusions that had not been possible previously. However, the paper describing his work was never published, and Shannon was too busy with new ideas to continue to pursue it.

Shortly after graduation from MIT, Shannon joined Bell Telephone Laboratories, where he worked for 15 years. It was here that he developed his theory of communication, carried out his study of the structure of the English language, and developed his theory of encryption. These theories are presented within the chapters of this text. The profound impact of this work on the field of information science is illustrated by the many applications and extensions of it that we shall highlight.

Shannon was somewhat shy, but he was also playful; and he was as creative in play as in technical work. It was not unusual to see him riding a unicycle in the hallways of Bell Labs. Juggling was one of his primary hobbies, and he was quite accomplished at it. He wrote a paper on the scientific aspects of juggling and built automatic juggling machines (one using a stream of air to propel objects upward). He also wrote papers on game-playing machines of various types, including one on chess.

Shannon's Approach to Problem Solving

Shannon's playful hobbies and his technical work shared the common attribute of reducing issues to their simple essence. He discussed this approach to problem solving in a talk that he gave in 1953:

> The first one [method] I might speak about is simplification. Suppose that you are given a problem to solve, I don't care what kind of problem—a machine to design, or a physical theory to develop, or a mathematical theorem to prove or something of that kind—probably a very powerful approach to this is to attempt to eliminate everything from the problem except the essentials; that is, cut it down to size. Almost every problem that you come across is befuddled with all kinds of extraneous data of one sort or another; and if you can bring this problem down into the main issues, you can see more clearly what you are trying to do and perhaps find a solution. Now in so doing you may have stripped away the problem you're after. You may have simplified it to the point that it doesn't even resemble the problem that you started with; but very often if you can solve this simple problem, you can add refinements to the solution of this until you get back to the solution of the one you started with.

Shannon's approach of abstraction to an essence should become clear as we study his contributions throughout this text. His work is a testament to the power of the method.

2.7 EXERCISES

1. (Four-event source) Consider a source with four events having probabilities 1/5, 1/5, 1/5, 2/5.
    (a) What is the information in bits conveyed by a report that the first event occurred?
    (b) What is the entropy of the source?

2. (Change of base) What is the general formula for entropy $H_b(S)$ using base-b logarithms in terms of entropy $H_a(S)$ using base-a logarithms?

3. (A counterfeit coin*) A certain counterfeit half-dollar has a probability p of being heads and $1 - p$ of being tails, where $p \neq 1/2$. John flips the coin and tells Jane the outcome.
    (a) What is the entropy associated with the statement that John makes to Jane?
    (b) On the next flip, Jane realizes that there is a probability q that after the flip John will reverse the coin before reporting the (altered) outcome. Is the new entropy of John's statement less than, equal to, or greater than that of part (a)? Prove your answer. This says something about the effect of mixing two sources.

4. (Maximum entropy) Show explicitly that the maximum possible entropy of a source of n events is $\log n$ bits and is attained when the events have equal probabilities.

5. (Source mixing) Prof. Babble is writing a mathematical paper that is a combination of English and mathematics. The entropy per symbol of his English is H_E and the entropy of his mathematics (using mathematical symbols) is H_M. His paper consists of a fraction λ of English letters and a fraction $1 - \lambda$ of mathematical symbols.
    (a) Show that the per-symbol entropy of his paper is

$$H_P = \lambda H_E + (1 - \lambda)H_M + H(\lambda).$$

    (b) The professor is proud of the fact that he mixes English and mathematics in such a way that his papers have maximum per-symbol entropy. Find the value of λ that he uses.

6. (Playing cards)
    (a) What is the amount of information in bits transmitted by announcing the name of a chosen card from a deck of 52 playing cards?
    (b) What is the total number of ways that a deck can be ordered? Hint: Find the logarithm of the number first.
    (c) What is the entropy in bits of a source consisting of a random deck of cards?

7. (Amoeba smarts) The DNA string of an amoeba holds roughly 10^9 bits of information. This tells the amoeba how to make its enzymes and indeed how to carry out all other functions for its life. If this information were translated into a written instruction manual for amoebas, about how many volumes would be required?

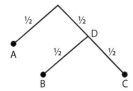

8. (Tree combination) Consider the three-event source with labels A, B, C and corresponding probabilities 1/2, 1/4, 1/4. By introducing an intermediate event D, this source can be constructed from the tree shown in figure 2.2. Let S be the source with events A and D and let P be the source with events B and C as seen from D (that is, the source P occurs only if D occurs).

FIGURE 2.2 A decomposition of a source into two sources.

   Find the entropy of the original source in terms of the entropies of S and P. Compare with the direct calculation of the entropy of the original source.

2.8 Bibliography

The classic paper on information theory is Shannon's original paper of 1949 [1]. Two basic textbook references are [2] and [3]. Quantitative estimates of the amount of data in various media are presented in [4]. An interesting study of the role of information theory in the study of biological systems is the book [5]. Shannon's vast collected works and a brief biography are found in [6]. A good survey of his work and philosophy is in the final project paper [7]. Shannon's talk on creativity was published in [8].

References

[1] Shannon, Claude E. *The Mathematical Theory of Communication*. Urbana: University of Illinois Press, 1949.

[2] Abramson, Norman. *Information Theory and Coding,* New York: McGraw-Hill, 1963.

[3] Cover, Thomas M., and Joy A. Thomas. *Elements of Information Theory*. New York: Wiley, 1991.

[4] Lyman, Peter, Hal R. Varian, James Dunn, Aleksey Strygin, and Kirsten Swearingen. "How Much Information?" 2000. www.sims.berkeley.edu/how-much-info.

[5] Loewenstein, Werner R. *The Touchstone of Life*. Oxford: Oxford University Press, 1999.

[6] Shannon, Claude E. *Collected Papers*. Ed. N.J.A. Sloane and D. Wynar. Piscataway, N.J.: IEEE Press, 1993.

[7] Chui, Eugene, Jocelyn Lin, Brok Mcferron, Noshirwan Petigara, and Satwiksai Seshasai. "Mathematical Theory of Claude Shannon." Final project paper in MIT course The Structure of Engineering Revolutions, 2001. http://mit.edu/6.933/www/Fall 2001/Shannon1.pdf.

[8] Shannon, Claude E. "Creative Thinking." Mathematical Sciences Research Center, AT&T, 1993.

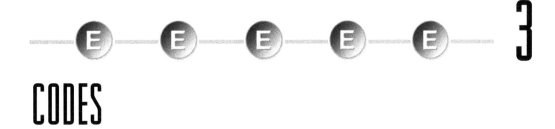

CODES

Alternative definitions of information and entropy, different from those presented in the previous chapter, could be reasonably postulated. The ones given by Shannon have intuitive appeal, but their true merit must rest on what new insights and actual results they help discover. This chapter shows that entropy is in fact closely related to the design of efficient codes for random sources. It seems unlikely that these results could be obtained without the concept of entropy.

Some people might think that the study of codes is somewhat arcane—simply a specialized branch of information theory. But actually, the ideas developed in the context of coding theory provide the foundation for techniques of compression, data search, encryption, organization of huge libraries, and correction of communication errors. Indirectly, they provide a pattern for the study of the economic value of information and for extraction of value from data. Shannon's development of information theory was originally motivated by issues related to efficient coding of English, binary sequences, and samples of continuous waveforms. The theory rapidly spread to other areas. For example, coding theory is essential to the study of some life sciences, including especially the study of DNA. And today coding theory has a continuing influence in many directions.

The main result of coding theory, **Shannon's first theorem**, is based on two principal ideas. The first is to use short codes for events that are highly likely. This is a common idea, for it is clear that it tends to shorten the average length of coded messages. The second idea is to code several events at a time, treating the group of these events as a package or metaevent, for this provides greater flexibility in code design. These two simple ideas are the essence of Shannon's approach presented in this chapter. It will be found that the implementation of these ideas leads back to entropy.

3.1 The Coding Problem

In the context of coding, the events of an information source are symbols to be transmitted, say $s_1, s_2, \ldots, s_m$. There is one symbol for each source event. The source symbols may be letters of the alphabet $a, b, \ldots, z$, they may be the digits 0 through 9, or they may be abstract symbols representing events, like s_1 for sunny weather and s_2 for cloudy.

A **code** consists of **codewords** made up of characters from a **code alphabet**. The code alphabet may be the binary alphabet consisting of zeros and ones, and this is in fact frequently the case in digital information systems, but it may also be larger. The number of characters in a code alphabet is denoted r. The sequence 011 is an example of a possible binary codeword of three characters. The word is read from left to right, just like English words are read. So in this example, 0 is the first character of the word.

A **code** is an assignment of codewords to source symbols. It can be thought of as a list, with codewords placed next to their corresponding source symbol.

As a specific example, suppose the source has symbols A, B, C and the code alphabet consists of 0 and 1. The assignment

$$
\begin{array}{ccc}
A & \longrightarrow & 0 \\
B & \longrightarrow & 0\,1 \\
C & \longrightarrow & 0\,1\,0
\end{array}
$$

is a code, mapping source symbols into codewords.

Some codes are used commonly and are familiar to many people. The Morse code of telegraphy uses dots, dashes, and spaces to encode the alphabet and the digits 0–9. For example, the letter e is encoded as a single dot. The letter a is a dot followed by a dash. Each letter has its own unique pattern. Another encoding of the alphabet is the ASCII code, in which each codeword is a string of seven binary digits. For example, the codewords for the letters A and B are 1000001 and 1000011, respectively. Items of merchandise are often marked with an identifying barcode whose character set is thin and thick vertical lines.

Code Length

An important feature of a code is the lengths of its codewords. Generally, short codewords are preferred to long ones, everything else being equal. A code in which all words have the same length is termed a **block code**. However, as we shall see, it is often advantageous to use words of various lengths.

A measure associated with a code is its **average word length**. If there are m source symbols with probabilities $p_1, p_2, \ldots, p_m$, respectively, and the corresponding codewords have lengths $l_1, l_2, \ldots, l_m$, the average word length is the weighted average

$$
L = \sum_{i=1}^{m} p_i \, l_i. \tag{3.1}
$$

A code is said to be **efficient** if it has the smallest possible average word length. We shall spend a good deal of effort making clear how to achieve this kind of efficiency, and we shall find that the entropy of the source is the proper benchmark for comparison.

Example 3.1 (Telephone area codes). As the telephone system expanded, the seven-digit phone numbers did not provide enough numbers to accommodate all customers. The Bell system then instituted three-digit **area codes** to be used as prefixes to calls directed outside one's immediate area.

In those days, telephones used the rotary dial system. Dialing a 9 took longer and caused more wear on the phone and the switching system than dialing a 1 or 2. Hence, to keep the average area code length small (as measured by dialing time), the lower digits were assigned to highly populated areas: 212 for New York, and 213 for Los Angeles, for example.

Code Properties

The general definition of a code allows for any assignment of codewords to source symbols. However, not all assignments are useful. For example, all source symbols might be assigned identical codewords, rendering the code useless. We say that a code is **nonsingular** if every codeword corresponds to a unique source symbol; otherwise it is **singular**. If the code is nonsingular, we can uniquely determine the corresponding source symbol from its codeword. Examples of singular and nonsingular codes are shown in figure 3.1.

Unfortunately, nonsingularity by itself is not a strong enough condition for a code to be useful. The reason is that codewords for several source symbols normally will be sent one after the other, like sending letters of the alphabet to transmit English words, and it may be impossible for the recipient to know where one word ends and the next begins. The code breaks down in such a case. To illustrate the problem, suppose that using the nonsingular code of figure 3.1 we receive the code characters 0010. When attempting to decode this string of characters, we find that they may correspond to the symbol sequences $s_1 s_4 s_1$ or $s_3 s_2$ or $s_1 s_1 s_2$. The nonsingular code becomes singular when applied to sequences of source symbols.

Codes that can be decoded uniquely even when arbitrary numbers of source symbols are coded in sequence are termed **uniquely decodable**, and it is these that we must use.

Instantaneous Codes

The easiest way to insure that a code is uniquely decodable is to guarantee that every word can be decoded as soon as it is received. A code with this property is called an **instantaneous code**.

Source Symbol	Singular Code	Nonsingular Code
s_1	00	0
s_2	10	10
s_3	01	00
s_4	10	01

FIGURE 3.1 Singular and nonsingular codes. A code is singular if there is not a unique mapping from codewords back to symbols.

A nonsingular block code (with all codewords of equal length) is an instantaneous code because as soon as the proper number of code characters is received, the corresponding source symbol can be determined. The block code below is an example.

Source Symbol	Codeword
s_1	00
s_2	01
s_3	10
s_4	11

If we receive the sequence 01101100, we decode it two characters at a time and find the message $s_2s_3s_4s_1$. Furthermore, as soon as two characters are received, the corresponding source symbol can be determined right away—instantaneously.

Block codes are simple to decode, but they are not always efficient. In general, one wants to assign short codewords to highly probable source symbols. For example, the Morse code assigns a single dot to the most common letter, e, whereas the less common letter q is assigned the relatively long codeword "dash dash dot dash."

Consider now the two codes shown in figure 3.2. We term these the **comma code** and the **capital code** respectively. Clearly both codes are nonsingular, and they are both uniquely decodable. The first one is uniquely decodable because once a zero is received, we know that it is the end of a codeword and we can decode the word. For instance, if we receive 01011100, we can decode it as $s_1s_2s_4s_1$. The zero acts like a comma, signaling that the word has ended.

The second code is also uniquely decodable because as soon as a zero is received, we know that it begins a new word. So if we receive 00101110, we again decode it as $s_1s_2s_4s_1$. Here the zero acts like a capital letter at the beginning of each word, showing the word separation.

However, the comma code is instantaneous, while the capital code is not. The zero in the comma code indicates the end of a word, and hence we can decode that word right away. The zero in the capital code indicates the beginning of a new word, and hence we can go back one character and decode the previous word. But decoding lags receipt of the codeword by one character. This code is therefore not instantaneous.

The criterion for a code to be instantaneous is that no codeword be the **prefix** of another codeword. If a word were a prefix of another word, a decoder receiving that prefix would have to wait to see if the subsequent characters form the extended word. On the other hand, if no word is a prefix of another word, then once the word is received, it can be decoded. In the capital code, for example, every word except the last is a prefix of another word.

Source Symbol	Comma Code	Capital Code
s_1	0	0
s_2	10	01
s_3	110	011
s_4	1110	0111

FIGURE 3.2 Comma and capital codes. Both are uniquely decodable, but the capital code is not instantaneous.

Code Trees

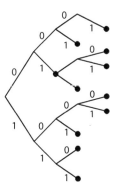

FIGURE 3.3 A code tree. Each heavy dot represents a codeword. The word itself is constructed by following the branches from the beginning of the tree and recording the 0's and 1's of the branches traveled. For example, the top dot corresponds to the word 0001. The code represented by this particular tree is not instantaneous.

The nature of instantaneous codes is best illustrated with a **code tree**. We shall consider trees for a binary code alphabet consisting of the two characters 0 and 1. Such a tree is shown in figure 3.3. The construction can be generalized to code alphabets of more than two characters.

A code tree has an origin that is a single node. From this node, one or two branches emanate, each leading to another node. Then each of these new nodes produces one or two more branches, and so on until an end node is reached.

The two branches emanating from a node are labeled 0 and 1 for up or down, respectively. Each heavy dot in a particular tree represents a codeword, defined by the sequence of zeros and ones on the path from the initial node to that node. For example, the top node in the tree of figure 3.3 corresponds to the codeword 0001. The tree of the figure contains eleven codewords, and hence it can serve as a code for eleven source symbols.

However, the code defined by the tree in the figure is not instantaneous. If the sequence 01 is received, we do not know if that is a complete word or if it is the beginning of 011 or 0101 or 0100, all of which are valid codewords. Since the codeword 01 is a prefix of another (in fact three other) codewords, the overall code is not instantaneous.

The tree of figure 3.3 can be altered to represent an instantaneous code by omitting the codeword 01 (corresponding to the leftmost codeword node in the tree). With that node left blank, no codeword is part of a path that leads to another word, and each remaining codeword is represented by an end node (but this code can only represent ten rather than eleven source symbols).

The code tree diagram should make clear that the condition that no codeword is the prefix of another codeword is equivalent to the condition that all codewords correspond to end nodes. A prefix is, after all, simply a node part way along a longer path.

The code trees corresponding to the comma and capital codes are shown in figure 3.4. The difference in terms of the criterion that words should be end nodes is apparent.

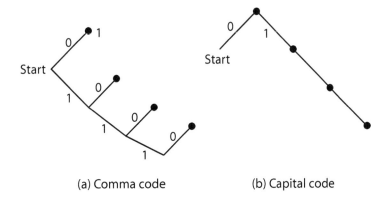

(a) Comma code   (b) Capital code

FIGURE 3.4 Two code trees. (a) The tree on the left represents the comma code with words 0, 10, 110, 1110. (b) The tree on the right represents the capital code with words 0, 01, 011, 0111.

The Kraft Inequality

There is a neat mathematical test that completely settles the question of whether a given set of codeword lengths $l_1, l_2, \ldots, l_m$ corresponds to some instantaneous code for m symbols. The result is known as the Kraft inequality.

Most often the result is applied to binary codes, but the general result stated below is applicable to codes made up of an arbitrary number r of code alphabet characters.

Kraft inequality. An instantaneous code can be constructed with given lengths $l_1, l_2, \ldots, l_m$ if and only if

$$\sum_{i=1}^{m} r^{-l_i} \leq 1, \tag{3.2}$$

where r is the number of code alphabet characters and m is the number of source symbols.

A special case of this inequality is that of $r = 2$, in which case the Kraft inequality is

$$\sum_{i=1}^{m} 2^{-l_i} \leq 1.$$

Let us consider some examples before looking at the proof of the inequality.

Example 3.2 (Block codes). Suppose we wish to construct an instantaneous code for m symbols using a binary $(0, 1)$ code that has all codewords of equal length l. The Kraft inequality requires

$$\sum_{i=1}^{m} 2^{-l} \equiv m \, 2^{-l} \leq 1.$$

Hence $m \leq 2^l$. It follows that the maximum number of symbols that can be coded with an instantaneous code using words of length l is $m = 2^l$. For example, for $l = 5$ we can code at most 32 symbols. Of course, $m = 2^l$ is also the maximum number of distinct words that can be made that are of length l, so in this case the Kraft inequality simply gives the inequality on the number of distinct codewords.

Example 3.3 (Two codes). Consider the two binary codes shown below.

Source Symbol	Code 1	Code 2
s_1	0	0
s_2	10	10
s_3	110	110
s_4	111	11

The first is the comma code but with the final 0 dropped from the fourth codeword. It is easy to see that it is still an instantaneous code. Its lengths are 1, 2, 3, 3. The Kraft inequality is

$$\frac{1}{2} + \frac{1}{4} + \frac{1}{8} + \frac{1}{8} \leq 1.$$

In this case the left side is exactly equal to 1, so (as expected) the Kraft inequality is satisfied.

Code 2 is a valid nonsingular code. However, the left side of the Kraft inequality is

$$\frac{1}{2} + \frac{1}{4} + \frac{1}{8} + \frac{1}{4} = 1\frac{1}{8}.$$

Since this is greater than 1, the Kraft inequality is not satisfied. We know that no instantaneous code can be constructed with these lengths, so in particular code 2 is not instantaneous. (In fact, it is not uniquely decodable because $110 = s_3 = s_4 s_1$.)

Proof of the Kraft Inequality*

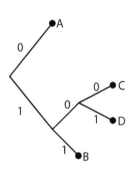

FIGURE   3.5 Illustration for Kraft inequality. The code for A uses one-half of the available codewords.

The Kraft inequality is proved by considering the code tree. We know that an instantaneous code has the property that each codeword is an end node of the tree. Consider the instantaneous binary code for the four symbols A, B, C, D shown in figure 3.5. The longest word is three characters long, and so for simplicity, let us impose that as the maximum length of codewords. There are therefore $2^3 = 8$ possible codewords. The code of the figure has only four words. Consider the codeword 0 for the symbol A. That assignment effectively excludes from the code all words with paths that continue through node A. Indeed, putting A where it is uses up a full one-half of the eight available words, since it excludes all other words that begin with 0. One says that the usage factor of A is 1/2, or equivalently 2^{-1}; and we note that 1 is the length of this word.

Now consider the word for the symbol B. The placement of B shown in the figure uses up one-fourth of the possible nodes, since it excludes all other words beginning with 11. Its usage factor is 1/4, which is equivalent to 2^{-2}; and we note that 2 is the length of that word.

Finally, the words for C and D each use up one-eighth of the available words. The usage factors are each 1/8, which is 2^{-3}. It is clear from these cases that the usage factor of a word of length l is 2^{-l}.

The total usage of all nodes in an instantaneous code must be less than what is available (which is 1). For the code of the example, the sum of the usages is $\frac{1}{2} + \frac{1}{4} + \frac{1}{8} + \frac{1}{8} = 1$. Some instantaneous codes may have total usage factors less than 1. To complete the proof, one recognizes that the usage factor is the left side of the Kraft inequality, and this must be less than 1.

Although this proof applies to binary codes, it easily extends to the general case of r code characters. Furthermore, although the Kraft inequality was originally developed to determine whether a code is instantaneous or not, McMillan extended the result, proving that the same test determines whether or not a code is uniquely decodable.

3.2 Average Code Length and Entropy

The concept of entropy can now be connected to the average code length of efficient codes. This leads to Shannon's important first theorem.

We first establish a special mathematical inequality that is used in the proof of the main result of this section.

Lemma 3.1. Let p_i, $i = 1, 2, \ldots, m$ and q_i, $i = 1, 2, \ldots, m$ satisfy $\sum_{i=1}^{m} p_i = \sum_{i=1}^{m} q_i = 1$ with all $p_i > 0$, $q_i > 0$. Then

$$\sum_{i=1}^{m} p_i \log p_i \geq \sum_{i=1}^{m} p_i \log q_i$$

with equality if and only if $p_i = q_i$ for each i.

Proof: Suppose the p_i's are fixed. We will use calculus to find the q_i's that

$$\text{maximize} \sum_{i=1}^{m} p_i \log q_i$$

$$\text{subject to} \sum_{i=1}^{m} q_i = 1.$$

For this purpose, we introduce a **Lagrange multiplier** λ for the constraint and form the **Lagrangian**

$$\mathcal{L} = \sum_{i=1}^{m} p_i \log q_i - \lambda \left[\sum_{i=1}^{m} q_i - 1 \right]$$

$$= \log e \sum_{i=1}^{m} p_i \ln q_i - \lambda \left[\sum_{i=1}^{m} q_i - 1 \right].$$

A necessary condition for the maximum is found by setting the derivative of $\mathcal{L}$ with respect to each q_i equal to zero. This gives

$$\frac{\partial \mathcal{L}}{\partial q_i} = \frac{(\log e) p_i}{q_i} - \lambda = 0, \qquad \text{for each } i.$$

Equivalently, $p_i = \lambda q_i / \log e$ for all i. Since the p_i's sum to 1 and the q_i's are required also to sum to 1, we conclude that $q_i = p_i$ for all i. Hence, the inequality of the lemma statement is true. ∎

We now come to the first relation between entropy and average code length.

Code length inequality. The average length L of a binary instantaneous code satisfies

$$L \geq H,$$

where H is the entropy of the source.

Proof: Let $l_1, l_2, \ldots, l_m$ be the word lengths of an instantaneous code. Consider the numbers

$$q_i = \frac{2^{-l_i}}{\sum_{j=1}^{m} 2^{-l_j}}.$$

These q_i's are positive and sum to 1. Application of the inequality of the lemma gives

$$H = -\sum_{i=1}^{m} p_i \log p_i \leq -\sum_{i=1}^{m} p_i \log q_i$$

$$= -\sum_{i=1}^{m} p_i \left[\log 2^{-l_i} - \log \sum_{j=1}^{m} 2^{-l_j} \right]$$

$$= \sum_{i=1}^{m} p_i \left[l_i + \log \sum_{j=1}^{m} 2^{-l_j} \right] \leq \sum_{i=1}^{m} p_i l_i = L.$$

The last step uses the Kraft inequality $\sum_{j=1}^{n} 2^{-l_j} \leq 1$, which means that $\log \sum_{j=1}^{n} 2^{-l_j} \leq 0$. Hence $H \leq L$. ∎

Thus entropy sets a lower bound on the average code length. It is possible to achieve this bound in certain cases, as shown by the next important example.

Example 3.4 (Power one-half probabilities). Suppose that the source symbol probabilities are all of the form $p_i = \left(\frac{1}{2} \right)^{k_i}$ for various integers k_i. Of course, the probabilities must sum to 1. For example, the probabilities might be $\frac{1}{2}, \frac{1}{4}, \frac{1}{4}$ or $\frac{1}{4}, \frac{1}{4}, \frac{1}{8}, \frac{1}{8}, \frac{1}{8}, \frac{1}{16}, \frac{1}{32}, \frac{1}{32}$. In such cases one can set $l_i = k_i$. These lengths are valid choices because the sum on the left of the Kraft inequality becomes $\sum_{i=1}^{m} 2^{-l_i} = \sum_{i=1}^{m} 2^{-k_i} = 1$.

The entropy of the source is

$$H = \sum_{i=1}^{m} 2^{-k_i} \log 2^{k_i} = \sum_{i=1}^{m} 2^{-k_i} k_i.$$

On the other hand, the average word length is

$$L = \sum_{i=1}^{n} 2^{-k_i} l_i.$$

Since $k_i = l_i$, we have $H = L$. Hence, for sources with probabilities that are powers of one-half, instantaneous codes exist with $L = H$.

Example 3.5 (A classic example). A specific example of the preceding situation is the classic one shown below.

Source Symbol	Probability	Codeword
s_1	1/2	0
s_2	1/4	10
s_3	1/8	110
s_4	1/8	111

The entropy of the source is

$$H = \frac{1}{2} + \frac{1}{4}2 + \frac{1}{8}3 + \frac{1}{8}3 = \frac{7}{4}.$$

On the other hand, the average code length is exactly the same:

$$L = \frac{1}{2} + \frac{1}{4}2 + \frac{1}{8}3 + \frac{1}{8}3 = \frac{7}{4}.$$

Shannon Coding

If the source probabilities are of the form $p_i = \left(\frac{1}{2}\right)^{k_i}$, the word lengths that achieve the lower bound are $l_i = k_i$, as shown above. Another way to write this is $l_i = \log\frac{1}{p_i}$.

If the source probabilities are not powers of one-half, one can still evaluate $l_i' = \log\frac{1}{p_i}$ for each i. These numbers may not be integers, and hence cannot serve as actual word lengths, but they do satisfy the Kraft inequality. Evaluation of the appropriate sum gives

$$\sum_{i=1}^{m} 2^{-l_i'} = \sum_{i=1}^{m} 2^{\log p_i} = \sum_{i=1}^{m} p_i = 1.$$

Following an idea of Shannon's, each of these l_i''s is increased to the next highest integer (that is, the l_i''s are rounded up) and the new values are denoted as l_i's. Since each of these l_i's is greater than the corresponding l_i', this new set of word lengths also satisfies the Kraft inequality. Hence, there is an instantaneous code with these code lengths.

It follows that $l_i < \log\frac{1}{p_i} + 1$ for each i. Hence the average word length satisfies

$$L < \sum_{i=1}^{n} p_i \left(\log\frac{1}{p_i} + 1\right) = H + 1.$$

Combining the lower bound and this upper bound leads to the following conclusion about bounds.

Average length bounds. A source with entropy H can be coded with an instantaneous binary code of average length L satisfying

$$H \leq L < H + 1.$$

The next chapter explains how to find instantaneous codes with L as small as possible. For now, the bound is all that is required.

3.3 Shannon's First Theorem

As stated earlier, there are two main ideas leading to Shannon's first theorem. The first is to design codes with various word lengths, assigning short words to highly probable symbols and longer words to rarely occurring symbols. This idea was explored in the past few sections. The second idea is to group symbols together, forming metasymbols. Coding these metasymbols provides additional flexibility.

The composite (or meta) symbols are sequences of the original symbols. One might consider two coin flips at a time, or even 10 at a time. An entire month of California weather reports might be sent bundled together. English words rather than individual letters might be encoded.

The formalism for forming the composite symbols was presented in chapter 2. The square of a source S consists of pairs of symbols, taken as if they were independent, and this new source is denoted S^2. Likewise, a source made up of independent groups of n symbols from S is denoted S^n.

According to the additive property of section 2.4, the entropy of the source S^n is $H(S^n) = n H(S)$.

Now suppose S^n is coded instead of S. This is the "big package" concept discussed in the introduction to this book. In the case of codes, sequences of symbols form packages that can be coded as a unit. Specifically, an instantaneous code for S^n can be found that has an average codeword length L satisfying

$$H(S^n) \leq L \leq H(S^n) + 1.$$

Equivalently,

$$n H(S) \leq L \leq n H(S) + 1.$$

Dividing by n produces

$$H(S) \leq L/n \leq H(S) + 1/n.$$

Note that L is the average codeword length for a symbol from S^n and hence is the average codeword length for a sequence of n independent symbols from S. Therefore, L/n is the average length per symbol of S. The upper bound on this length has been reduced from $H(S) + 1$ to $H(S) + 1/n$.

Note, too, that the code is instantaneous with respect to S^n, not necessarily with respect to S. Hence, an entire sequence of n original symbols is decoded after receipt of the whole sequence, not one at a time.

Finally, using both of the ideas for designing efficient codes, the average word length can be made essentially equal to the entropy of the source by increasing n. This is the content of Shannon's first theorem.

Shannon's first theorem. By coding sequences of independent symbols (in S^n), it is possible to construct decodable codes such that

$$\lim_{n \to \infty} \frac{L_n}{n} = H,$$

where H is the entropy of the source S, n is the length of the symbol sequences, and L_n is the average length of the codewords corresponding to S^n.

Hence, it is possible to get an average codeword length L_n/n as close as desired to H. The price paid for such improvement is increased coding complexity due to the increased dimension of the source and increased delay in the decoding process, since an entire sequence must be processed at once. Nevertheless, this is a remarkable result, and establishes the importance of the concept of entropy in information theory.

Example 3.6 (Two days of California weather). Suppose that California weather is sunny (S) with probability 7/8 and cloudy (C) with probability 1/8. Suppose also that the weather conditions on successive days are independent.

To send one day's weather report with a binary code would require a code of at least one character per day, since the best that can be done is use a single symbol for each weather condition. Hence $L = 1$.

If two days are sent together, the modified comma code might be used as shown below:

Source Symbol	Probability	Code
SS	49/64	0
SC	7/64	10
CS	7/64	110
CC	1/64	11

The average length of this code is

$$L = \frac{49}{64} + 2 \times \frac{7}{64} + 3 \times \frac{7}{64} + 3 \times \frac{1}{64} = \frac{87}{64}.$$

The average length per daily report is therefore

$$L/2 = \frac{87}{128} = .68.$$

The added flexibility has considerably reduced the average length of the code. By considering longer sequences, an average length close to the entropy $H = .54$ found for this source in chapter 2 can be achieved.

3.4 EXERCISES

1. (Code lengths). Which of the following code lengths are feasible for constructing an instantaneous binary code for five symbols?
   (a) 2 2 2 3 3
   (b) 1 2 2 4 5
   (c) 1 2 3 4 4
   (d) 1 2 3 3 8

2. (Decodable?) Is the following code uniquely decodable?

A	0 1
B	1 0
C	0 1 1
D	1 0 1

3. (Explicit) Show explicitly that the code represented by the tree in figure 3.3 is not uniquely decodable.

4. (Braille) In Braille, each character consists of a pattern of raised dots. There are six positions in a character, each of which can be either flat or raised. This gives a total of 2^6 or 64 possible letters that can be described by a single Braille character.

   For standard English there are more than 64 symbols to be described; a to z in both lower and upper case, 0 to 9, and standard punctuation (space, comma, etc.). Thus, some symbols require more than one Braille character.

   Suppose 8 percent of letters are capitalized and 12 percent of letters are digits.
   (a) In Grade 1 Braille, lowercase letters and standard punctuation all require a single character. There is also a special character that indicates that the following character will describe a capital letter. Similarly, there is a special character to indicate that the following character will describe a digit.

   What is the expected number of Braille characters used to describe a standard English letter?
   (b) In Grade 2 Braille, in addition to the characters denoting capitals and digits, there are single characters that are used to denote common groups of letters (for example, *the*, *and*, *ing*). Suppose that each group of letters is three long and that 20 percent of standard English text comes from one of the common letter groups used in Grade 2 Braille.

   What is the expected number of Braille characters used to describe a standard English letter?

5. (Double code) Suppose that as a small measure of security against eavesdroppers, the symbols of a source S are each assigned two codewords, and during transmission the word to send is chosen randomly, each with a 50 percent probability. Of course the code must still be uniquely decodable. In terms of the entropy $H(S)$, what is the lower bound on the average codeword length of the code?

6. (Classic code) For the classic example code of example 3.5, show that the probability of a 0 being transmitted at any time is .5.

7. (Three-word characters) Consider the source shown below:

s_1	1/3	s_5	1/27
s_2	1/3	s_6	1/27
s_3	1/9	s_7	1/27
s_4	1/9		

(a) Using base-3 logarithms, compute the entropy of this source.

(b) Assuming that there are three characters 0,1,2 in the code alphabet, find a code for the source that has average length equal to its base-3 entropy.

3.5 Bibliography

Basic coding theory is covered in all introductory texts on information theory. In addition to the references of chapter 2, the three below are excellent.

References

[1] Ash, Robert B. *Information Theory*. New York: Dover, 1990.

[2] Welsh, Dominic. *Codes and Cryptography*. Oxford: Oxford University Press, 1988.

[3] Roman, Steven. *Introduction to Coding and Information Theory*. New York: Springer, 1997.

4

COMPRESSION

Compression is any procedure that reduces data requirements of a file (or message) without seriously degrading the integrity of that file. Compression is an important part of modern information technology and is commonly used to reduce the file size of text, music, graphics, video, and large data sets. There are dozens of compression techniques, each with its advantages and areas of application, but many can be best understood by reference to information theory.

4.1 Huffman Coding

One of the first university courses in information theory was offered by Prof. Fano at MIT in 1951. He gave each student the option of writing a term paper or taking a final exam. As the subject for the term paper he suggested the problem of finding instantaneous codes of minimum average length L (which was known to satisfy $H \leq L \leq H+1$). David A. Huffman elected to write the term paper. Unaware that both Shannon and Fano had themselves attempted to solve the problem, he worked diligently for several months, apparently getting nowhere. Just before the course was about to end, he discovered the simple method leading to what has ever since been termed **Huffman coding**.

Given a number of source symbols and their probabilities, the Huffman procedure produces a code with average length L as close as possible to the source entropy H. The result of the procedure is not always unique, for there may be several codes with the same minimum average length, but any one of these is termed a Huffman code. Today, Huffman coding forms the basis of many compression methods, and even when it is not used, it provides a numerical benchmark to evaluate other methods.

The basic idea of the Huffman coding algorithm for binary codes is to start with a small number of symbols and work up. It is easy to code a source with two symbols no matter what their respective probabilities: the two codewords must simply be 0 and 1, giving an average length of 1. No better code exists. For three symbols, the situation is slightly more complicated, but the best way to proceed is to temporarily combine the two symbols of lowest probability into a single composite symbol with probability equal to the sum of the individual symbols. This composite together with the remaining symbol defines a new two-symbol source, which is easily coded. Then the composite symbol is split into its components and coded by adjoining 0 and 1, respectively, to the code character that has already been assigned to the composite. Hence if 0 is the composite, the final three codewords are 00, 01, 1. In general, when there are some number m of symbols, they are reduced by forming composites over and over again until only two symbols remain. Coding is then carried out sequentially by working backward, splitting the composites and adding code characters two at a time. When all composites have been split, the original source symbols are obtained, together with assigned codewords. The procedure is simple, but it is best understood in the context of an example.

Example 4.1 (Five symbols). Consider the source of five symbols, with probabilities .3, .2, .2, .2, .1. The entropy of this source is $H = -[.3 \log .3 + (3 \times .2) \log .2 + .1 \log .1] = 2.246$. The first step of the Huffman procedure is to order the probabilities by decreasing magnitude. In this example, they are already so ordered (ties do not matter). The symbol names and their probabilities make up the first two columns of figure 4.1.

The procedure has two phases. The first part is the **reduction phase**. The bottom two symbols are combined and their probabilities summed. This composite symbol is then treated like any other symbol, and a new ordered list is constructed. In the example, the two bottom symbols, with probabilities .2 and .1, are combined to form a symbol of probability .3, and this is placed in the new ordered list. In the example, it may be placed either first or second on the list, since there is another symbol of probability .3. We have elected to place it second. The lines in the figure keep track of the composites. In the example, three stages of reduction are required to obtain the final two composite nodes.

The second part of the procedure, the **splitting phase**, is shown in figure 4.2. This splits the combined symbols and assigns codewords. The two symbols in the final list

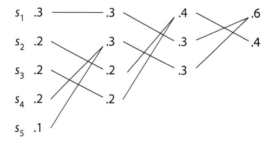

FIGURE 4.1 Huffman reduction phase. At each stage the two symbols of lowest probability are combined to form a single composite symbol. This is continued until only two symbols remain.

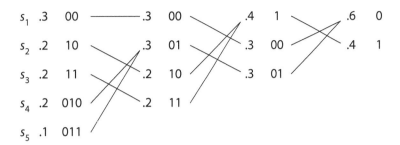

FIGURE 4.2 Huffman splitting phase. The two symbols in the last list are coded with 0 and 1. These are then taken backward to the previous list. One of the symbols must be split, and the existing codeword is appended with a 0 and a 1, respectively, to separate the symbols. In the example, the symbol with .6 probability is coded with 0. When it is split, the two components are 00 and 01.

are assigned the codewords 0 and 1 respectively. These are carried back to the next previous list of three symbols. The two symbols that were combined are now split, and the codeword previously assigned to the composite is now appended with a 0 for one of these components and a 1 for the other. In the example, the final node with a .6 probability was assigned the codeword 0. In the next list (going backward) this symbol is split, and the two components are coded as 00 and 01. The procedure moves back one list at a time until the original symbol list is obtained. The final codewords represent the Huffman code.

In this example, the final word lengths are 2, 2, 2, 3, 3. The average length is therefore

$$L = (2 \times .3) + (2 \times .2) + (2 \times .2) + (3 \times .2) + (3 \times .1) = 2.3,$$

which can be compared with the entropy H, which is 2.246.

It is often helpful to represent the code by its code tree, as shown in figure 4.3.

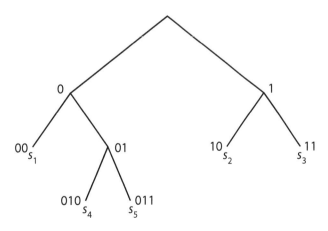

FIGURE 4.3 Huffman code tree. The code for the example can be shown on a tree.

Proof That Huffman Codes Are Efficient*

As mentioned in section 3.1, a code is said to be efficient if its average length is as short as possible. Huffman codes are efficient in this sense.

How can we prove that Huffman coding provides the minimum possible average length L? Suppose that a different efficient code for the same source has average length L' less than the value L for a Huffman code. We consider the code trees for both. It is clear, first, that in an efficient code there can be no dead branches, as illustrated in figure 4.4 (that is, there can be no final nodes that could be pushed back toward the beginning of the tree). This implies that codewords having the maximum length must appear in pairs, each pair emanating from a common predecessor node. Furthermore, these codewords must correspond to symbols of minimum probability or else the average length could be reduced by exchanging any one of these for a symbol of lower probability. It can be assumed that in the efficient code, the two symbols of lowest probability are pairs with the same predecessor, for if not, symbols having maximum length could be exchanged so that this is true.

Now imagine reducing the code trees of both the efficient code and Huffman code by combining the two lowest probability nodes, taking them back to their predecessor nodes. The word length of each of the nodes is reduced by 1. Hence the average word length of the reduced set of symbols in the efficient code is $L' - (p_q + p_{q-1})$, where p_q and p_{q-1} are the two lowest probabilities. The Huffman code tree will be reduced in a similar way by combining nodes of probability p_q and p_{q-1}, although, in the case of ties, these nodes may be different from those of the efficient code. Hence, the average word length of the Huffman code is reduced to $L - (p_q + p_{q-1})$. This argument applies at every stage of the reduction, with both the efficient code and the Huffman code being reduced by the same amount at each step. The final result for the Huffman code is a code for two symbols, and an average length of 1. The corresponding value for the efficient code would be less than 1, since $L' < L$. This is impossible, so we must have $L' = L$.

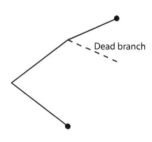

FIGURE 4.4 Dead branch. An efficient code can have no dead branches.

Nonuniqueness of Huffman Codes

There is flexibility in the Huffman coding algorithm whenever there are ties among symbol (or composite symbol) probabilities because these symbols can be interchanged in the ordered lists. Alternative orderings generally produce different code assignments, and often produce different word length combinations, but all resulting codes have the same average word length. This is illustrated by the following example. The reader is encouraged to actually go through the procedure to construct the codes.

Example 4.2 (Two ways). Two efficient codes are shown below for a five-symbol source.

Source Symbol	Probability	Code 1	Code 2
s_1	.4	1	00
s_2	.2	01	10
s_3	.2	000	11
s_4	.1	0010	010
s_5	.1	0011	011

These are both instantaneous codes constructed by the Huffman procedure. In constructing code 1, composite nodes were always placed at the lowest position in the list consistent with proper ordering, while in constructing code 2, composite nodes were always placed at the highest possible position. (For example, at the first step, the composite will have probability of $.1 + .1 = .2$, and hence there are actually three possible positions for it in the next list.) The average lengths of the two codes given in the table are, respectively,

$$L_1 = (.4 \times 1) + (.2 \times 2) + (.2 \times 3) + (.1 \times 4) + (.1 \times 4) = 2.2$$
$$L_2 = (.4 \times 2) + (.2 \times 2) + (.2 \times 2) + (.1 \times 3) + (.1 \times 3) = 2.2.$$

Thus, although the word lengths are different, both codes have the same (minimum) average word length.

Example 4.3 (Powers of A,B). Consider the source S consisting of two symbols A and B with probabilities $p_A = 3/4$, $p_B = 1/4$. The entropy of this source is

$$
\begin{aligned}
H &= -\left[\frac{3}{4}\log\frac{3}{4} + \frac{1}{4}\log\frac{1}{4}\right] \\
&= \log 4 - (3/4)\log 3 \\
&= 2 - (3/4)\log 3 = 2 - .75 \times 1.5849 = .8118.
\end{aligned}
$$

The Huffman code for this simple source assigns 0 to A and 1 to B, giving a minimum average length of $L = 1$.

The source S^2 consists of the four pairs AA, AB, BA, and BB with probabilities found by multiplication. The source S^3 consists of eight triples AAA, AAB, ABA, and so forth. The Huffman codes for each of these sources can be constructed. The resulting codes and their average lengths are shown in table 4.1. As the number of symbols in the compound source is increased, the average (per symbol) length of the Huffman code approaches the entropy of the source.

TABLE 4.1
Huffman Codes for Powers of a Source.   As the number of combinations is increased, the average per-symbol length of the Huffman code approaches the entropy of the source, $H = .8118$.

Symbol	Probability	Code	Symbol	Probability	Code	Symbol	Probability	Code
A	3/4	0	AA	9/16	0	AAA	27/64	1
B	1/4	1	AB	3/16	10	AAB	9/64	001
			BA	3/16	110	ABA	9/64	010
			BB	1/16	111	BAA	9/64	100
						ABB	3/64	00000
						BAB	3/64	00001
						BBA	3/64	00010
						BBB	1/64	00011
	$L = 1$			$L/2 = .84375$			$L/3 = .8229167$	

Universal Coding

The Huffman coding procedure requires knowledge of the symbol probabilities. This can be a limitation when dealing with specialized material, such as technical documents, where the actual frequencies of symbol occurrence may differ from standard probability tables. In such cases the resulting Huffman code does not achieve the maximum possible compression. Motivated by this concern, various adaptive techniques have been proposed that modify the list of probabilities according to the actual occurrences in the data. In a sense, these methods learn the frequencies. If an adaptive process is continued indefinitely, on an infinite stream of symbol occurrences, eventually the estimated frequencies will converge to the actual frequencies and the compression rate will be maximal. Such methods, which do not require that probabilities be given in advance but act in the limit as if they did have them, are termed **universal**.

4.2 Intersymbol Dependency

Successive symbols from a source are not always independent. Instead current symbol probabilities often depend on what symbols occurred before the current one. For example, in English, an *h* is more likely to occur if a *t* was the previous letter than if a *d* was the previous letter. This intersymbol dependency must be accounted for in an accurate measure of entropy. Typically, intersymbol dependency reduces entropy, and this implies that codes with shorter average word lengths can be constructed. Indeed, this idea is the basis for many modern compression methods.

English Probabilities

It is not surprising that considerable attention has been devoted to the possibility of encoding English text efficiently to achieve compression of text documents. An estimate of the actual entropy of English, accounting for its intersymbol dependencies, is necessary to predict the possible advantage of such compression.

For convenience it is often assumed that the alphabet consists of just 27 symbols (26 letters and a space or punctuation). Since $2^5 = 32$, all 27 symbols could be encoded with a binary block code with each word having length 5. However, 27 is less than 32, so the actual entropy is less than five bits.

The simplest estimate of the entropy of English (termed the **zero-th order estimate**) is based on the assumption that all letters of the alphabet are equally likely, with a probability of 1/27. The corresponding zero-th order entropy, denoted H_0, is therefore

$$H_0 = \log 27 = 4.755 \text{ bits/letter.}$$

This is an upper bound, since if the probabilities are not uniform, the entropy will be reduced. However, this bound is an important reference point.

The probabilities of letters of the alphabet as they occur in English text vary somewhat with the type of document, but these differences tend to be slight. The probabilities shown in table 4.2 are typical. The table shows, for instance, that the

TABLE 4.2
Alphabet Probabilities for English.
Probabilities are computed over thousands of words taken from various sources.

A	.064	N	.056
B	.014	O	.056
C	.027	P	.017
D	.035	Q	.004
E	.100	R	.049
F	.020	S	.056
G	.014	T	.071
H	.042	U	.031
I	.063	V	.010
J	.003	W	.018
K	.006	X	.003
L	.035	Y	.018
M	.020	Z	.002
		Space/punctuation	.166

letter E is apparently everybody's favorite, since it is much more common than any other letter except punctuation and space.[1]

The entropy of English using letter probabilities is termed the first-order estimate and denoted H_1. For the probabilities of table 4.2, $H_1 = 4.194$, representing a modest reduction in entropy over the case of uniform probabilities. Based on this estimate, one expects that a Huffman code for the alphabet would have an average length closer to four than to five. Huffman coding is indeed used as a method of compression of English text, and as expected, documents require about 20 percent less storage than when the standard ASCII code is used.

It is possible to go further and consider the frequencies of pairs of letters. Indeed, the $27^2 = 729$ probabilities of pairs have been estimated and these imply an entropy H_2 (per letter, not per pair) of approximately 3.3. Going further to the $27^3 = 19,683$ triples yields an entropy estimate H_3 of about 3.1, and since this is only slightly less than H_2, it seems that it is not worthwhile to go further.

Zipf's Law

Intuitively, the redundancy in English text is deeper and more complex than revealed by mere letter frequency variations. For example, most people have little trouble interpreting the corrupted word *scho_l* as *school*, and they do not accomplish this interpretation through use of letter frequencies.

To get to the more subtle redundancies, Shannon proposed looking at word frequencies instead of letter frequencies. He could then determine the entropy H_W of

---

[1] A notable exception is the 267-page novel *Gadsby* by Ernest Vincent Wright, which by design contains no occurrences of the letter *e*.

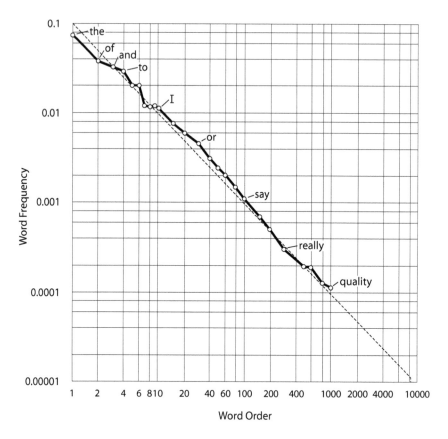

FIGURE 4.5 Zipf's law. The frequency of a word is approximately proportional to the reciprocal of its word rank.

English words, and from that estimate the per letter entropy of English as $H = H_W/\bar{w}$, where $\bar{w}$ is the average length of English words. To carry out this program, Shannon used an empirically derived relation among word frequencies proposed by George Zipf in 1949. **Zipf's law** holds in a surprisingly broad set of languages, including Yiddish, Old German, Plains Cree, Norwegian, and English.[2] Specifically, Zipf's law states that if the words of a language are listed in decreasing order of frequency (with word 1 being the most frequent and word m being m-th on the list), then the probabilities of these words approximately satisfy the relation

$$p_m = \frac{A}{m},$$

where A is a constant that depends on the number of active words in the language.

Zipf's law is usually illustrated by a plot on a log-log graph, using the form $\log_{10} p_m = \log_{10} A - \log_{10} m$. Such a plot for English is shown in figure 4.5. Shannon used a value of $A = .01$, which gives the approximation shown in the figure. This

---

[2]The law also holds for many other rankings, including city populations, energy consumption of animals, and some financial market data.

requires that the number of words be $M = 12{,}366$, so that the sum of the probabilities is 1. That is, $\sum_{m=1}^{M} p_m = \sum_{m=1}^{12{,}366} \frac{.01}{m} = 1$. Then

$$H_W = \sum_{m=1}^{12{,}366} \frac{.01}{m} \log \frac{m}{.01} = 9.72 \text{ bits/word.}$$

Using $\bar{w} = 4.5$ letters/word as the average word length then gives $H = 9.72/4.5 = 2.16$ bits/letter. This is a reduction of more than 50 percent over the raw entropy based on 27 equally probable and independent letters. However, even this figure is likely still too high because English words are themselves interdependent.

Redundancy

If the letters in English were not interdependent, the entropy of a string of n letters would be exactly n times the single-letter entropy. This is the additive property of independent copies of a source, which is expressed as $H(S^n) = n H(S)$. When successive symbols are not independent, the source made up of n symbols from S is written as $(S_1, S_2, \ldots, S_n)$ and the corresponding entropy is written as $H_n = H(S_1, S_2, \ldots, S_n)$. The average entropy per symbol is then H_n/n. This leads to the definition of the average per-symbol entropy over long sequences as

$$\bar{H} = \lim_{n \to \infty} \frac{1}{n} H(S_1, S_2, \ldots, S_n),$$

assuming the limit exists. This is the appropriate definition of the entropy of English.

Perhaps the simplest (and best) way to measure entropy is to study fluent native speakers who process the language in deep ways developed by experience. A useful experiment is to show a volunteer a partial line of text and ask him or her to guess the letter that comes next. For example, the volunteer might be shown the partial line[3]

> She was so astonished and bewildered that sh_

Virtually everyone would predict (correctly) that the next letter is e. It is likely that the following symbol would also be correctly predicted to be a space. However, the letter after that might be more difficult to predict. It happens to be c, leading to the rest of the sentence: "could make no reply."

In this experiment, a subject is shown the last $n - 1$ letters of text and attempts to guess the n-th. The probabilities of the possible letters for position n depend on the previous $n - 1$ letters. The conditional entropy corresponding to this uncertainty is denoted $H(S_n | S_1, S_2, \ldots, S_{n-1})$, which is the entropy of S_n given knowledge of the symbols that appeared in $S_1, S_2, \ldots, S_{n-1}$. It is generally true (as shown in chapter 5) that

$$H(S_1, S_2, \ldots, S_n) = H(S_n | S_1, S_2, \ldots, S_{n-1}) + H(S_1, S_2, \ldots, S_{n-1}).$$

In words, the entropy of the n symbols is the entropy of $n - 1$ symbols plus the entropy of the next symbol given that the previous $n - 1$ symbols are known.

[3] From Cervantes' *Don Quixote*, chapter 27.

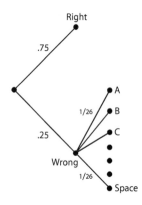

FIGURE 4.6 The probability structure after a guess. The upper branch denotes the letter guessed. The other 26 possibilities are assumed to have equal probabilities.

Suppose that when carrying out this experiment over many strings of text and using several volunteers, the predictions are correct 75 percent of the time. To be conservative we might assume that if a prediction is wrong, the volunteer has no idea what letter it might be, and hence assigns a probability of 1/26 to each of the remaining 26 possibilities. (See figure 4.6.) The entropy of the uncertainty is then

$$H = -.75 \log .75 - 26 \times (.25/26) \log (.25/26) = 1.99 \text{ bits.}$$

This is the lowest estimate for the entropy of English we have obtained so far. However, even it may be too large, since if a prediction of the n-th letter is incorrect, the volunteer is likely to assign unequal probabilities to the remaining choices. In fact, a reasonable estimate for the entropy of English based on various reported experiments is approximately 1.5 bits per letter.

Shannon introduced a measure of redundancy to express the degree by which the entropy of a language differs from the maximum value that its alphabet could have. He defined redundancy as

$$R = 1 - H/\log M,$$

where M is the size of the alphabet and H is the true entropy value. Hence for English

$$R \approx 1 - \frac{1.5}{\log 27} = 1 - 1.5/4.755 = 67 \text{ percent.}$$

This implies that there is tremendous opportunity for compression of English text. A text file can, in theory, be compressed to about one-third of its original size.

Our discussion implies, however, that to achieve results close to those that are theoretically possible, it is necessary to use a complex process of coding and decoding. Huffman coding alone will not get us far, since even coding the $27^2 = 729$ pairs of letters gives only modest compression compared to what is theoretically possible.

4.3 Lempel–Ziv Coding

Compression practice was advanced significantly by the innovative and clever method published by Ziv and Lempel in 1977.[4] The method is based on the fact that sequences of letters in English text are not entirely random, but repeat patterns from time to time. These patterns make up words or even phrases. The Lempel–Ziv method essentially constructs a dictionary of these common patterns.

The method may at first sight appear to be simply a clever way to take advantage of patterns, a method almost trivial in its design. In fact, in practice it is quite powerful and has deep theoretical underpinnings. It can be shown that in its ideal form it produces, in the limit of long text, a compression factor equal to that predicted by the entropy of the text. It is therefore a universal compression algorithm, based on a simple recording scheme.

[4]Due to a quirk of history, the now usual ordering of the names of the authors associated with the method differs from the order in the original publication.

In the Lempel–Ziv method, both sender and receiver keep a record of what has already been sent. Then, when preparing to send additional text, the sender looks back at previously sent text to find a maximum-length duplication of what needs to be sent next. Then a reference to the past duplicating string is sent instead of the string itself. For example, if the next portion of text is a word that was sent before, the sender merely sends a reference to the position of that word in the record of past letters.

In the ideal method, the entire history of previously transmitted text is kept by both sender and receiver, but in practice the stored history is restricted to a given size, located in a search buffer. Likewise, the possible length of string to be sent is restricted by a look-ahead buffer.

The reference to the past string is transmitted by sending a triple (x, y, z) where the first entry x is the number of places back in the buffer where the duplicate string begins, y is the length of the duplicate string, and z is the next letter in the look-ahead buffer after the duplicated string. For example, the message $(5, 3, F)$ means that the upcoming text begins with a string that is identical to one that begins 5 letters back, has length 3, and is then followed (in the new text) by the letter F.

The process is illustrated in figure 4.7 for sending the message, THIS-THESIS-IS-THE-THESIS. Initially, the search buffer, made up of text previously sent, is empty. Hence the first letter is sent directly by the message $(0, 0, T)$. Additional single-letter transmissions continue for a few steps as the search buffer begins to fill up. Finally, after five steps, the sequence THIS- has been sent, and the sender is preparing to transmit the next string. The look-ahead buffer begins with TH, which duplicates a TH sent earlier. Since the beginning T of that two-letter string is five places back in the search buffer, the message $(5, 2, E)$ is sent, indicating also that E is the next letter. This message has effectively transmitted the three letters THE. Next, the single letter S is sent. After that, there is a good deal of duplication, and letters are sent in groups of three, six, and seven in the final three steps.

It is possible for the duplicated string to be long enough that it spills over onto the look-ahead buffer. That means that a message such as $(5, 9, R)$ might be sent, with the second entry being larger than the first. That is fine, for the receiver can figure out the entire duplicated sequence.

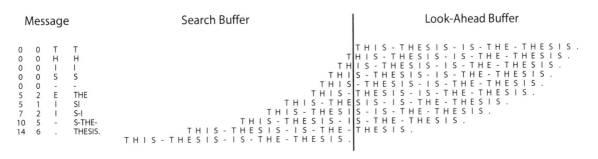

Message				Search Buffer	Look-Ahead Buffer
0	0	T	T		THIS-THESIS-IS-THE-THESIS.
0	0	H	H		THIS-THESIS-IS-THE-THESIS.
0	0	I	I		THIS-THESIS-IS-THE-THESIS.
0	0	S	S		THIS-THESIS-IS-THE-THESIS.
0	0	-	-		THIS-THESIS-IS-THE-THESIS.
5	2	E	THE		THIS-THESIS-IS-THE-THESIS.
5	1	I	SI		THIS-THESIS-IS-THE-THESIS.
7	2	I	S-I		THIS-THESIS-IS-THE-THESIS.
10	5	-	S-THE-		THIS-THESIS-IS-THE-THESIS.
14	6	.	THESIS.		THIS-THESIS-IS-THE-THESIS.

FIGURE 4.7 Lempel–Ziv (LZ77) coding. The sender looks back to find a maximum-length duplication of a string that is about to be sent.

Variations

There are several variations of the basic Lempel–Ziv procedure. In the original paper and method, referred to as LZ77, the buffers are each of finite size. Hence if a pattern repeats but the previous instance of it is more distant in the past than the length of the search buffer, it cannot be referenced. There is a trade-off between the size of the buffers (which use storage and require search and processing time) and the efficiency of the procedure in terms of the number of patterns that can be referenced.

Another consideration regards the sending of the triple (x, y, z). With large buffers, both x and y take on values in a large range, and hence may require a large number of bits to transmit. Frequently, these are coded with another compression technique, such as Huffman coding, to achieve greater overall efficiency.

Several popular compression algorithms found on personal computers, including ZIP, PKSip, LHarc, PNG, gzip, and ARJ, are based on the Lempel–Ziv LZ77 method together with compression coding of the reference information. Good compression ratios are achieved by these packages. For example, the LaTeX file of this chapter when compressed with ZIP is less than one-third the size of the uncompressed version.

The LZW Method

To get around the disadvantage inherent in the use of a finite buffer in the LZ77 method, Ziv and Lempel proposed a variation in 1978 (now termed LZ78) that builds an explicit dictionary of previously sent letter patterns. The LZ78 method was itself modified by other researchers, the most popular of these modifications being that proposed by Terry Welch and now referred to as LZW.

To begin the LZW method, a dictionary is loaded with a set of basic letters and words. When about to send a symbol, the sender looks ahead for the longest possible sequence that starts with that symbol and which is in the dictionary. The sender then transmits the entire sequence by referencing its dictionary index. In many cases, however, only a single symbol can be sent. An example is shown in figure 4.8, where again the message is THIS-THESIS-IS-THE-THESIS.

In this example, the alphabet consists of the symbols E, H, I, S, T, and –. These are entered into the dictionary initially. At the first step, the sender sees that the first symbol T can be sent by reference to the library but the longer string TH cannot. Hence, the sender sends the message 5 (to send the T) and records the string TH in the dictionary as entry number 7. Next, a 2 is sent to reference H in the dictionary and the string HI is recorded as entry number 8. These single-letter transmissions continue for a few steps while the dictionary is built up. Finally, when reaching the beginning of the word THESIS, the sender can send a reference to entry 7, to send TH, followed by the letter E. The sender will also append THE to the dictionary as entry 12. As the dictionary expands, longer strings in the dictionary are frequently sent.

The receiver likewise carries out the dictionary building as transmissions are received. Hence, the receiver can decode the message as it is received.

Optimality*

Remarkably, it can be shown that the Lempel–Ziv method achieves the optimal rate of transmission in the long run. To see what this means, suppose a binary alphabet of

Index	Entry		Index	Entry
1	E		11	-T
2	H		12	THE
3	I		13	ES
4	S		14	SI
5	T		15	IS-
6	-		16	-T
7	TH		17	IS-T
8	HI		18	THE-
9	IS		19	-TH
10	S-		20	HE
			21	ESI

T	H	I	S	-	TH	E	S	IS	-	IS-	THE	-T	H	E S	I S
5	2	3	4	6	7	1	4	9	6	15	12	11	2	13	9

FIGURE 4.8 LZW example. The Lempel–Ziv–Welch method builds a dictionary as it transmits, so that if a pattern in the dictionary is later encountered, only its dictionary index need be sent. The receiver is able to build the dictionary as it receives messages, and hence is able to decode the transmission. The top part of the figure shows the dictionary as it is constructed from the message. The bottom shows how the message is compressed by reference to dictionary entries.

0's and 1's is used and suppose there are at least n past message symbols. We look at the future sequence beginning with the current symbol and find the maximum length of that sequence that is duplicated in the past n symbols. The length of this longest duplicating sequence is denoted L_n. For example, consider the data below, showing k as the position step and x_k as the message symbol.

$k =$	−4	−3	−2	−1	0	1	2	3	4	5	6	7	...
$x_k =$	0	1	0	1	1	0	1	0	1	1	0	0	...

At $k = 0$, we have $L_4 = 3$ because within the last $n = 4$ steps there is a sequence of length 3 that duplicates the three message symbols x_0, x_1, x_2 and that is the longest duplication. Likewise, at that position $L_3 = 3$ and $L_2 = 1$. One step later, at $k = 1$, we have $L_4 = 2$ because there is a duplicating sequence of length 2 starting at $k = -2$; the symbol at $k = -4$ cannot be considered since it is five symbols back. Likewise, $L_3 = 2$ and $L_2 = 0$. However, at $k = 1$, we have $L_5 = 6$, because there is a duplicating sequence of length 6 (which spills over into the current and future symbols, and that is allowed).

It can be shown that

$$\frac{\log n}{L_n} \longrightarrow H \qquad \text{as } n \longrightarrow \infty, \tag{4.1}$$

where H is the entropy of the source defined by

$$H = \lim_{n \to \infty} \frac{1}{n} H(S_1, S_2, \dots, S_n),$$

which accounts for intersymbol dependencies. Expression (4.1) means that the lengths of the duplicating strings increase substantially as the span of history is increased. It also implies that the compression factor is the one predicted by entropy.

To show how (4.1) translates into a compression ratio, consider the following simple example. Suppose the 0's and 1's of the message are generated completely at random, so that $H = 1$. And suppose there is an accumulated history of $n = 1,024$ symbols. Then (4.1) says that $(\log 1024)/L_n \approx 1$. In the upcoming sequence one can expect to find, using (4.1), that about the next ten symbols (since $\log 1024 = 10$) will form a sequence that is somewhere duplicated symbol by symbol in the history.[5] It is necessary only to reference the starting position and length of that duplicating string, rather than the string itself. However, reference to the duplicating sequence requires sending a number between 1 and 1,024, and this takes $\log 1024 = 10$ bits.[6] Thus 10 bits of position information must be sent to reference a 10-symbol string. That averages to one bit per symbol, which is what could be done by directly sending the symbols one by one.

However, suppose that the 0's and 1's in the sequence are not completely random, but are such that the entropy is $H = 1/2$. Then, from (4.1), with $n = 1,024$ one expects to find a duplicating sequence of length $L_n \approx (\log n)/H = 20$. Still only 10 bits must be sent to locate the duplication, but now the 10 bits of position information convey 20 message symbols. Hence, on average, each symbol requires only one-half a bit. Perfect compression has been achieved, reducing the required bits per symbol to the entropy.

Note also that the process itself makes no reference to the statistics of the sequence. It is therefore a universal process, in that it achieves the optimum performance even if the entropy and probabilities are not known.

4.4 Other Forms of Compression

In addition to text, compression is also applied to graphs, pictures, speech, and video, where the data generally are expressed as numerical values rather than abstract symbols. A simple graph may represent an intensity variable over one dimension, characteristic of speech or music patterns. A color photograph has two spatial dimensions and two intensity dimensions corresponding to brightness and color. Video has several intensity levels defined over time and space.

It is common to process these kinds of data by first discretizing the time and spatial dimensions as well as the intensity levels. For example, speech levels may be recorded every millisecond to within eight-bit accuracy, and picture intensities may be recorded at a few million pixels, each giving 24 bits of color data.

Consider, for example, the graph in figure 4.9. The continuous curve is approximated by first discretizing the x-axis into a number of points. Next, the intensity (the y value) at each of those points is quantized by a rounding operation, using the closest vertical point on a grid. The resulting approximation is shown in the figure as the series of dots, and each dot can be encoded with a few bits. Obviously, both

---

[5] Here is an intuitive (nonrigorous) way to see that the maximum duplicated length is about 10. A particular sequence of length 10 will occur with probability of $(1/2)^{10}$ since each symbol has probability $1/2$. Hence if we look at 2^{10} different starting points of history, it is quite likely that we will find a duplication because the number of sequences we are looking at times their probabilities is 1; that is, $(1/2)^{10}2^{10} = 1$. In actuality, the result of (4.1) is much stronger, for it says that in the limit of large n, L_n will be exactly 10.

[6] The length L_n must also be sent, but since this value is around 10, it does not add much to the requirement.

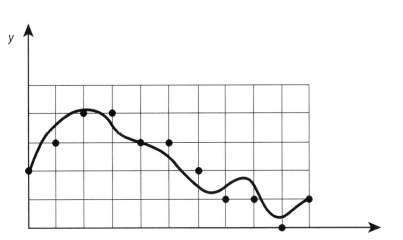

FIGURE 4.9 Quantization of a graph. At regular intervals on the x-axis, the y values of the graph are approximated by the nearest grid point.

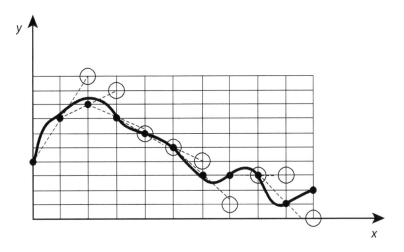

FIGURE 4.10 Quantization with prediction. A prediction of y, represented by an open circle, is made by drawing a dotted line between the two previous points and extending it to the next x value. Then it is only necessary to transmit the prediction error, which in the figure is often zero, and never more than two units (plus or minus).

the degree of approximation and the associated number of bits increase as the grid is made finer in either dimension.

Prediction Methods

It is possible to take advantage of the continuous nature of the data in graphs, pictures, voice, and so forth by predicting new points on the basis of points already known. For example, consider the graph used before, but now plotted in figure 4.10

14	12	10	18	20	24	30
12	12	14	16	18	26	32
12	14	12	16	20	24	30
18	10	10	12	16	40	50
12	12	14	35	40	55	60
14	45	48	50	56	114	80
50	52	46	55	110	116	92
55	56	50	58	112	108	94

A	B
C	X

(a) Luminance matrix   (b) Prediction elements

FIGURE 4.11 Luminance matrix for a picture. Part (a) shows the luminance values. Part (b) shows the elements A, B, C used to predict the value of X.

with finer y divisions. Reading from left to right, past points can be used to predict the next ones. One method, illustrated in the figure, computes a straight line between the two most recent points and extends this line to the next x-axis point; it then computes the nearest y grid value. This value is the prediction of the next point, and it can be determined by both sender and receiver. Thus the sender must merely send a message confirming or correcting the prediction, by say, sending one of the values -4, -3, -2, -1, 0, 1, 2, 3, 4, indicating the number of vertical grid points by which the actual value differs from the prediction. The small values, -1, 0, 1, will be most frequent and hence can be sent efficiently with short codewords, while the larger, less frequent, numbers can be assigned longer words. Overall, the average number of bits required per point will be smaller than without prediction.

The method is easily adapted to two-dimensional information, such as that derived from photographs. A small version of a luminance array is shown in figure 4.11. The values in the array can be transmitted one at a time starting at the upper left corner and moving across the rows. Efficiency is improved if past values are used to predict successive values. For example, as shown in part b of the figure, the values of A, B, C can be used to predict the value in X, and then only the error in the prediction must be sent. A suitable prediction is $X = B + C - A$.

Standards for compression were established by the committee called the Joint Photographic Experts Group (JPEG), and these standards are used in the popular JPEG image compression packages. The committee developed two principal methods: lossless and lossy. The lossless method faithfully transmits the luminance and chrominance arrays using the prediction method discussed above. Modest compression is obtained by this lossless scheme, but substantial compression is obtained by the more popular lossy method briefly discussed below.

Approximation

Another way to compress continuous data is to approximate it by a simple pattern. For example, a general curve might be approximated by a polynomial, which is defined by its coefficients. Figure 4.12 shows a curve approximated by a third-order polynomial

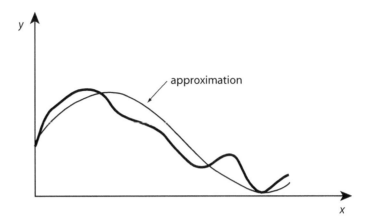

FIGURE 4.12 Polynomial approximation. The original curve is approximated by a low-order polynomial. It is then only necessary to send the coefficients of this polynomial to transmit an approximation of the original curve.

of the form $a_3x^3 + a_2x^2 + a_1x + a_0$. This approximating curve can be reconstructed from the four coefficients.

The popular lossy version of JPEG approximates the luminance matrix by a series of two-dimensional cosine functions, and records the coefficients of that series. The method is designed so that that even at a 10:1 compression ratio only minor image degradation is apparent to the human eye.

As an outgrowth of the JPEG committee, the Moving Picture Experts Group (MPEG) was formed to develop compression standards for efficient encoding of full motion pictures and high-quality audio. The result was the MPEG1 standard that can compress both video and audio. The popular music compression standard MP3 is actually the audio portion of the MPEG1 standard, and it provides music compression ratios of about 10:1, making the storage and playback of music easily adaptable to modern computer use. The concept behind these compression standards is sometimes called **perceptual coding** because although there is a significant loss of detail as measured by the number of raw bits used to record the original, the reduction is made in a way that does not significantly degrade one's perception of the recording.

4.5 EXERCISES

1. (Six-symbol source) A source has six symbols with the probabilities indicated below.

s_1	0.3	s_4	0.1
s_2	0.2	s_5	0.1
s_3	0.2	s_6	0.1

   (a) Find a Huffman code for this source.
   (b) What is the average length of the Huffman code?
   (c) What is the entropy of the source?

2. (Five-symbol source) Find three different Huffman codes for the source

s_1	0.4
s_2	0.2
s_3	0.2
s_4	0.1
s_5	0.1

3. (Huffman advantage) The compression advantage of Huffman coding S^2 rather than S is greatest when there are large differences in the source symbol probabilities. For each of the two cases below, find the source entropy and the average Huffman code length for S and for S^2.

   (a)
s_1	.60
s_2	.40

   (b)
s_1	.90
s_2	.10

4. (Huffman deduction) Consider the source with associated symbols and probabilities

s_1	.2	s_5	.1
s_2	.2	s_6	.1
s_3	.2	s_7	.1
s_4	.1		

   One Huffman code for this source has word lengths of $2, 3, 3, 3, 3, 3, 3$. Without using the Huffman procedure, find an instantaneous minimum average-length binary code for this source that has these word lengths.

5. (Run-length coding) Consider a source consisting of the symbols 0 and 1, where the probability p of a 0 is very close to 1. In this case, it might be considered advantageous to assign symbols to various runs of 0's. That is, one lets $s_1 = 1$, $s_2 = 01$, $s_3 = 001$ and so forth for all possible run lengths. Then, for example, the sequence 001010001 would be represented as s_3, s_2, s_4.
   (a) Let H_0 be the entropy of the original source, and let H_R be the entropy of the run-length source of symbols. Find the ratio H_R/H_0.
   (b) Find the average length L of the runs, and then the ratio of the entropy per unit length for the two methods.
   Hint: For $0 < r < 1$, $\sum_{k=0}^{\infty} r^k = \frac{1}{1-r}$ and $\sum_{k=1}^{\infty} kr^k = \frac{r}{(1-r)^2}$.

6. (Card determination) Mary selects a card from a pack of 52 playing cards. John attempts to determine the card by asking a series of yes-no questions that Mary answers correctly. In general, how many questions must John ask to be sure he can determine the card? What questions should he ask to minimize the average number of questions that are required? (Hint: the first question might be, "Is the card black or red?" Do not try to work out the Huffman code. Just think about how to ask the questions.)

7. (A special word) A certain language has the property that all letters occur independently with various probabilities. The Huffman code for this language has average word length L per letter. One special word of length l occurs quite frequently, with probability p. (That is, of 100 words, about 100 Pl will be from the special word.) It is suggested that this word be assigned the symbol 0 and that all other letters be coded with 1 appended by a Huffman code for this group of letters (also of average length L).
   (a) For what values of p is this proposal worthwhile?
   (b) Does this seem to be a useful strategy for coding English with the special word *the*, (with a space included) assuming that the Huffman code has length 4 per letter, the word *the* (with space included) occurs about once every 20 words, and the average word length is five letters?

8. (Refined estimate) Suppose the volunteer of section 4.2 gets 75 percent of the letters correct and assigns standard English letter probabilities to the others. What is the revised estimate of H?

9. (LZ77 example) Decode the message

   (0,0,I) (0,0,-) (0,0,M) (3,1,S) (1,1,-) (5,5,L) (5,3,Y).

10. (LZW case*) A special circumstance arises in the LZW procedure when the dictionary item sent is the last one entered into the dictionary; this item is not yet fully formed by the receiver. Consider the message *abababab*. The initial dictionary consists of 1:*a*, 2:*b*. The message will be sent as $1, 2, 3, 5, \ldots$, and the dictionary will be 1:*a*, 2:*b*, 3:*ab*, 4:*ba*, 5:*aba*.... However, when 5 is sent, the receiver will not yet have completed the construction of that dictionary entry. After receiving 3, the receiver will know that item 5 is *abX*, where X is unknown. So when the next transmission is 5, the receiver can translate it as *abX* and realize that the X in the dictionary item must be *a* because it is the first symbol of the transmission after receiving 3. This special case is part of the LZW method.
    Suppose the initial alphabet is 1:*a*, 2:*d*, 3:–. Translate the sequence 2 1 2 3 1 3 4 10 9 5 4.

11. (JPEG) Show that the formula $B + C - A = X$ is a reasonable one for JPEG by forming linear approximations at A to the partial derivatives of luminance with respect to x and y coordinates.

4.6 Bibliography

The Huffman coding method was first published in [1], and the story of Huffman's term project is reported in [2]. Tables of English letter frequencies appear in many places. The one used here is from [3]. Zipf's law for word frequencies was published in [4]. The experimental way for determining the entropy of English was presented by Shannon in [5]. Another interesting approach is presented in [6]. The original LZ77 and LZ78 methods were described in [7] and [8]. For further analyses of the optimality properties of these algorithms see [9] and [10]. Good general references on compression methods are the texts [11] and [12].

References

[1] Huffman, David A. "A Method for the Construction of Minimum Redundancy Codes." *Proceedings of the Institute of Radio Engineers* 40 (1952): 1098–1101.

[2] "Profile: David A. Huffman." *Scientific American*, September 1991, 54–58.

[3] Welsh, Dominic. *Codes and Cryptography*. Oxford: Oxford University Press, 1988.

[4] Zipf, George K. *Human Behavior and the Principle of Least Effort*. Cambridge, Mass.: Addison-Wesley, 1949.

[5] Shannon, Claude E. "Prediction and Entropy of Printed English." *Bell System Technical Journal* 30 (January 1951): pp 50–64.

[6] Cover, Thomas M., and R. King. "A Convergent Gambling Estimate of the Entropy of English." *IEEE Transactions on Information Theory* 24 (1978): 413–21.

[7] Ziv, J., and A. Lempel. "A Universal Algorithm for Data Compression." *IEEE Transactions on Information Theory* 23 (1977): 337–43.

[8] ———. "Compression of Individual Sequences via Variable-Rate Coding." *IEEE Transactions on Information Theory* 24 (1978): 530–36.

[9] Wyner, Aaron D., and Jacob Ziv. "Some Asymptotic Properties of a Stationary Ergodic Data Source with Applications to Data Compression." *IEEE Transactions on Information Theory* 35 (1989): 1250–58.

[10] Ornstein, Donald S., and Benjamin Weiss. "Entropy and Data Compression Schemes." *IEEE Transactions on Information Theory* 39 (1993): 78–83.

[11] Wayner, Peter. *Compression Algorithms for Real Programmers*. San Diego: Academic Press, 2000.

[12] Sayood, Khalid. *Introduction to Data Compression*. 2nd ed. San Diego: Academic Press, 2000.

CHANNELS

It is amazing how much has been accomplished using only Shannon's basic definition of entropy. The formalization and implementation of compression, the understanding of redundancy, and creative approaches to coding of discrete sources are all derived from this basic notion. Yet much more can be done by considering the structural connection between different random variables. In a physical or operational sense this structure takes the form of an information **channel**, as illustrated in figure 5.1. In mathematical terms, the structure is defined by **conditional probabilities** relating the output probabilities to the input. This probabilistic structure is a useful model of many real situations, and it is the structure addressed by Shannon's powerful **second theorem** stating that information can pass through a channel almost without error at rates bounded by the channel's capacity.

The notion of a channel is more general than a traditional communication system. It can represent any situation where observation of one random variable reveals something about another. Medical tests are a good example. An underlying random variable may be the condition of a patient's arteries. A cholesterol measurement reveals something (probabilistic) about that condition. The link from artery condition to cholesterol level is an information channel. Likewise some geologic tests are information channels probabilistically linking geologic structure to the presence of an oil deposit. A household smoke alarm is a channel from the presence or absence of fire to the possible sounding of a loud mechanical squeal. Opinion surveys can also be regarded as channels, with a survey result giving probabilistic information about

FIGURE 5.1 An information channel. The channel transmits information about the random variable X to the random variable Y.

55

attitude of the general population. The concept of a channel is a prefound construct, and it is used in several chapters of this book.

5.1 Discrete Channel

A **discrete channel** is defined with reference to a finite number of input and output events or symbol sets X and Y, respectively. These might be, for example, the binary alphabet 0, 1, the normal English alphabet, medical conditions and test results, or the amount that the price of a stock changed during the week.

The channel is defined by a set of conditional probabilities $p(y|x)$ for all $x \in X$ and $y \in Y$. The conditional probability $p(y|x)$ is interpreted as the probability of y occurring at the output when x is the input to the channel. The conditional probabilities are, in the context of a channel, called **transition probabilities**. A channel with r input symbols and s output symbols is characterized by $s \times r$ transition probabilities.

A channel defined as above is called a **discrete memoryless channel** because there are a finite number of possible inputs and outputs and the probability $p(y|x)$ is assumed not to depend on previous input symbols.

A simple, but important channel is the **binary symmetric channel** (BSC), which is characterized by the graph shown in figure 5.2. This channel has binary 0, 1 input and output symbols, and the transition probabilities are $1 - p$ for an input symbol to pass directly through the channel to its corresponding output symbol, and p for an input symbol to be diverted to its opposite symbol. The channel is said to have a probability of error equal to p.

For a general discrete channel with symbol alphabets of large size, the graph showing the transition probabilities is more complex. A general channel with four input and four output symbols is shown in figure 5.3. The diagram implies that any input symbol may with some positive probability be received as any output symbol. In practice, the set of input and output symbols may be identical, such as A, B, C, D;

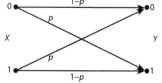

FIGURE 5.2 Binary symmetric channel (BSC). There are two input symbols and two output symbols. Proper transmission corresponds to horizontal progress across the graph. There is a probability p of an error, where a symbol is misdirected diagonally across the graph. This error probability is the same for each symbol.

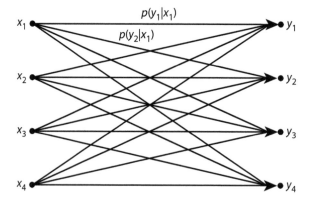

FIGURE 5.3 Discrete channel. Each input symbol may be received as any one of the output symbols. The channel is characterized by its set of transmission (or conditional) probabilities. In general, the number of output symbols may be less than, equal to, or more than the number of input symbols.

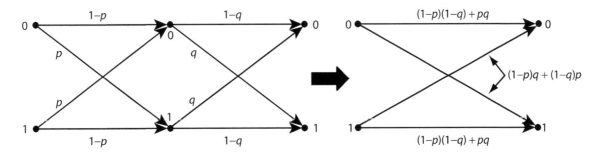

FIGURE 5.4 A channel with internal structure. The two BSCs combine to form a single BSC.

they may be different; they may differ in number; and the error rates may differ among symbol combinations.

A channel may have a complex inner structure with internal or intermediate variables. For example, one might transmit a code to a first station that then relays it to the final destination, each transmission being governed by a binary symmetric channel, but with different error probabilities. The net result is an overall BSC as shown in figure 5.4. Even more complex structures may arise when describing the path between an underlying medical condition and its several consequences before influencing an observed symptom. However, intermediate stages can be collapsed to produce net transition probabilities from input to output.

5.2 Conditional and Joint Entropies

Conditional entropy is defined relative to two random variables X and Y, which may or may not be related to a channel. The conditional entropy of Y given X is written $H(Y|X)$ and is interpreted as the average entropy in Y given knowledge of X.

The formal definition is built in stages. First, suppose that the specific value x_i of X is known. The entropy of Y given this knowledge is, according to the basic definition of entropy,

$$H(Y|x_i) = \sum_{y_j} p(y_j|x_i) \log \frac{1}{p(y_j|x_i)}.$$

For example, the entropy associated with the toss of a die Y is $H(Y) = \log 6 = 2.59$ bits. If we are told that the outcome is a high number (either 5 or 6), then the conditional entropy given that knowledge is $H(Y|\text{high}) = \log 2 = 1$ bit.

The overall conditional entropy is obtained by averaging with respect to all possibilities x_i in X according to their probabilities. The formal definition is as follows.

Conditional entropy. The conditional entropy $H(Y|X)$ is

$$H(Y|X) = \sum_{x_i \in X} H(Y|x_i)p(x_i). \tag{5.1}$$

In detail,

$$H(Y|X) = \sum_{x_i \in X} H(Y|x_i)p(x_i)$$

$$= \sum_{x_i \in X} \sum_{y_j \in Y} p(y_j|x_i)p(x_i) \log \frac{1}{p(y_j|x_i)}$$

$$= \sum_{x_i \in X} \sum_{y_j \in Y} p(x_i, y_j) \log \frac{1}{p(y_j|x_i)}. \tag{5.2}$$

The last line of the definition uses the **joint probability** $p(x_i, y_j)$, the probability that both x_i and y_j occur, which satisfies $p(x_i, y_j) = p(y_j|x_i)p(x_i)$.

Example 5.1 (The toss of a die). Let Y be the outcome of the toss of a die with six possible outcomes, and let X be the two events High (5 or 6) and Not high (1, 2, 3, or 4). The conditional entropy is

$$H(Y|X) = \frac{1}{3}\log 2 + \frac{2}{3}\log 4 = \frac{5}{3}$$

because after X is known there is a one-third chance that either 5 or 6 will remain as possible values of Y, and a two-thirds chance that 1, 2, 3, or 4 will remain. This conditional entropy is lower than the entropy of 2.58 bits of Y itself, and this is a general property, as stated by the following lemma.

Lemma 5.1 Entropy reduction. $0 \le H(Y|X) \le H(Y)$.

This lemma is established in section 5.4. It states that the entropy of a variable Y is, on average, never increased by knowledge of another variable X.

Example 5.2 (Two dice). According to lemma 5.1, knowledge of X tends to decrease the entropy of $H(Y)$ to a lower value $H(Y|X)$. Suppose, however, that I try to guess the outcome of a toss of a die Y by observing the outcome of a different die X. We can safely assume that X and Y are independent; neither influences the other. This means that $p(y_j|x_i) = p(y_j) = 1/6$ for all x_i and y_j. It follows that $p(x_i, y_j) = p(x_i)p(y_j) = 1/36$.

We have

$$H(Y|X) = \sum_{i,j=1}^{6} \frac{1}{36} \log 6 = \log 6$$

$$= H(Y).$$

In other words, the entropy of Y conditional on a variable that is independent of Y is the entropy of Y itself. Knowledge of irrelevant events does not change entropy.

Example 5.3 (Binary symmetric channel). Suppose that for the BSC the input X and output Y both consist of the symbols 0 and 1. The conditional entropies of Y

given $x = 0$ or given $x = 1$ are

$$H(Y|0) = H(p) \equiv -[p \log p + (1 - p) \log (1 - p)]$$
$$H(Y|1) = H(p).$$

In this case, the overall conditional entropy is

$$H(Y|X) = H(Y|0)p(0) + H(Y|1)p(1) = H(p),$$

which is independent of the input probabilities.

Joint entropy. The joint entropy of the random variables X and Y is the entropy of the pair (X, Y).

If there are r possible x_i's and s possible y_j's, there will be $r \times s$ possible pairs in (X, Y). These pairs are characterized by the joint probabilities $p(x_i, y_j)$. Such pairs might consist of two consecutive letters of the alphabet in English prose, as discussed in chapters 3 and 4. Or in the present context, they might be all possible input and output symbol pairs in a channel. The joint entropy can similarly be defined for any finite number of random variables $X_1, X_2, \ldots, X_n$.

There is an important relation between joint and conditional entropy.

Lemma 5.2. $H(X, Y) = H(Y|X) + H(X)$.

In words, this lemma says that the joint entropy of X and Y is the entropy of X plus the entropy of Y given X. Stated this way, it makes intuitive sense.

The proof of the formula is obtained by mechanical manipulation of the formulas for individual terms.

Proof:

$$\begin{aligned}
H(Y|X) + H(X) &= \sum_{(x_i, y_j) \in (X, Y)} p(x_i, y_j) \log \frac{1}{p(y_j|x_i)} + \sum_{x_i \in X} p(x_i) \log \frac{1}{p(x_i)} \\
&= \sum_{(x_i, y_j) \in (X, Y)} p(x_i, y_j) \log \frac{1}{p(y_j|x_i)} + \sum_{(x_i, y_j) \in (X, Y)} p(x_i, y_j) \log \frac{1}{p(x_i)} \\
&= \sum_{(x_i, y_j) \in (X, Y)} p(x_i, y_j) \log \frac{1}{p(x_i, y_j)} \\
&= H(X, Y). \quad \blacksquare
\end{aligned}$$

Example 5.4 (Conditional die). Consider again example 5.1. Y is the outcome of a die toss that can be any of the six numbers $(1, 2, 3, 4, 5, 6)$. X has the two possible values High (5 or 6) or Not high. $H(X, Y)$ can be computed two ways, using either the conditional entropy $H(Y|X)$ or $H(X|Y)$. According to the first way

$$H(Y|X) = \frac{1}{3} \log 2 + \frac{2}{3} \log 4 = \frac{5}{3}$$
$$H(X) = \frac{1}{3} \log 3 + \frac{2}{3} \log \frac{3}{2} = \log 3 - \frac{2}{3}$$
$$H(X, Y) = H(Y|X) + H(X) = \frac{5}{3} + \log 3 - \frac{2}{3} = 1 + \log 3.$$

According to the second way

$$H(X|Y) = 0$$
$$H(Y) = \log 6$$
$$H(X, Y) = H(X|Y) + H(Y) = 0 + \log 6 = 1 + \log 3.$$

5.3 Flipping a Channel

Perhaps surprisingly, the direction of a channel can be reversed (or flipped), as illustrated in figure 5.5. We hasten to remark that it is the probabilistic structure of information that is reversed, not the underlying physical process if there is one. For example, a small radio receiver cannot transmit back to the radio station. It is the information inferences that can be reversed. However, the channel can be flipped only when the probabilities of the input events or symbols are known.

An example (see exercise 2) is the relation between a person's height and weight. If we know someone's height, then on average we know something about his or her weight. There is a channel between height and weight. On the other hand, knowledge of a person's weight tells us something, on average, about his or her height, and this is the reverse channel from weight to height.

Once the input probabilities are specified, it is possible to calculate the corresponding output probabilities and treat them as input probabilities for the reverse direction. If the input is the random variable X and the output is the random variable Y, the output probabilities are found from the transition probabilities as

$$p(y_j) = \sum_{i=1}^{n} p(y_j|x_i)p(x_i).$$

Once the probabilities of X and Y are known, one may compute the **reverse probabilities** of the channel—the conditional probabilities $p(x_i|y_j)$. The appropriate formula, known as **Bayes' rule**, is

$$p(x_i|y_j) = \frac{p(y_j|x_i)p(x_i)}{p(y_j)}. \tag{5.3}$$

One way to remember this formula is start with $p(x_i, y_j) = p(x_i|y_j)p(y_j) = p(y_j|x_i)p(x_i)$, giving two ways to express $p(x_i, y_j)$. Dividing by $p(y_j)$ gives Bayes' rule.

These reverse probabilities are important from the receiver's viewpoint, for they allow the receiver to deduce the probability of input symbols based on the received output symbol. This is true of communication systems where the receiver wishes to deduce the original message, it is true of medical tests, where a physician wishes to

FIGURE 5.5 A channel and its flipped version.

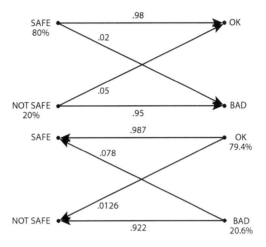

FIGURE 5.6 A water tester channel. As shown at the top, it has a 5 percent probability of reporting OK when actually the water is not safe. However, the flipped channel (shown at the bottom) shows that there is only a 1 percent chance that if the tester reports OK, the water is not safe.

make a diagnosis, and it is also true of investors in oil wells who from geologic and other data wish to estimate the probability that there is oil at a site.

Example 5.5 (Safe water). Suppose you are going out into the wild where the quality of the water is suspect. You have an inexpensive water tester that indicates either OK or BAD to inform you if the water is potable. However, the tester is not 100 percent reliable. The water itself is either SAFE or UNSAFE and the tester may give false readings of either kind: indicating BAD when the water is actually SAFE, and OK when the water is actually UNSAFE. Research on the tester has quantified the probabilities of these errors to be 2 percent and 5 percent respectively. You have learned that there is a 20 percent chance that water in this region is not safe. The water tester channel is shown in figure 5.6. The input is the actual condition either SAFE or UNSAFE and the output is the tester result OK or BAD. You are concerned about the 5 percent probability of one of the errors.

Your real interest, however, is the probabilities of the flipped channel, for you want to know the probability of the water being safe if the test says OK. You first calculate the output probabilities:

$$p_{OK} = .98 * .80 + .05 * .20 = .794$$
$$p_{BAD} = .02 * .80 + .95 * .20 = .206.$$

Then you calculate the reverse transition probabilities:

$$p(SAFE|OK) = .80 * .98/.794 = .987$$
$$p(UNSAFE|OK) = 1 - .987 = .0126$$
$$p(UNSAFE|BAD) = .95 * .20/.206 = .922$$
$$p(SAFE|BAD) = 1 - .922 = .078.$$

Hence there is only(!) about a 1 percent chance of water being UNSAFE when the test says OK.

5.4 Mutual Information

Mutual information is the information about one variable revealed by knowledge of another.

Mutual information. The mutual information of a random variable X given the random variable Y is

$$I(X; Y) = H(X) - H(X|Y). \tag{5.4}$$

The formula of the definition has a simple interpretation. The original uncertainty in X is embodied in the entropy $H(X)$. Observation of Y reveals something more about X. On average, the new entropy of X given knowledge of Y is the conditional entropy $H(X|Y)$. Knowledge of Y causes (on average) the entropy of X to drop from $H(X)$ to $H(X|Y)$, and hence the entropy is reduced by $H(X) - H(X|Y)$. This reduction in entropy measures, on average, the information supplied by knowledge of Y. Thus $I(X; Y)$ is the information about X transmitted by Y.

Lemma 5.3 (Properties of mutual information).

1. $I(X; Y) = H(X, Y) - H(X|Y) - H(Y|X)$
2. $I(X; Y) = H(X) - H(X|Y)$
3. $I(X; Y) = H(Y) - H(Y|X)$
4. $I(X; Y)$ is symmetric in X and Y
5. $I(X; Y) = \sum_x \sum_y p(x, y) \log \dfrac{p(x, y)}{p(x)p(y)}$
6. $I(X; Y) \geq 0$
7. $I(X; X) = H(X)$.

Proof: Item 2 is the definition. It is used to prove item 1.

$$
\begin{aligned}
I(X; Y) &= H(X) - H(X|Y) \\
&= H(X) + H(Y|X) - H(Y|X) - H(X|Y) \\
&= H(X, Y) - H(X|Y) - H(Y|X) \quad \text{from lemma 5.2.}
\end{aligned}
$$

Items 3 and 4 follow immediately from 1. Item 5 is proved by using 1 to write

$$
\begin{aligned}
I(X; Y) &= \sum_{x \in X} \sum_{y \in Y} p(x, y) \left[\log \frac{1}{p(x, y)} - \log \frac{1}{p(x|y)} - \log \frac{1}{p(y|x)} \right] \\
&= \sum_{x \in X} \sum_{y \in Y} p(x, y) \left[\log \frac{1}{p(x, y)} - \log \frac{p(y)}{p(x, y)} - \log \frac{p(x)}{p(y, x)} \right] \\
&= \sum_{x \in X} \sum_{y \in Y} p(x, y) \log \frac{p(x, y)}{p(x)p(y)}.
\end{aligned}
$$

To prove 6, item 5 is used to write

$$I(X;Y) = \sum_{x \in X} \sum_{y \in Y} p(x,y) \log p(x,y) - \sum_{x \in X} \sum_{y \in Y} p(x,y) \log p(x)p(y).$$

Notice that both $p(x,y)$ and $p(x)p(y)$ are probability densities on (X,Y). They both sum to 1 when summed over all x and y. Thus, by lemma 3.1 (chapter 3) $I(X;Y) \geq 0$. Item 7 follows from 2 and $H(X|X) = 0$. ∎

The symmetry of mutual information may not seem obvious at first. It says that knowledge of Y gives as much information about X as knowledge of X gives about Y. Both are $I(X;Y)$. However, studying examples will make this more intuitive.

Example 5.6 (Mutual die information). If I do not know the outcome Y of a die, the original entropy is $H(Y) = \log 6 = 2.59$ bits. If I am told X, which is whether the outcome is High (5 or 6) or Not high (1, 2, 3, or 4), the new entropy is on average $H(Y|X) = 5/3 = 1.67$ bits from example 5.1. The mutual information is the difference: $I(Y;X) = H(Y) - H(Y|X) = \log 6 - 5/3 = \log 3 - 2/3 = .918$ bits, since the entropy dropped from $\log 6$ to $5/3$.

This value may also be computed by reversing the roles of X and Y as $I(X;Y) = H(X) - H(X|Y) = H(X) - 0 = \frac{1}{3}\log 3 + \frac{2}{3}\log 3/2 = \log 3 - 2/3 = .918$, the same as before.

The fact that mutual information is always nonnegative implies that knowledge of Y always gives nonnegative information about X on average (but not necessarily in every instance). This result, by the way, establishes lemma 5.1, stated in section 5.2, that $0 \leq H(Y|X) \leq H(Y)$, since $I(X;Y) = H(Y) - H(Y|X) \geq 0$.

The relations between various entropy concepts can be represented on a **Venn diagram** as shown in figure 5.7. The complete circles represent the entropies $H(X)$ and $H(Y)$. The portion of a circle that does not overlap the other circle is the conditional

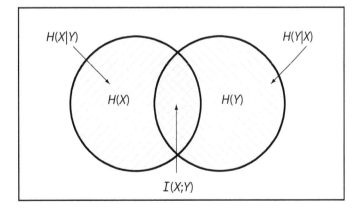

FIGURE 5.7 Venn diagram of entropy concepts. The circles represent the entropy of X and Y, respectively. The portion of a circle that does not overlap represents the conditional entropy. The region of overlap is the mutual information. Hence, according to the diagram, $I(X;Y) = H(Y) - H(Y|X) = H(X) - H(X|Y)$.

entropy. $H(Y|X)$, for example, is the part of the $H(Y)$ circle that does not overlap the $H(X)$ circle. The part that does overlap is the mutual information $H(Y) - H(Y|X)$. The figure implies that mutual information is symmetric, since the intersection can be computed also as $H(X) - H(X|Y)$.

Example 5.7 (The Oregon weather channel). In the state of Oregon the general weather condition G is either SUNNY, CLOUDY, or RAINY, with probabilities .5, .25, .25, respectively. The temperature condition T is either HOT, WARM, or COLD, and the probabilities of those temperatures are related to the general conditions as shown in figure 5.8.

Let us answer the following questions:

1. What is the entropy of the general weather condition G?

2. What is the entropy of the temperature condition T?

3. On average, how much information about the general weather condition is conveyed by knowledge of the temperature condition?

4. What is the conditional entropy $H(T|G)$?

5. On average, how much information about the temperature condition is conveyed by the general weather condition?

Here are the solutions:

1. $H(G) = H(.5, .25, .25) \equiv .5 \log 2 + .25 \log 4 + .25 \log 4 = .5 + .25 \times 2 + .25 \times 2 = 1.5$ bits per symbol.

2. The conditional entropy of T given any general condition in G is seen to be 1. Hence, the weighted sum of these defined by the probabilities in G is also 1. Hence, $H(T|G) = 1$.

3. It is easy to calculate the probabilities: $P_{HOT} = .25$, $P_{WARM} = .25 + .25 \times (.5 + .5) = .5$, $P_{COLD} = .25$. Then $H(T) = H(.5, .25, .25) = 1.5$ bits per symbol.

4. $I(T; G) = H(T) - H(T|G) = 1.5 - 1.0 = .5$ bits per symbol.

5. By symmetry we know that $I(G; T) = I(T; G) = .5$ bits per symbol.

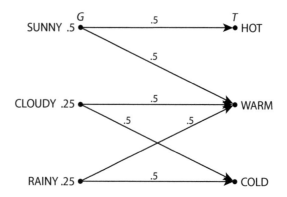

FIGURE 5.8 Oregon weather. The general weather G on the left is associated with temperature conditions T on the right.

5.5 Capacity*

Shannon's deepest and most surprising results explicitly address the problem of communicating reliably through channels subject to error. We recall that before his work, experts believed that reducing the error rate toward zero implied that the rate would go to zero as well. Shannon changed that conclusion. He showed that to every communication channel there corresponds a capacity, which is a rate at which reliable communication is possible. The definition of that capacity is straightforward, following from the concept of mutual information.

Capacity. The capacity of a channel is the maximum possible mutual information that can be achieved between input and output by varying the probabilities of the input symbols. Mathematically, if X is the input of the channel and Y is the output, the capacity C is

$$C = \max_{\text{input probabilities}} I(X, Y). \tag{5.5}$$

The mutual information about X given Y is the information transmitted by the channel per symbol on average, and this depends on the probability structure. The input symbol probabilities can be adjusted by suitable coding of the underlying sources; the transition probabilities are fixed by the properties of the channel; and the output probabilities are determined by the input and transition probabilities. Therefore it is the input probabilities that determine the mutual information, and they can be varied by coding. The maximum mutual information with respect to these input probabilities is the channel capacity.

Example 5.8 (Capacity of BSC). For the BSC we know from example 5.3 that $H(Y|X) = H(p)$. Hence $I(X, Y) = H(Y) - H(Y|X) = H(Y) - H(p)$, and thus

$$C = \max_{\text{X probabilities}} \{H(Y) - H(p)\}.$$

The entropy $H(p)$ is fixed but $H(Y)$ can be varied indirectly by changing the input probabilities. The maximum possible value of $H(Y)$ is 1, and this is achieved when the probabilities of the two input symbols are each $1/2$. Therefore, the capacity of the BSC is

$$C = 1 - H(p).$$

Let us check some special cases. If $p = 0$, then $H(0) = 0$, and $C = 1$ consistent with the ability to reliably send one bit of information with each symbol. Likewise, if $p = 1$, again $C = 1$. The worst case is $p = \frac{1}{2}$, since then $H(p) = 1$ and hence $C = 0$. It is impossible for the receiver to deduce whether a 0 or a 1 was sent.

Example 5.9 (The Oregon weather). Consider the structure of example 5.7. As originally stated, this is not really a communication channel, since we cannot vary the input probabilities. However, just for fun, suppose that they can be adjusted. For this structure we found $H(T|G) = 1$, independent of the input probabilities.

Hence

$$C = \max_{\text{input probabilities}} H(T) - 1.$$

The maximum possible value of $H(T)$ would be $H(T) = \log 3$ if we could achieve equal probabilities in the temperature conditions. However, this is impossible (as we shall see shortly). Suppose probabilities α, β, and $1 - \alpha - \beta$ are assigned to the three input probabilities in G. The corresponding output probabilities can be easily found to be $\alpha/2$, $1/2$, $(1-\alpha)/2$, which shows that it is impossible to achieve $1/3$, $1/3$, $1/3$. The capacity is therefore

$$\begin{aligned} C &= \max_{\alpha} \left\{ -\left(\frac{\alpha}{2}\right)\log\left(\frac{\alpha}{2}\right) - \frac{1}{2}\log\frac{1}{2} - \left(\frac{1-\alpha}{2}\right)\log\left(\frac{1-\alpha}{2}\right) \right\} - 1 \\ &= \max \tfrac{1}{2}[H(\alpha) + 2] - 1 = \max \tfrac{1}{2}H(\alpha) \\ &= 1/2, \end{aligned}$$

where the maximum is achieved by $\alpha = \frac{1}{2}$. Hence the channel as originally presented operates at maximum capacity.

5.6 Shannon's Second Theorem*

The definition of capacity presented in the previous section follows logically from the formal concept of mutual information. However, that definition does not say what information is transmitted through the channel. Conceivably, if the capacity is, say .5 bits per symbol, the message may be garbled so that only half of it gets through, on average. Shannon's second theorem states that, in fact, it is possible to communicate almost perfectly at a rate equal to the channel capacity. This theorem is considered to be Shannon's greatest achievement. It is the answer to the perplexing problem of whether reliable communication can be achieved through unreliable channels.

Theorem 5.1 (Shannon's second theorem). Suppose a discrete channel has capacity C and the source has entropy H. If $H < C$, there is a coding scheme such that the source can be transmitted over the channel with an arbitrarily small frequency of error. If $H > C$, it is not possible to achieve arbitrarily small error frequency.

Shannon presented a schematic characterization of the theorem that, although not the basis of a formal proof, shows the concept underlying the theorem in a simple way. A version of his schematic is shown in figure 5.9 for the BSC.

As with Shannon's first theorem, the key idea is to consider long blocks of message symbols to define "meta messages." We go through the argument for the BSC channel where the basic symbols are 0 and 1 and one symbol is transmitted per second. In this case the capacity is $C = 1 - H(p)$ bits per second, where p is the probability of error.

If the message blocks have length T, there are a total of 2^T possible blocks. These blocks are represented by the dots on the left side of the figure. Likewise there are 2^T possible output blocks, represented by the dots on the right side.

When a particular block is sent, it may be corrupted by error, which occurs with a probability p per symbol. In a long block of length T, it is highly likely that about pT errors will occur. The number of different blocks with this many errors is equal to the number of ways that pT positions can be chosen from the T available positions. The

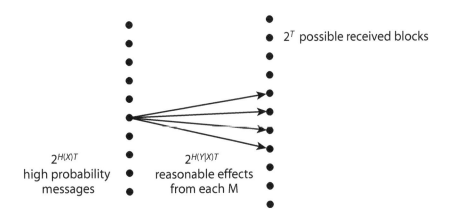

FIGURE 5.9 BSC channel representation. There are 2^T possible input blocks and 2^T output blocks. Associated with each input is a fan of $2^{TH(p)}$ highly probable output blocks. Hence for the fans to be disjoint we must use less than $2^{T-TH(p)}$ input blocks.

number of these is about $2^{TH(p)}$.[1] One can think of a message block as generating a fan of $2^{TH(p)}$ likely effects or output blocks, as shown in the lower part of the figure.

If the fans do not overlap, one can look backward from a received block and tell with certainty which message was sent. To insure that the fans do not overlap, the system must use fewer than the 2^T available message blocks. In fact, since there are 2^T possible output blocks and each message generates a fan of $2^{TH(p)}$ highly probable outputs, a condition for nonoverlap is that the number M of fans times the size of each fan must be less than 2^T. Thus $M2^{TH(p)} \leq 2^T$. Equivalently, $M \leq 2^{T(1-H(p))} = 2^{TC}$, where $C = 1 - H(p)$ is the capacity of the channel.

The rate R of information transmission in bits per second when M blocks are used every T seconds is $R = [\log M]/T$. If M satisfies the condition for nonoverlap, then $R \leq \log 2^{TC}/T = C$. If $R < C$, it is possible to arrange things so that the fans do not overlap, and hence it is possible to communicate reliably at this rate.

Random Codes

One of the most disquieting yet provocative aspects of Shannon's proof of the second theorem is his coding method. The schematic discussion of the second theorem, presented above, does not demonstrate that a suitable coding scheme exists; it only gives a condition on the number of codewords necessary to avoid overlap of the highly probable received blocks. Shannon showed that there is a suitable code, and he did so by selecting codes randomly and showing that at least one (actually many) of these would work. He did not show how to construct a specific code.

Shannon extended his theory to continuous channels, channels in which signals are functions of time rather than sequences of symbols. These results are every bit as important as those of this chapter, and are presented in chapter 21.

---

[1] The number of ways is actually $\binom{T}{pT} \equiv \frac{T!}{(T-Tp)!(Tp)!}$. By Stirling's approximation $T! \approx T^T$, and from this we find $\log \binom{T}{Tp} \approx -T[p \log p + (1-p) \log (1-p)] = TH(p)$. Hence, $\binom{T}{pT} \approx 2^{TH(p)}$.

5.7 EXERCISES

1. (Die rolls) A fair die is rolled. If the outcome is 1, 2, or 3, it is rolled once more; otherwise not. How much information about the number of rolls (one or two) is conveyed by telling whether the final outcome is odd or even?

2. (Height and weight channel) Consider the relation between height L and weight W shown in figure 5.10, indicating that tall people tend to be heavier than short people.
   (a) What is the entropy of L?
   (b) What is the conditional entropy $H(W|L)$?
   (c) Find the probabilities of the weight categories.
   (d) Flip the channel to find the reverse transition probabilities.
   (e) Find the mutual information $I(L; W)$, which is how much information, on average, about a person's height is given by his or her weight.

3. (Single-side error) A source X sends 0's and 1's with probabilities π and $1 - \pi$, respectively, to a receiver Y. The 1's are transmitted without error, but 0's are changed to 1's with probability p and remain as 0's with probability $1 - p$. What is the mutual information of the source and receiver?

4. (One error only) Consider a source X that has two symbols. These are coded in binary as 00 and 11 respectively. Each X symbol (two-bit word) has probability 1/2. When these two-bit words are sent through a certain channel, there is a probability p that one of the two bits will be changed by an error to the opposite binary bit before it is received at Y. For example, 00 may be received as 00 with probability $1 - p$, 01 with probability $p/2$, or 10 with probability $p/2$. There is no probability of two errors in a word. (Throughout the exercise use the notation $H(p) = -p \log p - (1 - p) \log (1 - p)$.)
   (a) Find $H(Y|X)$.
   (b) Find $H(Y)$.
   (c) What is the mutual information $I(X; Y)$?

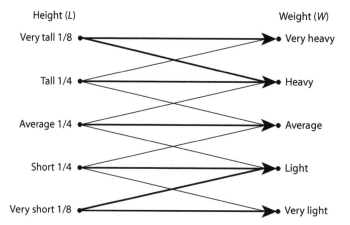

FIGURE 5.10 The transitions between categories are shown. The heavy lines correspond to probabilities of 1/2. The light lines correspond to probabilities of 1/4. The probabilities of the height categories are listed on the left.

5. (Function) Suppose that the source X is mapped into the source Y having values $y_j = g(x_i)$ for $x_i \in X$.

    (a) Show that $H(Y) \leq H(X)$. Under what conditions will there be equality?

    (b) Under what conditions is $H(X|Y) = 0$?

6. (Binary erasure channel) Consider a channel in which some bits are lost during transmission. The receiver knows which bits are lost but cannot recover them. The situation is shown in figure 5.11. Find the capacity of the channel.

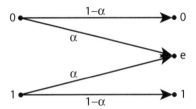

FIGURE 5.11 The binary erasure channel.

5.8 Bibliography

Shannon fully described the nature of channels, conditional entropy, mutual information, and channel capacity in his original paper [1]. This material is covered in the information theory textbooks cited in earlier chapters. The procedure for flipping a channel is widely used in other application areas (as illustrated in chapters 9 and 18).

Reference

[1] Shannon, Claude E. *The Mathematical Theory of Communication*. Urbana: University of Illinois Press, 1949.

6

ERROR-CORRECTING CODES

Errors inevitably occur during the transmission of information, arising in electronic circuits subject to heat or outside disturbance, during audio reception near background noise, while light waves pass through murky media, when blurry type is read, as automatic transcription fails, and in the course of many other modes of information transmission. Errors are ubiquitous, but it is possible to guard against them and to some extent even correct them automatically. How to most effectively accomplish this was the central issue that concerned Shannon when he embarked on the development of his theory of communication.

This chapter addresses the issue of code character errors, errors that cause a given character to be received as different from that which was sent. The frequency of these errors depends on the nature of the physical environment, and is often determined by the **signal-to-noise ratio**—the power that carries the signal divided by the power associated with noise. For example, if a binary message is sent electronically, with a 1 being represented by a voltage of one volt and a 0 being represented by zero volts, the receiver is likely to observe some intermediate value such as .82 volts. Using a nearest-neighbor criterion, the receiver might consider anything over .5 volts to be the binary signal 1 and anything less than .5 volts to be the binary signal 0. If the noise level is small on average, most transmissions will be interpreted correctly. However, there generally remains a chance that an intended 1 will be received as .47 volts and hence interpreted incorrectly as a 0. The probability of such an error is related to the magnitude of the signal-to-noise ratio.

There are two principal ways to increase the reliability of communication. One is to improve the reliability of the physical environment by increasing the signal-to-noise ratio. That is why, for example, audiophiles purchase high-power amplifiers, why noise filters are installed in telephone lines, why you want a quiet room when talking on the telephone—all to improve the signal-to-noise ratio.

The second way to improve communication reliability is to use a coding scheme that is tolerant of errors; a coding scheme that allows the receiver to automatically detect errors and correct them. This chapter is the study of such coding methods.

These methods are used in daily life, although usually we are unaware of their behind-the-scenes work. But nevertheless, we rely on them in data transmission, in codes for identifying or cataloging books, and in audio and visual equipment.

6.1 Simple Code Concepts

There are some simple techniques for improving the reliability of codes that are in common use. Their study introduces basic concepts that are used in more elaborate schemes.

Repetition Codes

If there are two symbols A and B in a source alphabet and binary coding is used, the straightforward way to send them is to assign, say, 0 to A and 1 to B. This is straightforward but has no error protection.

Protection can be provided easily with the **repetition code**, discussed briefly in chapter 2. In the simplest version, each code character is repeated, so that 0 is replaced by 00 and 1 by 11. If a single error occurs, the receiver might receive, for example, 01 and could detect the presence of an error. The receiver would, however, not be able to correct the error, since 01 could have been 00 or 11. More repetition improves reliability.

Check Sums

A more efficient means of error detection and correction is obtained by the use of parity check sums.

A **parity check bit** is a bit adjoined to a binary codeword that assures that the sum of the bits (in modulo 2 arithmetic) is a fixed value. For example, a bit can be adjoined to the codeword 0110 to make the sum zero. The new word is 0110$\underline{0}$, where the underlined bit is the parity bit. The sum of the individual bits is zero (in mod 2 arithmetic), or equivalently, there is an even number of 1's. If the codeword were 1110, the same scheme would produce the new word 1110$\underline{1}$ to keep the number of 1's even. This scheme is termed **even parity**. There is of course also the less often used **odd parity** that keeps the sum of the 1's odd.

Example 6.1 (Four symbols). Consider encoding four source symbols A, B, C, D with a binary code. These can be encoded by the simple binary code shown below. When the codewords are augmented by a parity bit, the code can detect a single error.

Symbol	Simple Binary Code	Code with Parity Check
A	00	000
B	01	011
C	10	101
D	11	110

The concept of a parity bit can be extended in several ways. The following example illustrates an extension to a code based on an alphabet of 10 characters.

Example 6.2 (ISBN numbers). The International Standard Book Number (ISBN) is printed on most books. Usually it is a 10-digit number assigned by the publisher. For example, the number may be

$$0\text{-}691\text{-}12418\text{-}3,$$

although the pattern of dashes differs somewhat from book to book. The first digit indicates the language in which the book is written (0 indicates English). The second three (or sometimes two) digits indicate the publisher (691 denotes Princeton University Press). The next six digits represent a book number assigned by the publisher. The final digit is a check digit, which can take the values 0, 1, 2, 3, 4, 5, 6, 7, 8, 9, X with X representing 10. The check digit is designed so that a certain linear combination of all digits is zero modulo 11 (or briefly, mod 11), meaning that the combination is a multiple of 11. Specifically,[1]

$$x_1 + 2x_2 + 3x_3 + 4x_4 + 5x_5 + 6x_6 + 7x_7 + 8x_8 + 9x_9 + 10x_{10} = 0 \quad \text{mod } 11.$$

That is, the sum on the left is an integral multiple of 11.

For example, for the ISBN number given above, we have

$$1 \cdot 0 + 2 \cdot 6 + 3 \cdot 9 + 4 \cdot 1 + 5 \cdot 1 + 6 \cdot 2 + 7 \cdot 4 + 8 \cdot 1 + 9 \cdot 8 + 10 \cdot 3$$
$$= 198 = 11 \cdot 18 = 0 \quad (\text{mod } 11).$$

Note that -10 is the same as 1 in mod 11 arithmetic (since $-10 = -1 \cdot 11 + 1$, and $1 = 0 \cdot 11 + 1$). Therefore one can solve for x_{10} as

$$x_{10} = -10x_{10} = x_1 + 2x_2 + 3x_3 + 4x_4 + 5x_5 + 6x_6 + 7x_7 + 8x_8 + 9x_9 \quad \text{mod } 11.$$

Hence for our example,

$$x_{10} = 1 \cdot 0 + 2 \cdot 6 + 3 \cdot 9 + 4 \cdot 1 + 5 \cdot 1 + 6 \cdot 2 + 7 \cdot 4 + 8 \cdot 1 + 9 \cdot 8$$
$$= 168 = 11 \cdot 15 + 3 = 3 \quad (\text{mod } 11),$$

as it should.

The remarkable property of the ISBN code is that it can detect a single error in any digit or an interchange of two digits. It is the coefficients in the check sum that permit the detection of an interchange. This is a practical code, for it has been found experimentally that single-digit errors and transpositions are the most common errors people make when transcribing ISBN numbers.

Rectangular and Triangular Codes

More than one check sum can be incorporated into a code, with each sum checking a different combination of code symbols, or applying different weights to the symbols. Generally, each check provides additional error detection and correction capability.

---

[1] In general, the term x mod n is equal to the remainder after x is divided by n. For example, 25 mod $11 = 3$.

```
1  0  0  1  1  1  0  0
0  0  0  1  1  1  1  0
0  0  1  1  0  0  1  1
1  0  0  1  0  0  1  1
0  1  1  1  1  1  0  1
1  0  1  0  1  1  1  1
0  0  1  1  0  0  1  1
1  1  0  0  0  0  1  1


1  0  1  1  1
1  1  0  1
1  0  0
0  1
1
```

FIGURE 6.1 A rectangular code. Each grayed bit provides a parity check for its respective row or column. The bit in the lower left corner provides a parity bit for both the first column and the last row.

An interesting class is made up of the **rectangular codes**, an example of which is shown in figure 6.1. The message symbols are the nongrayed symbols forming a rectangle. The grayed bits in the left column are parity bits assigned so that the sum of the corresponding row has even parity. Likewise, the grayed bits in the last row are assigned so that the sum of each column has even parity.

The rectangular code can correct one error no matter where it appears in the array. For example, if the error is in the third row and fourth column, the parity check of the third row and the fourth column will not be even, and hence these checks will indicate the position of the error. An error in one of the check bits can also be corrected.

A similar idea is contained in the triangular code illustrated below.

```
1  0  1  1  1
1  1  0  1
1  0  0
0  1
1
```

Here the parity bits are defined so that the sum of elements in both the row and column of the bit has even parity. For example, the bit at the end of the second row is 1, so that the sum of the second row and the fourth element of the first row is even. We leave it to the reader to see that any single error can be found and corrected.

6.2 Hamming Distance

Most early work on error-correcting codes concentrated on block codes (codes in which all codewords have the same length), and the basic foundation of this work is presented in the next few sections.

The **Hamming distance** between two codewords in a block code is the number of symbols by which the two words differ. For example, the words 11000010 and 10010010 are a distance two apart, since they differ in the second and fourth binary digit. Two words are identical if and only if the Hamming distance between them is zero.

Codewords for different source symbols must, of course, differ from each other by a distance of at least one; otherwise the code is singular. Codes that are resistant to errors consist of words separated by greater distances.

Most decoding procedures use a distance criterion to select the most likely intended codeword. Specifically, when receiving a word, the decoder searches for the valid codeword closest to the word received in the sense of having the smallest Hamming distance. If there is a single closest codeword, it is assumed to be the intended codeword. If there are ties for the minimum distance, the error is simply detected, but not corrected.

Example 6.3 (A short set of codewords). Consider the codewords for a four-symbol source shown below:

$$
\begin{array}{c|cccc}
s_1 & 0 & 0 & 0 & 0 \\
s_2 & 1 & 1 & 0 & 1 \\
s_3 & 0 & 1 & 1 & 0 \\
s_4 & 1 & 0 & 1 & 1
\end{array}
$$

Suppose that the sequence 0001 is received. To decode this we look for the word in the code table that is closest to it. This is the word 0000 for s_1, which is a distance of one from the received sequence. All other codewords are a greater distance than one from the received pattern.

Suppose, however, that 0100 is received. There are two closest codewords: 0000 (for s_1) and 0110 (for s_3), both at a distance of one. Hence, although an error can be detected, it cannot be corrected using the minimum-distance criterion.

Clearly, an important characteristic of a code is the **minimum distance** d between any two words. In the example above, this minimum distance is two, achieved by the distance between s_1 and s_3. In general, the minimum distance determines the degree to which a code is immune from errors.

To quantify the relationship between a code's minimum distance and its error tolerance, multiple error **detection** and multiple error **correction** are distinguished. For an integer e, the term **e-error detection** means that if at most e errors exist in a received word, the receiver can detect that the word has errors but cannot necessarily correct those errors. For an integer f the term **f-error correction** means that if at most f errors exist in a received word, the receiver can both detect and correct those errors using the closest distance criterion.

It is easy to see the difference between these two notions in the case of single errors. If a single error is made, it can be detected if the minimum distance between codewords is $d = 2$. This is because a single error causes the received word to be a distance one from the intended codeword, but there is no codeword that close. Hence the error is detected. However, this error cannot always be corrected when $d = 2$ because there may be more than one valid codeword at a distance one from the received word.

A minimum distance of $d = 3$ assures that the receiver can both detect and correct a single error. If the received codeword is a distance one from the intended word (due to the single error), it will be at least a distance two from every other valid codeword. Hence, there is a unique codeword closest to the received word and that indeed must be the word intended.

An example of a code where the distance between codewords is five is shown schematically in figure 6.2. It should be clear that up to four errors can be detected and up to two can be both detected and corrected.

This analysis is generalized by the following fundamental relationship.

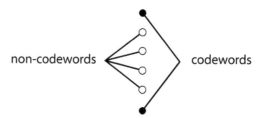

FIGURE 6.2 Codewords are separated by a Hamming distance of 5. An error in four or less binary digits cannot result in a new codeword, and hence such an error can be detected. However, at most two errors can be corrected by the majority rule procedure.

TABLE 6.1
Hamming Distance *d* Required for *e*-error Correction or *f*-error Correction.

d	2	3	4	5	6	7	8
e	1	2	3	4	5	6	7
f	0	1	1	2	2	3	3

Detection and correction distances. A code with minimum distance d is e-error detectable if and only if $d \geq e+1$. The code is f-error correctable only if $d \geq 2f+1$, as shown in the little chart in table 6.1.

Code Size

A binary block code of length n can consist of at most 2^n distinct codewords. However, because it is necessary to separate codewords by a distance greater than one to obtain error tolerance, practical codes have fewer words than this maximum. It is conventional to characterize a block code by the triple (n, M, d), where n is the length of the codewords, M is the number of words in the code, and d is the minimum distance of the code. For example, the code of example 6.3 in this notation is a (4, 4, 2) code. In some cases the d is dropped and the code is characterized by just the two numbers (n, M).

Good block codes are those that maximize M for given values of n and d. Much of coding theory is devoted to a development of methods for achieving this kind of efficiency.

6.3 Hamming Codes

A systematic treatment of parity checks for block codes was presented by Hamming in 1950.[2] Suppose we want a binary code of length n that will correct a single error in any codeword. We approach this by partitioning the n available bits into two groups: k are devoted to the message and q are devoted to parity checks. The structure is therefore as shown here:

$$\overbrace{\underbrace{x\ x\ x\ \cdots\ x\ x}_{k}\ \underbrace{x\ \cdots\ x}_{q}}^{n}.$$

If the code is able to correct a single error, the q parity checks must be able to locate it. Since the error can occur in any of the n places, or in no place, there must be a total of $n+1$ combinations of parity check combinations. In other words, a requirement is

$$2^q \geq n+1.$$

A popular combination is $q = 3$, $n = 7$, which leads to $k = 4$. The structure of a codeword is $x_1\ x_2\ x_3\ x_4\ x_5\ x_6\ x_7$, where x_1, x_2, x_3, x_4 are message bits and x_5, x_6, x_7 are devoted to parity checks.

---

[2] Independently proposed by Golay a year earlier.

A parity check is the sum of certain bits in the word, arranged so that the mod 2 sum is zero for an actual codeword. When applied to a received version of a word, a nonzero value of this sum signals that an error is present. If there are several checks, their pattern defines a **syndrome**[3] associated with the received version of the word, and this can be used to diagnose and correct the error.

A nice idea embodied in Hamming codes is that it is possible to construct the parity checks so that the syndrome, when its zeros and ones are interpreted as an integer in binary form, directly indicates the location of the error. The code with $n = 7$ will have three parity checks with values s_1, s_2, s_3 that define the syndrome, and this three-place binary number $s_1 s_2 s_3$, represents an integer between 0 and 7. Specifically, the syndrome s_3, s_2, s_1 defines the number $4s_1 + 2s_2 + s_3$. For example, the syndrome 110 is $4 + 2 + 0 = 6$, in the Hamming code this number gives the location of the error bit in the corrupted word or gives 0 if there is no error. For this to be true, it is required that

$$
\begin{array}{lll}
s_3 & \text{signals positions} & 1, 3, 5, 7 \\
s_2 & \text{signals positions} & 2, 3, 6, 7 \\
s_1 & \text{signals positions} & 4, 5, 6, 7.
\end{array}
$$

For example, 6 is signaled by $s_1 = 1, s_2 = 1, s_3 = 0$. Together the redundant bits x_5, x_6, x_7 must be defined to make

$$
s_3 = x_1 + x_3 + x_5 + x_7 = 0 \ \text{mod } 2
$$
$$
s_2 = x_2 + x_3 + x_6 + x_7 = 0 \ \text{mod } 2
$$
$$
s_1 = x_4 + x_5 + x_6 + x_7 = 0 \ \text{mod } 2.
$$

These represent three equations in the three unknowns x_5, x_6, x_7, and these equations can be solved using ordinary algebra modified to account for mod 2 arithmetic. The solution is

$$
\begin{array}{lll}
x_5 = x_2 + x_3 + x_4 & \text{mod } 2 & \text{(6.1a)} \\
x_6 = x_1 + x_3 + x_4 & \text{mod } 2 & \text{(6.1b)} \\
x_7 = x_1 + x_2 + x_4 & \text{mod } 2. & \text{(6.1c)}
\end{array}
$$

The top equation (for s_3) can be verified by the calculation

$$
\begin{aligned}
s_3 &= x_1 + x_3 + x_5 + x_7 \\
&= x_1 + x_3 + (x_2 + x_3 + x_4) + (x_1 + x_2 + x_4) \\
&= 2x_1 + 2x_2 + 2x_3 + 2x_4 = 0 \quad \text{mod } 2
\end{aligned}
$$

because $2x$ is identically zero (mod 2) for any binary number x.

There are 16 codewords in the resulting **Hamming code**, corresponding to the 16 words possible with four message bits. The minimum distance must be at least three since the code is single-error correcting. In fact the minimum distance is exactly three. Hence the resulting Hamming code is a $(7, 16, 3)$ code (often referred to as an $[n, k] = [7, 4]$ Hamming code). The 16 codewords are shown in the following table 6.2. To construct them, the first four bits (on the left) define the sequence of

---

[3] Syndrome, a medical term for a pattern of symptoms characteristic of a disease; here applied to a diseased word.

TABLE 6.2
The Sixteen Hamming Codewords.

0000000	0100101	1000011	1100110
0001111	0101010	1001100	1101001
0010110	0110011	1010101	1110000
0011001	0111100	1011010	1111111

binary numbers 0000, 0001, 0010, 0011, and so forth up to 1111. Three parity bits are adjoined on the right to each of these using the equations (6.1). For example, starting with 0001, equations (6.1a, b, c) give $x_5 = x_6 = x_7 = 1$. Hence the codeword is 0001111, the second word in the first column.

To see how syndrome decoding works, suppose the sequence 0101000 is received. The check sums (mod 2) are evaluated to find $s_1 = x_4 + x_5 + x_6 + x_7 = 1 + 0 + 0 + 0 = 1$, $s_2 = 1 + 0 + 0 + 0 = 1$, $s_3 = 0 + 0 + 0 + 0 = 0$. When arranged as a three-place binary number, the syndrome is $s_1 s_2 s_3 = 110$, which corresponds to six in binary. Hence there is an error, and it is in position six. The correct codeword is found by changing this bit, obtaining 0101010, the sixth word in the code.

There are other Hamming codes with larger values of n, all of which are single-error correcting. More importantly, the Hamming code structure suggests other directions for code development.

6.4 Linear Codes

The Hamming code discussed in the previous section can be viewed in terms of the **parity check matrix**

$$\mathbf{P} = \begin{bmatrix} 0 & 0 & 0 & 1 & 1 & 1 & 1 \\ 0 & 1 & 1 & 0 & 0 & 1 & 1 \\ 1 & 0 & 1 & 0 & 1 & 0 & 1 \end{bmatrix}. \tag{6.2}$$

The rows of this matrix correspond to the three parity checks in the Hamming code. For example, the top row defines the syndrome s_1. In matrix terms a valid codeword $\mathbf{c} = (c_1, c_2, \cdots, c_7)$ must satisfy

$$\mathbf{Pc}^T = \mathbf{0} \quad (\text{mod } 2),$$

where $\mathbf{c}^T$ is the column vector that is the transpose of $\mathbf{c}$ and $\mathbf{0}$ denotes a three-dimensional vector of zeros. Indeed, any $\mathbf{c}$ that satisfies that equation is a valid codeword since the parity checks are all satisfied.

This idea can be generalized to define other codes. Beginning with a matrix $\mathbf{P}$ having n columns and $n - k$ rows, one defines the code C as consisting of all binary words (rows) $\mathbf{c}$ of length n that satisfy $\mathbf{Pc}^T = \mathbf{0}$ mod 2. Note that if $\mathbf{c}_1$ and $\mathbf{c}_2$ are two words satisfying $\mathbf{Pc}^T = \mathbf{0}$ mod 2, then their sum (mod 2) also satisfies the equation and hence is a codeword. A code that satisfies this property is termed a **linear code**. Using a $\mathbf{P}$ matrix as above is one way to produce a linear code. Conversely, given a linear code, there is always some matrix $\mathbf{P}$ that generates it in this sense. Most of the important block codes are linear.

Example 6.4 (The short set of codewords). The codewords of example 6.3 are listed below.

$$
\begin{aligned}
c_1 &= 0 \quad 0 \quad 0 \quad 0 \\
c_2 &= 1 \quad 1 \quad 0 \quad 1 \\
c_3 &= 0 \quad 1 \quad 1 \quad 0 \\
c_4 &= 1 \quad 0 \quad 1 \quad 1
\end{aligned}
$$

These words comprise a linear code because, for example, $c_2 + c_3 = c_4$, $c_3 + c_4 = c_2$, $c_2 + c_2 = c_1$. By trial and error one finds the corresponding parity check matrix

$$
\mathbf{P} = \begin{bmatrix} 0 & 1 & 1 & 1 \\ 1 & 1 & 1 & 0 \end{bmatrix}
$$

that produces the code.

One important property of binary linear codes is that the minimum distance of the code can be determined relatively easily. The **weight** of a binary codeword is the sum of its characters (not in mod 2). Thus the weight of the codeword 1001010 is 3, since there are three ones. This definition is used in the following result.

Minimum distance criterion. The minimum distance of a binary linear code is equal to the weight of the nonzero codeword of least weight.

Proof: Suppose that the minimum distance is achieved by the distance between the codewords $\mathbf{c}_1$ and $\mathbf{c}_2$. Since the code is linear, the difference $\mathbf{c}_1 - \mathbf{c}_2$ (which is the same as $\mathbf{c}_1 + \mathbf{c}_2$ in binary) is a codeword. The weight of this word is the minimum distance. Clearly no other nonzero word can have smaller weight, for then its distance to zero would be less than d. ■

For example, now that we know that the short code of example 6.4 is linear, it follows that the minimum distance is two since c_3 is the nonzero codeword with minimum weight, and that weight is two.

6.5 Low-Density Parity Check Codes

These codes (abbreviated **LDPC codes**) were first introduced by Robert Gallager in his 1960 MIT Ph.D. thesis. However, because of the limits of computing power at that time, the codes were considered impractical, and they were all but forgotten. Now, however, with greater computing power available, LDPC codes have had a recent revival and are among the best-performing codes in low signal-to-noise situations.

LDPC codes are linear codes in which the parity check matrix is sparse, having very few nonzero elements. A **regular** (n, s, r) LDPC code is defined to be a block code of length n having a parity check matrix with exactly r 1's per column and s 1's per row and where r and s are small compared to n.

An example parity check matrix for an $(n, s, r) = (10, 4, 2)$ LDPC code is shown below, which happens to be an $(n, M, d) - (10, 2^6, 3)$ code:

$$\mathbf{P} = \begin{bmatrix} 1 & 1 & 1 & 1 & 0 & 0 & 0 & 0 & 0 & 0 \\ 1 & 0 & 0 & 0 & 1 & 1 & 1 & 0 & 0 & 0 \\ 0 & 1 & 0 & 0 & 1 & 0 & 0 & 1 & 1 & 0 \\ 0 & 0 & 1 & 0 & 0 & 1 & 0 & 1 & 0 & 1 \\ 0 & 0 & 0 & 1 & 0 & 0 & 1 & 0 & 1 & 1 \end{bmatrix}. \tag{6.3}$$

The special feature of Gallager LDPC codes is the manner in which they are decoded. The parity check matrix is viewed as involving each codeword bit in exactly r parity conditions. Likewise, each parity check involves exactly s codeword bits.

The simplest way to decode is by an iterative flipping method. First, the parity checks are all evaluated. Then each codeword bit is examined and is changed if it is involved in more than a fixed number of violated parity checks. Then the parity checks are reevaluated and the process continues until all parity checks are satisfied. Although this method is not guaranteed to correct all errors, it works well in many situations.

A more complex decoding procedure is based on the assignment of probabilities to each codeword bit being 0 or 1. These probabilities are determined iteratively by considering the information transmitted by other bits through the parity check equations.

Sparsity of the parity check matrix facilitates the decoding in both methods by limiting the computational chore of evaluating parity checks and of updating codeword bits associated with a parity check equation.

Gallager codes and the decoding methods have been extended to **irregular** LDPC codes where the parity check matrix is sparse but the number of nonzero elements in rows or columns may not be constant. The concept has also been extended to codes with alphabets other than binary. In some situations, the length n of a LDPC codeword is 1 million or more.

6.6 Interleaving

Errors sometimes occur in bursts, caused, for example, by sudden outside disturbances, defects in recording media, or outright interruptions. In such environments some codewords may suffer numerous errors while, still, on average the error rate is low. The effects of error bursts can be ameliorated by a process of interleaving that spreads out the characters of a particular codeword, sprinkling them between the characters of other words. The errors are then diffused among several codewords rather than concentrated in a few. Then a modest level of error protection in all codewords can work in concert to correct the burst. The necessary diffusion can be accomplished by an **interleaver**.

There are several versions of interleavers, but one of the simplest is illustrated in figure 6.3. In this interleaver, codewords of length four are read into a matrix row by row. The matrix is then read out column by column to produce new groups of length four. It is these groups that are transmitted and subjected to possible error. For decoding, the groups received are read into a corresponding matrix column by column

$$x_1 \ x_2 \ x_3 \ x_4$$
$$x_5 \ x_6 \ x_7 \ x_8$$
$$x_9 \ x_{10} \ x_{11} \ x_{12}$$
$$x_{13} \ x_{14} \ x_{15} \ x_{16}$$

$(x_1 \ x_5 \ x_9 \ x_{13}) \quad (x_2 \ x_6 \ x_{10} \ x_{14})$
$(x_3 \ x_7 \ x_{11} \ x_{15}) \quad (x_4 \ x_8 \ x_{12} \ x_{16})$

FIGURE 6.3 A simple interleaver. Codewords of length 4 are read in row-wise and read out column-wise. The receiver reverses this process. If the code has single-error correction capability, the interleaved system can correct a burst of up to four characters in length.

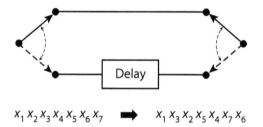

$x_1 \ x_2 \ x_3 \ x_4 \ x_5 \ x_6 \ x_7$ ➡ $x_1 \ x_3 \ x_2 \ x_5 \ x_4 \ x_7 \ x_6$

FIGURE 6.4 Delay interleaver. The switches change position with each character causing the characters to be interleaved.

and then read out row by row. If no error occurs, the original codewords appear. If there is a burst of four sequential errors and the code can correct one error in each codeword, the system employing the interleaver can correct the burst because the four errors will be spread evenly across the original codewords.

Many variations are possible. Larger matrices and more complex patterns of read-in and read-out can be devised. One popular method is to read the word characters one by one into a large matrix and then read them out in a more or less random order, an order known by both the sender and receiver.

Another type of interleaver employs a delay system; a simple version is shown in figure 6.4. Incoming characters are alternately directed to the upper branch or the lower branch, and likewise the output is alternatively taken from the upper or lower branch. The lower branch contains a delay of one-character duration so that a character entered at the left is not immediately available for output. The result is that the symbols become intermixed. The figure shows an example of how a stream entered on the left is transformed.

Actual delay interleavers are much more complex, involving several alternative paths with different delay lengths, some as long as 20 or more symbol periods.

6.7 Convolutional Codes

Convolutional codes are generated by **linear shift registers**, which are arrangements of delays, mod 2 summations, and feedback paths. An example of a third-order linear shift register is shown in figure 6.5. It has three registers, which delay the output from the input by one time unit (say a millisecond), and several taps and summations points. The value at each point is either 0 or 1, and the additions are carried out mod 2.

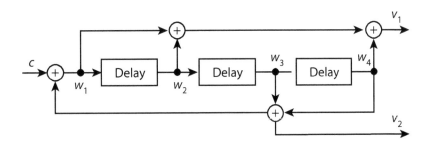

FIGURE 6.5 A convolutional encoder made from a third-order linear shift register.
Message bits cycle through the delay and feedback system, resulting in a complex output
pattern. Additions are carried out mod 2 so that all characters are either 0 or 1.

TABLE 6.3
Results of Convolutional Encoder.   As the binary bits move
through the delay and feedback system, a complex pattern of
outputs is generated.

Step	w_1	w_2	w_3	w_4	v_1	v_2
0	1	0	0	0	1	0
1	0	1	0	0	1	0
2	1	0	1	0	1	1
3	1	1	0	1	1	1
4	1	1	1	0	0	1
5	0	1	1	1	0	0
6	0	0	1	1	1	0
7	1	0	0	1	0	1
8	0	1	0	0	1	0
9	1	0	1	0	1	1

The arrangement in the figure serves as a rate 1/2 convolutional encoder because two
output symbols are generated for every one input symbol. These two output symbols
are sent side by side as (v_1, v_2). A convolutional encoder can be constructed to have
only a single output, but there is no net code redundancy in the resulting code. Hence,
single-output shift registers (or equivalently, convolutional encoders) are used only
as components of more complex encoding systems (as discussed in the next section,
for example).

The effect of a single input symbol in the encoder of figure 6.5 is easily traced.
Suppose specifically that a 1 enters the encoder at time zero, and that the 1 is followed
by a long sequence of 0's. The resulting sequence of values at the four points in the
center w_1, w_2, w_3, w_4 as well as the two outputs v_1, v_2 is shown in table 6.3.

There is a simple process for updating the table. The bits w_2, w_3, w_4 are equal to
w_1, w_2, w_3, respectively, from the previous step. The bit w_1 is the binary sum of w_3
and w_4. Then v_1 is the binary sum of w_1, w_2, and w_4; and v_2 is the binary sum of w_3
and w_4.

Notice that the line for step 8 is identical with that of step 1, and hence step 9 will be identical to step 2, and so on. Without new nonzero input, the system will cycle.

Convolutional codes can be decoded by using a similar convolutional process matched to the encoder. Today, convolutional encoders are used frequently because they have excellent error-correcting capabilities.

6.8 Turbo Codes

A tremendous advance in coding occurred in 1993 with the invention of turbo codes by Berrou, Glavieux, and Thitimajshima. The success of turbo codes stems from the fact that a given message symbol influences the transmission pattern over extremely wide intervals of symbols and in multiple ways. Subsequent decoding is complex, but for many applications the results are worth the effort.

A turbo encoder is formed by interleaving two convolutional encoders constructed as linear shift registers. One possible structure is shown in figure 6.6. The encoder of the figure is termed a rate $1/3$ turbo encoder because three code bits are formed for every input bit, so the redundancy is three. Turbo encoders can be constructed with other rates by using more or fewer convolutional encoders.

The two convolutional encoders may have relatively low dimension, but the interleaver typically has high dimension, which spreads the effect of a message character over many code positions and tends to separate the influence of the two convolutional encoders. The two convolutional encoders may or may not be identical in structure. As a simple example, they may both have the structure shown earlier in figure 6.5.

Decoding is carried out iteratively, where an approximation to the message is first formed and then gradually improved by subsequent passes of the result through a complex algorithm. The success of turbo codes seems to illustrate the fact that effective coding must, by nature, be extremely complex.

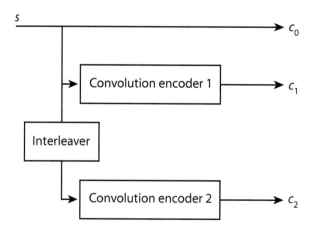

FIGURE 6.6 Turbo encoder structure. The raw bit symbols are sent through three paths. The three output symbols form three symbols of the code.

6.9 Applications

Error-correcting codes have had a great impact on modern technology, making possible communication over interplanetary distances, recording and playing of high-fidelity music, reliable wireless transmissions, and many other information services.

In many applications the transmission system first encodes the raw data with **source coding**. This is simply compression (such as Huffman coding) applied to the message data to reduce inherent redundancy. The resulting source-coded bit stream is then nearly random. This stream is next transformed by an error-correcting code that puts redundancy back in, but in the systematic way called for by the code.

Space Missions

Transmission of messages during space missions has been a major challenge. The source, on a small space probe for example, has low power, the transmission distances are enormous, and the signal must pass through the corrupting influence of the earth's atmosphere. Imagine trying to listen on Earth to a radio located on Mars powered by a small battery.

Photographs of another planet were first transmitted during the *Mariner 4* mission in 1965. Each photograph was divided into 200×200 pixels (as compared with the 400×525 pixels of commercial television). Each pixel was encoded into one of 64 possible brightness levels (equivalent to six bits). Therefore, the total number of bits per photograph was $200 \times 200 \times 6 = 240,000$. At a rate of 8.5 bits per second, it took eight hours to transmit a single photograph.

Subsequent Mariner missions obtained improved picture resolution. A **Reed–Muller** (32, 64, 16) code was used, which having $d = 16$ provides significant error correction capability. As a result, data transmission rates were increased to 16,200 bits per second.

Voyager missions began in 1977. A Voyager full-color image in digital form consisted of 15,360,000 bits. By this time, convolutional codes were found to be most effective at the low signal-to-noise ratios inherent in interplanetary communication. It was long recognized, however, that planetary images are highly redundant, and good compression algorithms (based on prediction) could achieve a compression factor of about 2.5. Use of compression was attractive, of course, since it could potentially improve transmission rates. However, decompression processes are highly susceptible to errors, a single error causing effects propagating over several pixels. The encoding system employed on Voyager missions therefore consisted of two parts: a convolutional code followed by a **Reed–Solomon code** that cleaned up errors introduced by the convolutional coder. This combination has proved to be highly effective and has since been used in many applications.

The *Cassini* was built in the mid-1990s and attained an orbit of Saturn in 2004. It also employed a convolutional code and sent beautiful pictures of Saturn to Earth. However, by the time *Cassini* reached Saturn orbit, turbo codes and low-density parity check codes were available. Some versions of LDPC codes achieve rates at near the Shannon capacity, with extremely low error rates. These codes are likely to be used in future space missions as well as in satellite television and mobile communication systems.

Compact Disc Players

The compact disc for music recordings is one of the most notable applications of coding theory because it is used by millions of people everyday, and the difference in quality compared to earlier techniques of phonograph records or tapes is strikingly apparent. Compact discs are widely known for their superb fidelity and lack of background noise. This clarity is largely due to error correction coding.

Two kinds of errors occur in a CD player: errors that occur randomly from time to time, and error bursts that occur in bundles caused by surface imperfections or errors in the optical reading device. Some surface imperfections are manufacturing defects. A standard CD holds about 650 megabytes, which is about 5 gigabits. So a manufacturing error rate of one per million results in about 5,000 errors. Other surface imperfections are due to fingerprints, scratches, and so forth.

The information on a CD is coded by interleaved $(n, M, d) = (32, 28, 5)$ and $(28, 24, 5)$ Reed–Solomon codes with interleaving delays of up to 27 characters. This interleaved code has excellent error-correction capabilities, for it is capable of correcting bursts of up to 4,000 sequential errors. If you experiment by scratching the surface of a CD with a sharp pointed object, you will find that there is no loss of fidelity. Indeed normally, after heavy scratching the player will stop altogether rather than deliver a scratchy sound. Similar codes are used in DVD players, bar codes, microwave links, and digital television.

Modems

Commercial voice band modems in 1962 operated at a rate of 2,400 bits per second. Gradually, the rate was increased 9,600 bits per second, which was then considered to be the maximum possible rate over standard telephone lines. However, the invention of trellis coding (a precursor to turbo coding) in the late 1970s made further gains possible, and rates of 14,400 and 19,200 were obtained. More recently, rates have climbed to 28,800, 31,200, 33,600, and 56,000 bits per second.

Frequency Hopping Systems

Some communication systems operate by periodically hopping from one base frequency to another (as discussed in chapter 21). Such systems allow several channels over the same waveband, for if a receiver is programmed to follow the hopping pattern of a certain sender, other patterns will cause little interference. The system is best if the hopping patterns have as little overlap as possible. An analogy with codes is clearly apparent. If the frequency bands are assigned numbers, say $0, 1, 2, \ldots, q - 1$, a hopping pattern is a sequence such as $(3, 6, 0, \ldots, 9)$ of q symbols. The separation between the patterns is their Hamming distance. Reed–Solomon codes have been used to construct efficient hopping patterns.

6.10 EXERCISES

1. (Rectangular code) Show that the parity bit in the lower-left-hand corner of the rectangular code is consistent in both directions. That is, if it should be a 1 to make the bottom row have correct parity, it should also be 1 to make the left column have correct parity.

2. (ISBN correction) Find the missing digit in the ISBN numbers below.
   (a) 0-385-49531-?
   (b) 0-201-1?794-2

3. (ISBN code)
   (a) Show that the ISBN code can detect an error that is either a transposition of two digits or a numerical error in a single digit.
   (b) Show that if the position of a single error is known, it can be corrected.
   (c) Show that the ISBN code is a (nonlinear) $(10, 10^9, 2)$ code.

4. (Syndrome parity bits) Derive the formulas for the parity bits x_5, x_6, x_7 of the Hamming [7, 4] code.

5. (Hamming code error) A message is sent using the [7, 4] Hamming code. The first word received is 0001111, which has no errors. Assuming that subsequent words are either correct or have at most one error, what are the correct versions of the following received words?
   (a) 0101101
   (b) 0100111

6. (Code shortening) Suppose a binary code of length n has M words and a minimum distance d—an (n, M, d) code. The set can be shortened by using only the words that agree in a certain position. For example, one may consider all the words in the code that have 0 in the second position. Taking this subset of words and dropping the selected position produces a set of words of length $n - 1$ but still with minimum distance no greater than d. Produce a set of eight words of length six by applying this technique to the first position of the Hamming [7, 4] code.

7. (Code lengthening) A code can be lengthened by one bit by adjoining a parity bit. If this is done, what happens to d and to the error-correcting and error-detecting properties of the code?

8. (Even-weight codes) An even-weight binary code of length n consists of all possible words of length n that have even weight (that is, have an even number of 1's).
   (a) How many codewords are there in an even-weight code of length n?
   (b) The **dual** of a code C whose words are of length n is the code $C^\perp$ consisting of all words of length n orthogonal to C. (A word w is in $C^\perp$ if and only if $w \cdot c = 0$ for all codewords c in C, where here $w \cdot c$ is defined as $\sum_{i=1}^{n} w_i c_i$. mod 2.) Find $C^\perp$ for the even-weight code C of length n.

9. (Linear code*) A certain linear code is defined by the parity matrix

$$\mathbf{P} = \begin{bmatrix} 1 & 1 & 0 & 0 & 0 & 1 \\ 1 & 0 & 0 & 0 & 1 & 1 \\ 0 & 1 & 1 & 1 & 0 & 1 \\ 1 & 0 & 1 & 0 & 0 & 1 \end{bmatrix}.$$

(a) Find the corresponding codewords. Hint: It may be useful to perform row operations on the matrix to put it in nearly triangular form.

(b) How many errors can this code detect?

(c) How many errors can this code correct?

10. (Team wagers) A series of four football games is to be played. Each game is played to conclusion, so that there is always a winner. You have the opportunity to participate in a betting exchange. To do so, you submit a list forecasting the winners of each of the four games.

(a) How many different lists must you submit to be sure that at least one list correctly forecasts at least three out of four winners?

(b) In each game one team is a home team and the other an away team. Hence a forecast can simply indicate H or A, and a complete list might be, for example, (H, A, A, H). Show a set of lists that satisfies part (a).

11. (Rate of LDPC) A binary block code that translates k message bits into codewords of length n is said to have rate $R = k/n$ (which is k bits of information for every n transmitted binary digits. Shannon's theorem states that for $R < C$ (where C is the capacity of the channel) and $\epsilon > 0$, there is a code of some finite length n that attains rate R with a probability of error less than ϵ.

(a) Show that parity check matrix (6.3) is singular (mod 2).

(b) What is the rate of the LDPC code defined by parity check matrix (6.3)?

12. (Dual code) Referring to exercise 8, find the dual code of the code with words 0000, 0110, 1001, 1111.

13. (The famous hat problem*) Recently there has been wide fascination among coding theorists with the **hat problem**: Suppose there are n people in a room, and each is given either a red or blue hat. The hat colors are determined by n separate coin flips. People can see everyone's hat except their own. Next, each person in turn must either guess the color of his or her hat or must pass. The group of n people will win $1 million if at least one person guesses correctly and no one guesses incorrectly. The group is able to agree on a strategy before the hats are assigned. What is the probability that the group wins, and how is that probability achieved? (The usual first reaction is that this probability is one-half—which is incorrect.)

(a) First consider $n = 3$. Suppose each person responds as follows: if the other two hats are the same color, then the person guesses the opposite color. If the other two hats are different, the person passes. Everyone passes after one person guesses. What is the probability of the team winning? Hint: Consider the number of possible hat color configurations, and the number of these for which the team will lose.

(b) What is the ratio the number of codewords in a Hamming code of length 3 to the total number of possible words of length 3?

(c) Suppose $n = 7$. Consider the strategy whereby each person determines whether there is a choice of his or her hat color that in conjunction with the others would form a codeword in an (n, M, d)-(7, 16, 3) Hamming code (using 0 for red and 1 for blue). If there is such a choice, the person guesses the opposite color; if not, he or she passes. What is the probability of winning in this case?

6.11 Bibliography

Excellent introductory treatments of error-correcting codes are found in [1], [2], [3], [4] (with a nice geometric interpretation of Hamming codes), and [5] (with a discussion of applications such as the Mariner missions). The original published paper on LDPC codes is [6]. Turbo codes were presented in [8]. Book treatments of modern convolutional, trellis, turbo, and low-density parity check codes are [7], [9], [10], and [11]. Reed–Solomon codes and their applications, including error correction coding in music CDs, are presented in [12].

References

[1] Roman, Steven. *Introduction to Coding and Information Theory*. New York: Springer, 1996.

[2] Cover, Thomas M., and Joy A. Thomas. *Elements of Information Theory*. New York: John Wiley and Sons, 1991.

[3] Welsh, Dominic. *Codes and Cryptography*. Oxford: Oxford University Press, 1988.

[4] Hamming, Richard W. *Coding and Information Theory*. Englewood Cliffs, N.J.: Prentice-Hall, 1980.

[5] Hill, Raymond. *A First Course in Coding Theory*. Oxford: Oxford University Press, 1986.

[6] Gallager, Robert. "Low-Density Parity-Check Codes." *IEEE Transactions on Information Theory* (1962): 21–28.

[7] Schlegel, Christian. *Trellis Coding*. New York: IEEE Press, 1997.

[8] Berrou, C., A. Glavieux, and P. Thitimajshima. "Near Shannon Limit Error-Correcting Coding and Decoding: Turbo Codes." *Proceedings of the International Conference on Communications* (1993): 1064–70.

[9] Vucetic, Branka, and Jinhong Yuan. *Turbo Codes: Principles and Applications*. Boston: Kluwer Academic Publishers, 2000.

[10] Blahut, Richard E. *Algebraic Codes for Data Transmission*. Cambridge: Cambridge University Press, 2003.

[11] Soleymani, M. R., Y. Gao, and U. Vilaipornsawai. *Turbo Coding for Satellite and Wireless Communications*. Dordrecht, The Netherlands: Kluwer, 2002.

[12] Wicker, Stephen, and Vijay K. Bhargava, eds. *Reed–Solomon Codes and Their Applications*. New York: IEEE Press, 1994.

SUMMARY OF PART I

Shannon introduced the concepts of **information** and **entropy** to express the idea that learning that a rare event occurred carries more information than learning that a common event occurred. The entropy of a source with events having probabilities $p_1, p_2, \ldots, p_n$ is defined as $H(p_1, p_2, \ldots, p_n) = -\sum_{i=1}^{n} p_i \log p_i$. The use of the base-2 logarithm is standard, and when it is used, the entropy has the units of **bits**. In many instances, the events are letters of the alphabet derived from a textual message.

Entropy plays an important role in a surprisingly large number of situations. It occurs naturally in each of the five parts of this text, motivating, characterizing, and setting fundamental limits on analytic and computational methods.

One application of the concept, and indeed the one that motivated Shannon's work, is the representation of events (such as letters in text) by codewords using zeros and ones as code symbols. **Shannon's first theorem** states that the average codeword length of a code, in symbols per event, must be greater than or equal to the entropy of the source.

The average codeword length can be as low as the source entropy, but this may require that coding be applied to long sequences rather than single instances of events. Coding and decoding may then become complex, and there may be a long delay between transmittal and final interpretation of corresponding codewords.

The minimum average codeword length for a source with a finite number of independent events is obtained by **Huffman coding**. However, this length is still greater than the entropy of the source except in special cases.

English text is highly **redundant** because letter frequencies are uneven, and there is a great deal of grammatical structure. Instead of the $4.755 = \log 27$ bits per letter that would apply if all letters (plus a space) were independent and of equal probability, the entropy of English is only about 1.5 bits per letter. This redundancy implies that lengthy bodies of English text can be compressed by a factor of about three by suitable coding. An especially elegant and practical class of compression methods work by building a dictionary of previously encoded short sequences of letters. When a sequence occurs a second time, a short dictionary reference is sent in place of the sequence itself. These **dictionary methods** are **universal** in the sense that the underlying probability structure need not be known in advance, and yet in the limit of long sequences they achieve the compression ratio predicted by entropy.

Shannon also introduced the notion of an information **channel**, which probabilistically transforms input events into output events. A given input event can produce one of several possible output events, each with its own probability. A channel might represent the probabilistic path from disease to medical symptom, from student ability to test results, from oil deposits to geological configurations, or from electrical signals to telephone output. An important method for analysis of channels is **flipping** with **Bayes's rule**. A flipped channel gives the probabilities of various inputs for each output.

Associated with any two random variables X and Y are the entropies $H(X)$ and $H(Y)$, and the **joint entropy** $H(X, Y)$. The **conditional entropy** is $H(X|Y) \equiv H(X, Y) - H(Y)$, interpreted as the entropy of X given Y. The **mutual information** of X and Y is $I(X; Y) = H(X) - H(X|Y)$, interpreted as the information about X revealed by knowledge of Y. For example, if X is the input of a channel and Y is the output, $I(X; Y)$ is the average amount of information about the input given by observation of the output. Mutual information is always nonnegative and symmetric; that is, $0 \leq I(X; Y) = I(Y; X)$.

Shannon defined the **capacity** of a channel as the maximum possible value of mutual information of input and output, where the maximum is taken with respect to the input probabilities. **Shannon's second theorem** states that it is possible to send information with arbitrarily good reliability at any rate less than the channel capacity. However, he did not supply a specific coding scheme that achieves this rate.

The general issue of designing codes for sending information reliably through channels that are subject to error is termed **error correction coding**, and a great deal of effort has been devoted to such design. Error-correcting codes are typically applied to messages that have first been coded efficiently with a compression algorithm such as Huffman coding. Error-correcting codes add back redundancy in a controlled manner so that the result is somewhat immune from errors. The simplest way to add redundancy is to repeat message symbols several times. A more effective method is to add **parity checks**. Several other sophisticated methods have been developed. The most effective methods, however, tend to be those, such as **convolutional**, **turbo**, and **low-density parity check codes**, that spread the influence of any particular input symbol over many disparate output symbols in complex ways. Decoding these codes is complex and must be delayed until the full effect of a particular symbol has been received. Effective **block codes** may be one million bits or more in length.

PART II

ECONOMICS

Strategies for Value

7

MARKETS

Information has both micro and macro manifestations. At the micro level are bits and bytes, codes and errors, capacity and compression. As shown in previous chapters, the concept of entropy provides a basis for analysis of many of these micro issues.

At the macro level are information products and services: books and movies, software and music, art and theater, telephone service and cable, consulting and research, insurance and guarantees. At this level, a high degree of organization and order is more important than randomness, and quality is not measured in bits, but more often in economic terms. It is products and services that people pay for.

From an economic perspective information products are fundamentally different from most other commodities because it is easy to produce them in large quantity from an initial version. In fact, in many cases the cost of producing additional copies is essentially zero. Economists say that the marginal cost of an additional copy is small.

Additionally, most information products can be used repeatedly. Books can be read more than once, music CDs can be played often, and software can be used everyday. Furthermore, most information products, including books, music, movies, and computer files, can be easily duplicated without authorization and passed on to other people. Yet the original creation of these products may entail considerable expense. How then, with all these difficulties, can such products survive in a free market? How can the creators be properly compensated?

We know that compensation does occur in practice, for these products and services are available in the market. Sometimes compensation rests on copyright protection, which essentially grants monopoly rights to the creator. There are other methods as well; but one may well ask whether existing methods of distribution and compensation are socially efficient. Perhaps the exclusive right granted by copyright leads to excessive prices, denying access to people who could benefit from the information. Or perhaps the opposite holds, and copyright laws do not provide enough incentive to produce some socially valuable materials.

Basic economic theory provides a framework for addressing these questions, allowing one to systematically trace economic value delivered to consumers, firms, and society as a whole. This chapter introduces this framework and applies it to some of the questions raised in the previous paragraphs. The analysis provides general conclusions, some of which may be surprising at first, and provides a stepping stone for analysis of the innovative methods that have evolved for marketing information products.

7.1 Demand

Consider a single individual contemplating the purchase of some item. Our interest is primarily directed toward information items, but here the item can be anything that is sold. Assume that the individual, referred to as a **consumer**, assigns a value w, termed the **willingness-to-pay**, to the item. This value is the maximum amount the consumer would pay to obtain the item. If the item is priced at or below w, the consumer will purchase the item. If the price is higher than w, he or she will not purchase it. This is the basic model of consumer behavior used in many economic analyses. We state it here.

Consumer rule. A consumer has a willingness-to-pay (WTP) w for an item. If the price p of the item satisfies $p \leq w$, then the consumer purchases the item. If $p > w$, the consumer does not purchase it.

Often it makes sense for a consumer to purchase more than one unit of the same item. This is true of tomatoes, cans of Coca-Cola, or bags of cement. In some cases it can be thought of as being true of books and CDs as well. For instance, someone may wish to get several identical CDs for gifts or several copies of a software package for a group.

When a consumer might possibly purchase more than one unit, he or she is considered to have a willingness-to-pay for each successive unit of that item. That is, there is a willingness-to-pay w_1 for the first unit, w_2 for the second, w_3 for the third, and so forth. Usually, successive values decrease since each one assumes that previous units have already been obtained.[1] Under this assumption, the willingness-to-pay for the second unit is less than the first, the third less than the second, and so forth. The consumer is therefore characterized by the decreasing sequence of willingness-to-pay values $w_1, w_2, \ldots$, as shown in figure 7.1. If the price per unit is p, the consumer will purchase the number i of units such that $w_i \geq p$ and $w_{i+1} < p$.

Now return to the case where a consumer will acquire at most one unit of the item, but consider several consumers, each with perhaps different willingness-to-pay values. If these willingness-to-pay values are rank-ordered from greatest to smallest and plotted, the result will again look something like that of figure 7.1. Hence the aggregate willingness to pay is similar to that of an individual who might purchase several units of the item.

---

[1] In some cases, the willingness-to-pay may increase after a consumer has experienced the product. This is why, for example, some firms offer low-cost trials.

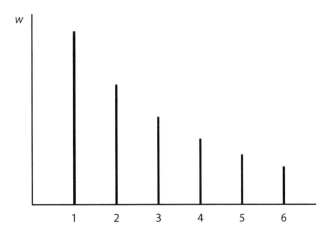

FIGURE 7.1 Willingness-to-pay. The willingness-to-pay usually decreases with each additional unit.

Continuous Approximation

When the item under consideration is a commodity, such as sugar, that can be divided into arbitrarily fine units, it is assumed that the consumer assigns a willingness-to-pay value for each tiny quantity increment Δq. Letting Δq go to zero, a **marginal willingness-to-pay function** $w(q)$ is defined. At quantity q, the willingness-to-pay for an additional amount Δq is $w(q)\Delta q$. Figure 7.2 shows a typical marginal willingness-to-pay function.

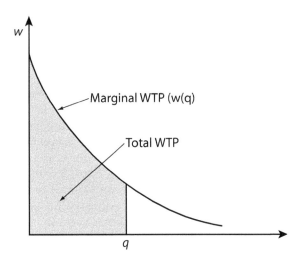

FIGURE 7.2 Marginal willingness-to-pay. The marginal willingness-to-pay function $w(q)$ is defined so that $w(q)\Delta q$ is the WTP for adding Δq to q. The total WTP for an amount q is the area under the $w(q)$ curve.

Again, a similar diagram applies when there are many consumers, even if each would acquire at most a single unit. In that case the scale of q is considered to be large compared to one unit, so that a single unit is essentially a tiny quantity. For example, the available quantity of a certain CD album might be measured in tens or hundreds of thousands, with each consumer buying at most one. The **total willingness-to-pay** for an amount q is the sum of the willingness-to-pay values for every unit acquired. In the continuous approximation, it is equal to the area under the marginal willingness-to-pay curve, as shown in the figure. In equation form,

$$\text{Total WTP}(q) = \int_0^q w(s)\,ds. \tag{7.1}$$

The relation between several individuals' marginal willingness-to-pay functions and that of the corresponding aggregate group is a bit complex, but there is a simple case. Suppose that n people in a group all have identical marginal willingness-to-pay functions $w_i(q)$. Then the marginal willingness-to-pay for the group of these n people is the function[2] $w_g(q) = w_i(q/n)$ because an amount q for the group translates into q/n for each person.

Demand Curve

If a fixed price p is established for all units of a commodity, then in aggregate the group of consumers will purchase the quantity q, satisfying $w(q) = p$, where $w(q)$ is the aggregate willingness-to-pay. If the price is varied from a high price downward, the total amount bought will vary as well, with more being purchased as the price is lowered. The relation $w(q) = p$ is termed the **demand curve**, since implicitly it gives the quantity purchased as a function of price.

Consumer Surplus

Suppose a fixed price p is set per unit and that $w(q)$ is the aggregate marginal willingness-to-pay function. As stated above, the amount q will be sold that satisfies $w(q) = p$. For most of the units purchased the marginal willingness-to-pay will be higher than p. Hence, consumers as a group pay less than it is worth to them. For instance, many consumers might be willing to pay $30 for a new CD; if it is priced at $12, those consumers get $18 in extra value. This extra value is termed **consumer surplus**. In terms of the marginal willingness-to-pay curve, the total consumer surplus is equal to the area under the curve and above the horizontal line at p, as illustrated in figure 7.3. As a general rule, **consumer surplus** is the difference between the total willingness-to-pay for a quantity and the amount paid.

---

[2]It is *not* equal to $nw_i(q/n)$. We have $\text{WTP}_g(q) = \int_0^q w_g(q')dq' = \int_0^{q/n} nw_i(q')dq'$. Differentiation gives $w_g(q) = w_i(q/n)$.

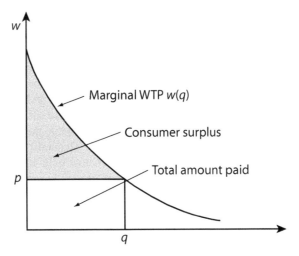

FIGURE 7.3 Consumer surplus. The extra value to consumers, above the price paid, is the consumer surplus.

7.2 Producers

Producers, too, are concerned with value. However, they are on the opposite side of the economic equation, seeking profit that depends on payments from consumers and on production costs. Accordingly, to analyze producers' actions, one must characterize their costs.

The cost of production of an item is a function of how many units are produced. A typical cost function for information products is shown in figure 7.4. There is **fixed cost** associated with merely getting ready for production. In book publication, for example, the fixed cost includes the cost of manuscript preparation, editing, art work, typographical composition, and press setup. It may also include marketing costs. Once production is begun, there are additional costs that depend on how many units

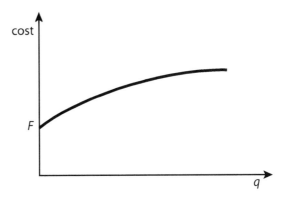

FIGURE 7.4 Cost function. Total cost increases with the quantity q. There is a fixed cost that must be paid at any nonzero production level. Then costs tend to rise smoothly.

are produced. For a book these are press charges, paper costs, bindery costs, and shipping. Total cost increases as additional units are produced.

The incremental cost associated with increasing production by a small amount is termed the **marginal cost**, and is usually denoted $m(q)$. Formally, if production is increased from the level q to $q + \Delta q$, the increase in cost is $m(q)\Delta q$.

The marginal cost function $m(q)$ can be viewed as defining the cost of each successive unit. In book production, the cost of the first copy is likely to be very large even after the fixed setup costs, while the cost of a second copy may be smaller; the cost of a third copy smaller yet. Generally the cost of an additional copy (the marginal cost) decreases as the number of units increases, finally reaching a relatively small value associated with large production runs and efficient distribution. Indeed, information products are typically characterized by high fixed cost followed ultimately by low marginal cost at high production levels. Decreasing marginal cost is reflected in figure 7.4 by the flattening of the cost curve as q is increased.

Marginal cost is the slope of the total cost curve $c(q)$. That is,

$$m(q) = c'(q). \tag{7.2}$$

This can be turned around to obtain

$$c(q) = F + \int_0^q m(s)\,\mathrm{d}s. \tag{7.3}$$

The fixed cost F is $c(0)$.

Decreasing marginal cost is shown more explicitly by the **marginal cost curve**, illustrated in figure 7.5. The total cost is the fixed cost plus the integral from 0 to q of the marginal cost, as expressed in (7.3).

Production processes for many traditional commodities, such as wheat, are characterized by marginal cost curves that increase rather than decrease, because at high levels of production less efficient resources (such as poor land) must be employed. This difference in the nature of marginal cost curves is one reason that information products are economically distinct from many traditional commodities.

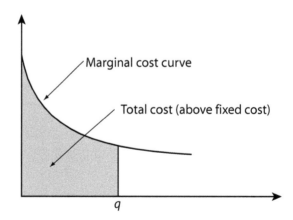

Marginal cost curve

Total cost (above fixed cost)

q

FIGURE 7.5 Cost. The marginal cost is the cost for an additional unit. The total cost is the sum, or integral, of all marginal costs plus the fixed cost.

Constant Marginal Cost

Cost functions of information products frequently can be approximated by **constant marginal cost**, written as $m(q) = m$. In that case the total cost function is

$$c(q) = F + m \cdot q. \tag{7.4}$$

Because production of many information products entails extremely low marginal cost, it is sometimes assumed as an approximation that $m = 0$. For example, the marginal cost of a CD is a fraction of a cent, and that of a hardback book is a few dollars. The marginal cost of an Internet message is practically zero.

Profit and Producer Surplus

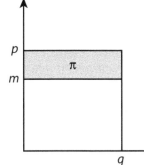

Producers are interested in profit: the amount of money received minus the cost of production. If the total quantity sold is q and all units are sold at price p, then the total amount received is $p \cdot q$. Hence the total profit is

$$\text{Profit} = p \cdot q - c(q).$$

If marginal cost is constant, as is typical of information products, then $c(q) = F + m \cdot q$, and profit is

$$p \cdot q - F - m \cdot q = (p - m) \cdot q - F.$$

The **producer surplus** π is the part of profit that does not include the fixed cost F. In the case of constant marginal cost the producer surplus is $\pi = (p - m)q$. This portion of the profit is shown as the shaded area in figure 7.6. It is the total amount received by producers beyond what they would receive at marginal cost.

Typically, both consumers and producers try to maximize their respective surplus measures, and of course these objectives are to a large extent conflicting.

FIGURE 7.6 Profit. Producer surplus is the shaded area representing the price minus the marginal cost times the quantity.

7.3 Social Surplus

Consumer surplus is value to consumers, and producer surplus is value to producers. The sum of these is termed **social surplus**; it is total value to all.[3] This definition of social surplus is general in the sense that it is independent of how transactions are carried out: through competition, monopoly, or individual negotiation. Social surplus is associated with each transaction and then summed, exactly the way the consumer and producer surpluses are associated with each transaction and summed.

The most important case is when all items are sold at a common price p. Consumer and producer surplus are then equal to the areas defined by the marginal willingness-to-pay curve and marginal cost curve. Figure 7.7 shows the situation when marginal cost is a constant m.

The quantity sold at price p is determined by the demand curve (identical to the marginal willingness-to-pay curve). Consumer surplus is therefore the area of the

[3] If payments are made to the government (as from taxes), these would be added to the definition of social surplus.

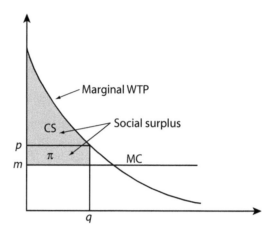

FIGURE 7.7 Social surplus. At a fixed price p, consumers purchase a quantity q such that p = marginal WTP; that is, p is determined by the demand curve. The associated consumer surplus is the area between that curve and the horizontal line at p. Producer surplus π is equal to the area of the rectangle of height $p - m$ and width q. Social surplus is the sum of these two.

shaded triangular-shaped region above the horizontal line at p and under the demand curve. Producer surplus $\pi = p \cdot q - m \cdot q$ is the area of the shaded rectangle. Hence social surplus is the total of the two shaded areas.

From the perspective of society, a transaction is valuable if the associated social surplus is positive. It does not matter whether the value accrues to consumers or to producers since (1) firms are owned by individuals and firms' profits are distributed to the owners, and (2) value to either consumers or producers can, at least in theory, be redistributed through tax mechanisms. On the other hand, from the same perspective, a product is socially valuable if the associated social surplus is greater than the fixed cost, for then the net gain to society is positive.

7.4 Competition

Most goods and services are sold through competition, but competition is an exceedingly complex process, taking many forms. In order to analyze economic issues, economists idealize the complex process, simplifying it while capturing its primary characteristics.

In an idealized form of competition, termed **perfect competition**, there is a single price p for all items of a given kind. This price is determined in ways that are difficult or impossible to model, but for purposes of analysis economists often simply assume that there a common price determined by market forces.

It is also assumed that each producer's and consumer's market participation is small compared to the size of the overall market, and there is no collusion. It follows that each party is a **price taker** in the sense that any individual's or firm's market action does not affect the price.

For our purposes, we also assume that all producers have identical cost structure, with constant marginal cost.

These idealizing assumptions are sufficient to draw a significant conclusion about the nature of the competitive price p.

Marginal cost pricing.   Under perfect competition, if a product is produced, its price is equal to its marginal cost.

Proof: As stated, marginal cost is assumed to be a constant m, although the conclusion is true more generally. Suppose first that $p < m$. Then no producer will produce because to do so would entail an immediate loss that could be avoided by not producing. Hence, it follows that $p \geq m$.

Suppose next that $p > m$. Then any producer could offer the item at a lower price $p' < p$ and virtually all consumers would move their business to that producer, increasing that producer's profit enormously. Other producers would react by lowering their price below p', and this would continue until $p = m$. ∎

Marginal cost pricing has important implications for information products. If price equals marginal cost, producer surplus is zero, which means that actual profit, accounting for the fixed cost, is negative. Hence, a typical information product sold in a competitive environment will garner strictly negative profit. Who would want to produce under these conditions?

Competition *is* viable for goods with increasing marginal cost, such as wheat, because profits can be made. But, when marginal cost is constant (or worse yet, decreasing), as is the case for most information products, the overall profit associated with marginal cost pricing is negative.

7.5 Optimality of Marginal Cost Pricing

Competition, with its attendant marginal cost pricing, is not desirable from a producer's perspective, but it is highly desirable from a social perspective. When a product is sold at a common fixed price p per unit, setting that price equal to marginal cost yields the maximum possible social surplus. This optimality result can be established algebraically or by the simple graphical analysis shown in figure 7.8.

The graphical analysis is explained in the figure caption. The algebraic derivation is also straightforward. If consumers select q such that $w(q) = p$, it follows that

$$\text{CS}(q) = \int_0^q (w(s) - w(q))\mathrm{d}s$$
$$\pi(q) = (w(q) - m)q$$
$$\text{SS}(q) = \int_0^q w(s)\mathrm{d}s - mq.$$

Setting the derivative of the social surplus with respect to q to zero produces

$$w(q) = m.$$

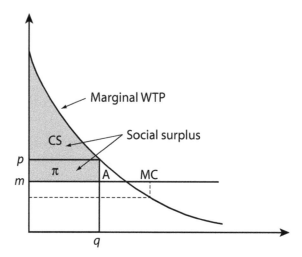

FIGURE 7.8 Maximum social surplus. If the price p in the figure is reduced to m, the associated social surplus will increase by the area of the small triangular region A. If the price is reduced further, to the level of dashed line, consumer surplus will increase, but producer surplus will be negative and equal in magnitude to the area defined by the dashed rectangle. The net effect is to reduce social surplus. The maximum social surplus is obtained at $p = m$, although producer surplus is zero at that point.

Since $p = w(q)$, the price is $p = m$, which is marginal cost pricing. We reach the following conclusion about optimal pricing.

Optimality of marginal cost pricing. In a single-price system, social surplus is maximized when price is equal to marginal cost.

It is clear that there is an economic dilemma associated with constant (or decreasing) marginal cost. Perfect competition is optimal from a social perspective, but yields negative profit to producers. How then are books, music, motion pictures, software, or the vast assortment of other information products to be sold?

7.6 Linear Demand Curves

For purposes of analysis, it is convenient to use demand curves—the marginal cost curves—that are downward sloping straight lines. A straight line is a reasonable approximation in many cases, and it occurs naturally in some situations.

Consider a product that is either purchased as a unit or not at all. Nobody purchases more than one unit. For instance, the product may be a book, a magazine subscription, or a software program. With reference to this product, a consumer is characterized by his or her single willingness-to-pay value. The collection of all consumers is accordingly characterized by the distribution of these WTP values.

Assume that there are many consumers; so many that it is reasonable to describe their WTP values as distributed continuously on a value axis. A special case of such a distribution is shown in figure 7.9(a). This figure implies that the WTP values are

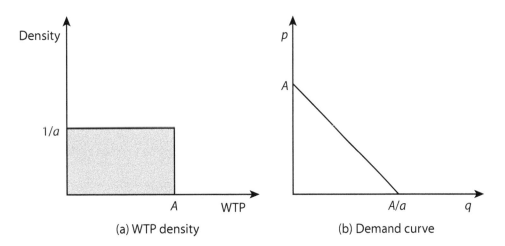

FIGURE 7.9 **Linear demand from uniform density.** A uniform density of WTP values implies a linear demand curve.

distributed uniformly between 0 and A. For instance, one-fourth of the people have WTP less than or equal to $A/4$. The total number of people in any ΔW range of WTP is $\Delta W/a$. The total number of people is A/a. For example, the WTP for a book may range between \$0 and $A = \$40$, and the total number of people considered might be $A/a = 40/a = 1$ million, and hence $a = 4 \times 10^{-5}$.

The uniform distribution of WTP values leads to the linear demand curve of figure 7.9(b). To see that, consider first the lowest price that can be set such that no one will buy the item. Clearly, that price is A, for no one has a WTP figure higher than that. Hence the demand curve intersects the price axis at $p = A$. Next, it is easy to see that the maximum quantity that can be sold (at $p = 0$) is A/a since that is the total number of people in the market.

The demand curve moves down linearly between the $p = A$ and $q = A/a$ units. This is because at, say $p = A/4$, one-fourth of the people will have WTP greater than or equal to p and hence, exactly $A/4a$ people will buy.

7.7 Copyright and Monopoly

Before the advent of the printing press, copying of manuscripts was tedious but rarely of legal concern. However, in 567 an Irish monk copied from a neighboring monastery the abbot's Psalter without permission. When the monk refused to return the copy, the abbot appealed to the king, who ordered the copy returned.

The printing press was introduced into England in 1476 and that same year a law was established that required printers to license books they printed. This law effectively granted a monopoly to printers, not to authors.

The first real copyright law was the Statute of Anne of 1710, which established the principles that the copyright belongs to the author and that the term of protection is limited.

These principles were embodied in the U.S. Constitution with the phrase, "the Congress shall have power...to promote the progress of science and useful arts, by securing for limited times to authors and inventors the exclusive right to their respective writings and discoveries." The term of copyright was originally set by Congress in 1790 as 14 years plus a possible renewal of 14 years. U.S. copyright law has subsequently been modified several times, but the latest version is embodied in the Sonny Bono Copyright Term Extension Act, which extended copyright protection to life of the author plus 70 years, and for works made for hire to 95 years (which many say was so that the Disney corporation could maintain control of Mickey Mouse).

Copyright law is intended to increase social welfare (that is, social surplus) by enabling authors and producers to make profit sufficient to cover the fixed costs of socially desirable products. It is fair to ask, however, to what extent such law resolves the dilemma inherent in information products: that competition leads to zero producer surplus and hence no incentive to expend initial resources for creation of an information product. Fortunately, there is a nice answer to this question (at least for the single-period case), which can be deduced by simple graphical reasoning or by elementary calculation.

A monopolist has the power to set the price of a product, rather than being forced to accept the price set by the market. An astute monopolist will therefore set price to maximize profit. To see what this entails, first assume, for simplicity, that demand is linear, as shown in figure 7.10. Also assume that marginal cost is a constant, m. The monopolist can select any price, and the corresponding quantity will be

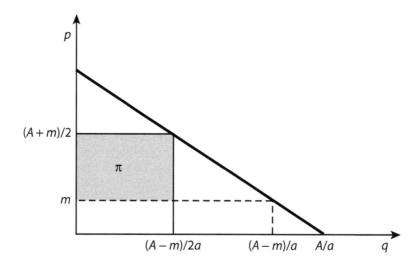

FIGURE 7.10 Monopoly pricing. To maximize profit when the demand curve is linear, a monopolist with constant marginal cost sets the quantity at one-half the quantity associated with marginal cost pricing. The producer surplus is then one-half of the maximum possible social surplus. The total social surplus at that price is three-fourths of the maximum possible. However, only half as many consumers are served as would be under perfect competition.

determined by the demand line. The associated producer surplus π is the area of the rectangular region bounded by the vertical axis, the marginal cost line, and the demand curve.

The quantity that maximizes profit is one-half the quantity associated with marginal cost pricing. This choice is shown in the figure.

Using calculus, it is easy to prove that the halfway point is best. Let the demand curve be $p = A - aq$, which means that A is the largest price point (with $q = 0$) and A/a is the largest quantity point (with $p = 0$). The quantity associated with perfect competition is the quantity where $p = m$, leading to $m = A - aq$, or equivalently, $q = (A - m)/a$.

At an arbitrary price p the producer surplus is

$$\pi = (p - m)q = (A - m - aq)q.$$

Setting the derivative of this to zero yields

$$A - m - 2aq = 0,$$

or equivalently

$$q = (A - m)/(2a),$$

which is the halfway point on the q axis to the quantity associated with marginal cost pricing. The corresponding p is $p = A - \frac{1}{2}(A - m) = \frac{1}{2}(A + m)$.

The producer surplus associated with this solution is exactly one-half the area under the demand curve and above the marginal cost line. Hence the producer surplus is one-half the maximum possible social surplus. The consumer surplus is half of the producer surplus. The total social surplus at this solution is therefore three-fourths of the maximum possible. The quantity produced is only one-half of what would be produced under perfect competition. In the case where each consumer buys at most one unit, this means that only half as many people are served as would be under perfect competition.

This result has important social implications. A product is socially desirable if the potential social surplus exceeds the fixed cost. From an economic perspective, society should encourage the production of such products. Competition does not provide that encouragement because marginal cost pricing yields zero producer surplus and hence negative net profit.

If producers are granted monopoly rights, they can earn producer surplus equal to one-half the potential social surplus. If fixed costs are less than this, producers will produce. However, if fixed costs are greater than one-half the potential social surplus, a monopolist will have no incentive to produce, even if the potential social surplus exceeds the fixed cost. Only products with fixed cost less than one-half the potential social surplus will be produced. The granting of monopoly rights is therefore only a partial solution to the problem of encouraging production of useful products. This result is summarized below.

Monopoly profit. When demand is linear and marginal cost is constant, a monopolist can collect producer surplus equal to one-half of the potential social surplus. Only one-half of the quantity that would be sold under marginal cost pricing will be sold under monopoly.

Nonlinear Demand

The results about monopoly can be generalized to demand curves that are nonlinear but convex as in figure 7.11: starting out nearly vertical and gradually becoming more horizontal. Many real demand curves are of this shape.

The optimal combination of price and quantity is some point on the curve, and the producer surplus is equal to the area of the rectangle defined by that and the marginal cost line. Suppose at that point a line tangent to the curve is constructed as shown in the figure. From the analysis for a linear demand curve, we know that the area of the rectangle is at most one-half the area of the triangle defined by the tangent line down to the marginal cost. It is easily seen that the area of this triangle is at most equal to the area under the curve above the marginal cost. Hence, producer surplus is no more than one-half the potential social surplus, just as in the linear case.

This general result implies that other methods should be sought to encourage the development of information products that are beneficial to society but which have large fixed cost.

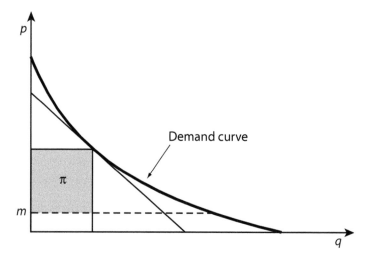

FIGURE 7.11 Monopoly profit. When the demand curve is convex, a monopolist can earn at most one-half the available consumer surplus, and the quantity sold is less than one-half that which would be sold under marginal cost pricing.

7.8 Other Pricing Methods

A variety of pricing methods have been proposed to encourage the production of socially beneficial products that could not otherwise withstand the pressure of competition. One alternative is for the government to fund the development of these products and make them available at their marginal cost. This funding can take the form of government laboratories, university research grants, or business tax credits for research and development. Two other useful methods are discussed in this section.

Regulated Monopolies

On the one hand, competition is infeasible as a practical matter when marginal costs are constant or decreasing. On the other hand, monopoly does not yield maximum social surplus. An alternative lying between these two extremes is the **regulated monopoly**. In this arrangement a regulating agency explicitly sets price to maximize social surplus subject to the constraint that fixed costs are recovered.

The regulated solution reduces the price to the point where the producer surplus is just equal to the fixed cost F. This maximizes consumer surplus while providing an incentive to produce.

Regulated monopolies are common arrangements for postal services and energy production, but are rare in information products, partly because it is difficult to determine demand curves and costs.

Voluntary Payment

Some information products such as public television, church services, street performances, and some teaching are supported by voluntary contributions. To see how this method of support fits into the general framework, suppose for simplicity that marginal cost is zero and that all consumers of a product voluntarily contribute an amount equal to one-half of their willingness-to-pay. The resulting producer surplus is then equal to the shaded triangular region indicated on the diagram of figure 7.12. It is easy to see that this producer surplus is one-half of the maximum possible social surplus. This is exactly the profit that a monopolist would obtain. However, with voluntary contributions, the total social surplus is the entire area below the demand curve. This is better than under monopoly, where the social surplus is only three-fourths this amount. Furthermore, everyone is served under a voluntary system. Hence, if the same profit can be generated under either system, the voluntary system is preferable to a monopoly.

In practice, of course, consumers may contribute less than half their consumer surplus, owing to the **free rider** effect: that is, without contributing themselves, consumers can obtain the benefit of everyone else's contributions. Low contributions considerably reduce the producer surplus, but the social surplus is still maximal provided that contributions are sufficient to cover fixed cost.

Other useful mechanisms for pricing information products are discussed in the next chapter.

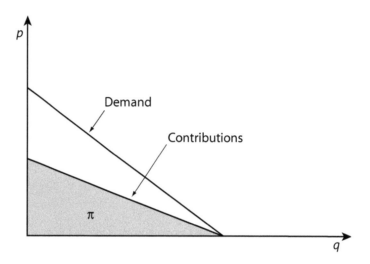

FIGURE 7.12 Voluntary contributions. The diagram assumes that marginal cost is zero. When contributions are voluntary, everyone is served, and total social surplus is equal to the maximum possible. If the average rate of contribution is one-half of a consumer's willingness-to-pay, the producer surplus is equal to that which would be obtained by a monopolist.

7.9 Oligopoly

There are situations intermediate between perfect competition (with many producers) and monopoly (with one producer). A market in which there are a limited number of producers is an **oligopoly**. The simplest is a **duopoly**, where there are two producers, and we shall study that form first.

Assume that there are two producers, each with marginal cost m. If they produce quantities q_1 and q_2, respectively, their corresponding producer surpluses are

$$\pi_1 = p(q_1 + q_2)q_1 - mq_1 \tag{7.5a}$$
$$\pi_2 = p(q_1 + q_2)q_2 - mq_2. \tag{7.5b}$$

The profit to each producer depends, therefore, not only on its own production level but also on that of the other producer. Neither producer can maximize its producer surplus without knowing what the other will do.

One way they might proceed is by **collusion**, agreeing on a total quantity and then dividing that among themselves. The best total quantity would be the monopoly quantity. The oil cartel OPEC attempts to operate this way. However, within a regulated economy such collusion is not allowed.

A more realistic assumption is that suggested by the economist Augustin Cournot in 1838. In a **Cournot equilibrium** each producer maximizes its own profit assuming that the quantity produced by its competitor is fixed, but fixed at what would be that competitor's optimal quantity. In other words, the producers maximize separately but simultaneously.

To work this out in detail, assume a linear demand function $p(q) = A - aq$. The producer surpluses (7.5) are then

$$\pi_1 = (A - a(q_1 + q_2))q_1 - mq_1 \tag{7.6a}$$
$$\pi_2 = (A - a(q_1 + q_2))q_2 - mq_2. \tag{7.6b}$$

The first of these is maximized by setting its derivative with respect to q_1 to zero. Likewise, the second is maximized by setting its derivative with respect to q_2 to zero. The result is the two simultaneous equations

$$A - m - 2aq_1 - aq_2 = 0 \tag{7.7a}$$
$$A - m - 2aq_2 - aq_1 = 0. \tag{7.7b}$$

These are easily solved by noting that symmetry implies $q_1 = q_2$. Thus

$$q_1 = \frac{A - m}{3a}, \qquad q_2 = \frac{A - m}{3a}.$$

The corresponding price can be found to be

$$p^* = \frac{A + 2m}{3}.$$

This solution is illustrated in figure 7.13.

Analysis of an oligopoly with n identical competing firms is similar to that for a duopoly. With the same demand function as above and marginal cost m, the optimal quantity for each firm is

$$q_i = \frac{A - m}{(n + 1)a},$$

with corresponding price

$$p^* = A - \frac{n(A - m)}{n + 1}.$$

Notice that $p^* \to m$ as $n \to \infty$, which is marginal cost pricing, corresponding to perfect competition.

The concept of a Cournot equilibrium was extended by the brilliant mathematician John Nash[4] to a basic result in game theory involving several players. The players each have payoff functions that may depend on the actions of all players. The Nash

---

[4]Nash became schizophrenic while quite young, but he was later awarded a Nobel Prize for his work in game theory. His life is reported in the best-selling biography and popular motion picture *A Beautiful Mind.*

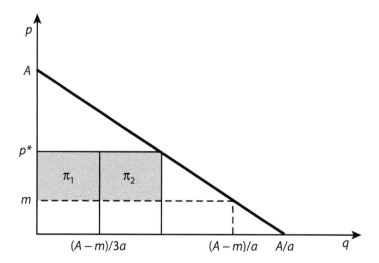

FIGURE 7.13 Duopoly. In a duopoly of identical producers, the equilibrium price p^* is lower than under monopoly. Hence the overall producer surplus is less than under monopoly, but social surplus is greater.

equilibrium, like its predecessor the Cournot equilibrium, is the result of simultaneous individual optimization.

Example 7.1 (Entry blocking). Suppose a monopolist in a certain product is challenged by a potential entrant to the market. How can the monopolist discourage such entry? One strategy is for the monopolist to lower price, so that entry will be less attractive.

The monopolist will not want to reduce his or her profit below that which would hold under duopoly, but that is a good benchmark for the strategy. Assume a linear demand function $p(q) = A - aq$ and zero marginal cost. The price p^* under duopoly would be $p^* = A/3$, and the quantity produced by each producer would be $A/3a$. Hence the producer surplus to the firm under duopoly would be $\pi_1 = A^2/9a$.

As a monopolist, the firm can achieve this same surplus at a q satisfying

$$\pi = p(q)q = (A - aq)q = Aq - aq^2 = A^2/9a.$$

The solution to this quadratic equation is

$$q = \frac{1}{2a}\left[A \pm \sqrt{A^2 - 4A^2/9}\right] = \frac{A[3 \pm \sqrt{5}]}{6a}.$$

Of the two solutions, the one with a plus sign is the one needed for blocking (why?). This gives $q = .873A/a$. The corresponding price is $p = .127A$ as compared to $p^* = .33A$ under duopoly. This low price may discourage the potential entrant.

7.10 EXERCISES

1. (Sharing the cost*) Two people with identical preferences for e-services decide that they will each purchase individual levels of service, then place a joint order for the total and share the cost equally (as is often done when dining out with friends). Specifically, each person has marginal willingness-to-pay of $w(x) = 10 - x$ for service level x. There is a fixed unit cost of $6.00 per unit level of service. The marginal production cost of the service is $5.00. (By symmetry, each person purchases the same amount x.)
   (a) If the two people did not share the cost, but each paid for his or her own order, how much would each purchase? What is the consumer surplus for each person? What is the profit to the service provider for each person? What is the total social surplus per person?
   (b) Under the sharing arrangement, how much will each purchase?
   (c) What is the consumer surplus of each person in this arrangement?
   (d) What is the profit to the service provider for each person, and what is the total social surplus per person?

2. (Superior product) Currently, the industry for delivering complex financial data is highly competitive with a marginal cost of M. A firm has just devised a new technology for delivering this data at a marginal cost of $m < M$. The firm can either enter the market directly on its own, or license the technology by charging other firms a fixed fee for each sale they make. Show that the firm can make at least as much by licensing as by entering the market on its own.

3. (Two markets) A book publisher sells in two distinct markets: A and B. The marginal cost of books is essentially zero. The demand functions in the two markets are different. They are $p_A = 600 - 3q_A$ and $p_B = 400 - 2q_B$, respectively.
   (a) If the publisher uses the same price in each market, what is the effective demand function for the total? Restrict attention to $p_T < 400$. (Hint: it is of the form $p_T = T - tq_T$.)
   (b) Under the conditions of part (a), what is the total profit of the publisher?
   (c) Now assume that the publisher is able to discriminate by charging different prices in the two markets A and B. What is the maximum profit to the publisher?

4. (Taxes) Suppose that an information product is produced with zero marginal cost and sold with perfect monopoly power. Each consumer purchases either one or zero units of the product. The overall demand curve is $p = A - aq$. However, the government taxes the product an amount t dollars per unit. The tax is paid by the producer, so that p is the actual price paid by consumers.
   (a) Find the amount sold and the selling price.
   (b) Find the producer surplus, consumer surplus, and total tax revenue received by the government. Call the sum of these T.
   (c) How does T compare with the total social surplus S that would be obtained without taxes?

5. (Copyright term*) Recently the Supreme Court upheld Congress's right to extend the term of basic copyright protection to 95 years. There has been a great deal of debate about whether this serves the public welfare. This exercise suggests a (highly simplified) model of the issue.

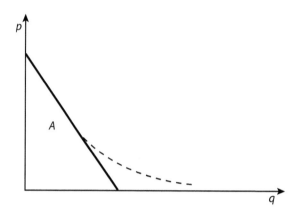

FIGURE 7.14 Yearly rate of demand for copyrighted material.

Suppose that the yearly rate of demand for a creative work is constant and follows the linear curve shown in figure 7.14. The area in the triangular region is A. The marginal cost of production is zero.

It follows that the yearly rate of producer surplus, if the product is actually produced, is $\pi = A/2$, and if the producer values future surpluses by discounting at the rate r, the total surplus to time T is

$$\pi = \int_0^T \pi(t)e^{-rt}dt = \frac{A}{2r}\left(1 - e^{-rT}\right).$$

The originator will produce the item only if the producer surplus exceeds the fixed cost F of creation. Assume that over the vast assortment of information products with the demand rate of the figure, the cost F is uniformly distributed between 0 and some maximum M. Hence the probability that a given work will be created and marketed is proportional to the total available producer surplus π. Likewise, the expected value $\overline{F}$ of the fixed cost given that this cost is less than π is $\pi/2$. Argue that the expected net benefit to society is proportional to

$$B = \{(\text{SS over } 0 \le t \le T) + (\text{SS over } t > T) - \overline{F}\} \cdot \text{prob}(\pi \ge F).$$

(a) Assuming that society uses the same discount rate r, give an explicit expression for B (to within a constant multiple).

(b) Show that there is no finite time that maximizes the social objective.

(c) Suppose that the actual demand curve slopes outward at the lower end as indicated by the dashed curve so that the total area under the curve is ρA where $\rho > 1$. Find the optimal value of T.

(d) Find T when $\rho = 1.2$ and $r = .05$.

(e) What is the limit of T as $\rho \to \infty$ in terms of r? Compare with the original 14-year term set by Congress in 1790.

6. (Asymmetric duopoly) The demand curve for a certain product is

$$p = 10 - q.$$

Two firms operate as a duopoly in this market. Firm 1 has constant marginal cost $m_1 = 2$ and firm 2 has constant marginal cost $m_2 = 3$. If the firms operate as a Cournot equilibrium (each firm maximizing profit while assuming the other firm's output is fixed), what are their respective output levels q_1 and q_2?

7. (Cartel cheating) Suppose that a product has a demand curve $p = A - aq$ and the marginal cost is zero. Suppose also that a cartel of two identical firms controls this product.

    (a) If the two firms act together and divide the market equally, what is the maximum profit (ignoring fixed costs) that each firm can obtain, and the corresponding production quantities?

    (b) Suppose that one firm faithfully produces the quantity of part (a). Show that the other firm has an incentive to break the agreement by producing more than agreed to. What is the maximum profit this second firm can obtain?

8. (Convex demand*) Suppose the demand function for a certain information product is

$$p = 10 - 7q + q^2.$$

However, there is a maximum quantity $q = 2$ that the market can absorb. The product is manufactured with zero marginal cost.

    (a) What is the quantity that would be sold under perfect competition?

    (b) What is the quantity that would be sold if the firm had a monopoly in that product?

7.11 Bibliography

Most of the material in this chapter is included in the subject of intermediate microeconomics. Three texts on the subject (in order of increasing difficulty) are [1], [2], [3]. A theory of the economics of information, including the material on the advantage of copyrights, is presented in [4]. [5] is a biography of John Nash.

References

[1] Pindyke, Robert S., and Daniel L. Rubinfeld. *Microeconomics*. 2nd ed., New York: Macmillan, 1992.

[2] Varian, Hal R. *Intermediate Microeconomics*. New York: Norton, 1987.

[3] Luenberger, David G. *Microeconomic Theory*. New York: McGraw-Hill, 1995.

[4] Bell, Hanan S. "A New Approach to Incentives for Information Creation and Distribution." Ph.D. diss. Department of Engineering-Economic Systems, Stanford University, May 1989.

[5] Nasar, Sylvia. *A Beautiful Mind*. New York: Touchstone, 1998.

8

PRICING SCHEMES

reators of information products face considerable challenges when seeking compensation for their work. A free competitive market, coupled with low marginal (or copying) cost, leads to vanishingly small transaction profits that may not cover the fixed cost of production nor be sufficient to fairly reward the artists, writers, and designers who created the work. Yet these products are often valuable as gauged by potential social surplus. The challenge faced by the producer of an information product is to participate creatively in the market in order to extract a significant share of the social surplus potentially available.

Copyright and patent laws and voluntary contributions provide possible mechanisms for appropriate recompense, but as described in the previous chapter, these are partial solutions, suitable for some products but not for all. If creators relied solely on them, many socially worthwhile products would not be produced.

These observations invite exploration of sophisticated marketing procedures—special schemes—to increase social surplus and arrange for producers to garner much of this social surplus as producer surplus. This chapter discusses some methods that are used for this purpose.

8.1 Discrimination

From a producer's viewpoint the ideal selling technique is to discriminate by negotiating each transaction separately with knowledge of each customer's willingness-to-pay. The producer will then charge each customer the full willingness-to-pay, provided only that this is greater than the marginal cost. This procedure yields the maximum possible social surplus, and all of this surplus accrues to the producer. Figure 8.1 illustrates the rewards of this perfect discrimination procedure.

Perfect discrimination is sometimes approximated by public auctions or by perspicacious street vendors, but these methods are usually impractical for selling large quantities. Furthermore, some forms of discrimination are illegal. Nevertheless, perfect discrimination is a useful benchmark, and its consideration motivates a search for some, perhaps imperfect, method that captures significant surplus.

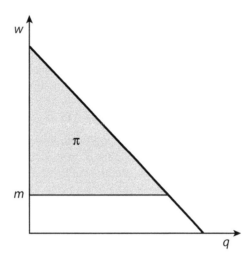

FIGURE 8.1 Perfect discrimination. By charging customers their full willingness-to-pay, a producer obtains maximum surplus.

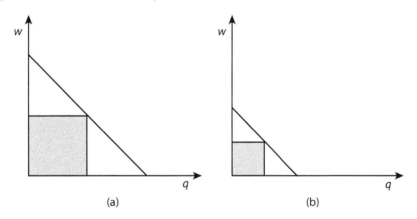

FIGURE 8.2 Serving two types. (a) Customers of type A with high WTP; (b) customers of type B with low WTP. Both are served if the producer can discriminate.

One important and sometimes practical form of discrimination is **discrimination by an observable factor**. An example is the offering of discounts to senior citizens, with "age" being the observable factor in that case. Many other factors are commonly employed. Individuals are often categorized into **types**, with each type being defined by a specific willingness-to-pay (WTP) level or interval. An observable factor (or factors) may reveal information about an individual's type.

The situation depicted in figure 8.2 illustrates the potential of this kind of discrimination. There are two types of individuals, A and B, and we suppose that their type can be easily identified (perhaps by their age or home address). Type As tend to have higher WTP than type Bs. The overall demand curves of the two types are the linear ones shown in the two parts of the figure. If a monopoly producer can discriminate between these types, establishing separate prices for each type and setting each as a monopolist would, the resulting producer surplus will be the sum of the two shaded areas in the figure. If, however, the producer cannot discriminate, the best single price will be close to that of that of type A, and most type Bs would not buy. In this simple example, discrimination leads to greater social surplus, greater profit, greater consumer surplus, and greater numbers of individuals served. In other words, it is a win–win–win procedure.[1]

Examples of factors used to discriminate include the following:

1. Location. Prices for ordinary commodities (such as toothpaste or soda) are often higher at resorts, and part of the reason is that most people at resorts are more affluent than the average population, and hence have greater willingness-to-pay. Prices also vary in different sections of the world according to general cost-of-living indices.

2. Sex. Although it is not legal to explicitly discriminate by sex, it is commonly done implicitly. Women's clothing is frequently priced higher than men's for items of comparable cost of manufacture, for example.

3. Age. We mentioned senior discounts, but there are also child discounts for food, entertainment, and clothing. Student discounts are also available for many information goods such as software, theater and museum attendance, and books.

4. Business structure. Nonprofit organizations are often accorded discounts for goods and services in business-to-business transactions.

5. Saturday stay-over. Airlines offer heavy discounts to travelers whose travel includes a Saturday night stay at a destination. The stay-over is an observable factor, which is significantly correlated with nonbusiness travel.

As the last example illustrates, it can be highly profitable to find an observable factor that is correlated with an underlying characteristic distinguishing the willingness-to-pay of different customers.

8.2 Versions

Although discrimination by observable factors can be effective, it is more subtle, more practical, and often even more profitable to arrange things so that prospective consumers reveal their type through their own market actions. This general idea goes under the heading of the **revelation principle**. A marketeer using this principle realizes that customers seeking to maximize consumer surplus will act in a manner characteristic of their type, and the marketeer accounts for this when establishing the price system—designing it so as to maximize producer surplus. That is, the marketeer accounts for customer maximization when solving his or her own maximization problem.

One class of schemes based on this general idea is **versioning**, where the availability of two or more versions of a product allows customers of different types to select the version that is best for them and also provides more profit.

[1] However, type A consumers face a higher price than when there is no discrimination.

Examples of versions include the following:

1. Hardcover and softcover books. Usually there is a delay in release of the latter version of a book, so customers anxious for the book will purchase the hardcover edition, while others will wait for the lower-priced softcover version.

2. Real-time stock data and 20-minute delayed data. Serious day-traders are willing to pay a premium for real-time data, while lower prices attract other customers.

3. Movies in theaters versus videotapes or DVDs. This is similar to hardcover and softcover books.

4. First-class and coach air fares.

5. Professional and academic software. Some companies issue a discounted academic version of their software, which is actually more costly to man-ufacture because certain functions must be disabled to justify the lower academic price.

6. High- versus low-resolution photographs. People wishing professional pictures buy more expensive high-resolution pictures than casual users.

Example 8.1 (Two versions). Suppose there are two types of people, A and B, and they are equal in number (normalized to 1 each). Type As have a high willingness-to-pay for a certain product, and type Bs have a low WTP. Specifically the WTP values are $w_A = 9$, $w_B = 5$. Assume that the product is produced by a monopolist with zero marginal cost.

If the price is set high, at $p = 9$, only the type As will purchase it. Consumer surplus CS is zero, producer surplus π is 9, social surplus SS is 9, and only type As are served.

Suppose instead that the price is set at $p = 5$. Then As and Bs will both buy, and as a result CS $= 4$, $\pi = 10$, SS $= 14$, and everyone is served. This is better for everyone, and is a case where monopoly yields maximum possible social surplus.

Now suppose that a lower-quality version of the same product is introduced. The high- and low-quality versions are labeled H and L, respectively. Suppose that the willingness-to-pay values are $w_A^H = 9$, $w_A^L = 6$, $w_B^H = 5$, and $w_B^L = 4$, as shown schematically in figure 8.3.

As a first attempt at assigning prices to the two versions, consider $p^H = 9$, $p^L = 4$. If As buy H and Bs buy L, the producer surplus is $\pi = 9 + 4 = 13$, which is an improvement over the $\pi = 10$ achieved with one version. However, this solution is not in accord with the revelation principle because As will choose to buy L instead of H and obtain a corresponding consumer surplus of $6 - 4 = 2$ instead of 0.

The better solution is to set p^H low enough so that the type As will not migrate to L. The price scheme $p^H = 7$, $p^L = 4$ will in fact induce self-selection and maximize profit. (In the case where two CS values are the same, it is assumed the better product is chosen, since a small change in price would insure that.) At this solution CS $= 2$, $\pi = 11$, SS $= 13$, and everyone is served. Compared to the single-version situation, producers are better off, but consumers lose some surplus.

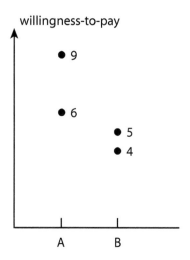

FIGURE 8.3 Versions. The willingness-to-pay figures are shown for two types of consumers and two versions of a single product. The upper values are for the high-quality version.

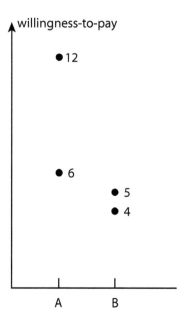

FIGURE 8.4 Revised versions. New willingness-to-pay figures are shown in which the high-end product is very attractive to type A consumers.

Example 8.2 (Two other versions). Suppose everything is the same as in the previous example except that the willingness-to-pay values are $w_A = 12$, $w_B = 5$. The monopolist will set the price at $p = 12$. In this case, CS = 0, $\pi = 12$, SS = 12, and only type A customers are served.

Now suppose that a lower-quality version is introduced, so that the willingness-to-pay values are $w_A^H = 12$, $w_A^L = 6$, $w_B^H = 5$, $w_B^L = 4$, as shown in figure 8.4.

Analysis similar to that of the previous example shows that the prices that yield the most profit are $p^H = 10$, $p^L = 4$. The two types will self-select, with As choosing H and Bs choosing L. The result is CS = 2, $\pi = 14$, SS = 16. Thus, CS, π, and SS have all increased.

8.3 Bundling

One of the most important pricing techniques is **bundling**, where several items or services are sold as a bundle at a fixed price. There are many familiar examples of bundling:

1. Software suites. Some software firms (such as Microsoft) sell individual products separately and also bundled together as a "suite" that is priced at a substantial discount over the sum of the prices for the individual components.

2. Nuts and bolts. These are frequently sold in packages of fixed amounts.

3. Automobile options. It is common to bundle together several accessories into an option package that is sold as a unit.

4. Travel packages. Air travel, hotel, and sightseeing are offered at a considerable discount relative to the sum of individual prices.

5. Round-trip airfares. The price of a round-trip ticket is usually less than the price of two one-way tickets. In fact, sometimes the round-trip fare is even less than the one-way fare.

6. Season tickets. Opera, sports events, and theater offer season tickets at prices that are well discounted over the same number of individual tickets.

7. Subscriptions. Subscriptions to magazines, newspapers, book clubs, premium television channels, and music clubs are all forms of bundling.

8. Economy-size packages. Large packages of laundry soap, large tubes of toothpaste, and economy sizes of all sorts of things are a form of bundling.

9. Amusement park tickets. Disneyland originally sold tickets to individual amusements, but later charged a fixed park entry fee that essentially bundles all amusements into a single package.

Given the right circumstances, bundling can significantly increase a firm's revenue, and as the preceding examples indicate, these circumstances are commonplace. Three prototypical situations are discussed in this section.

Complementary Preferences

Traditional bundling theory is based on the existence of a diversity of preferences among consumers. The idea is this: Suppose there are two types of consumers, A and B, having different preferences for two goods, 1 and 2. If the products are sold separately, the consumers may separate by their type and each purchase a single product. But if the two products are bundled together both types may purchase the complete package. An example is cable TV packages. Some people may not care for all of the channels in a package, but they buy the package if it contains enough channels that they do want and it is priced attractively.

Example 8.3 (Goods bundle). Suppose type A individuals have willingness-to-pay values $w_A^1 = 5$, $w_A^2 = 1$, for goods 1 and 2 respectively, while type B individuals have values $w_B^1 = 1$, $w_B^2 = 5$. These values are represented graphically in figure 8.5(a).

The two axes in the figure are the WTP values for good 1 and good 2, respectively. The dots represent the two potential customer types. Type A has WTP values of 5 and 1 and hence is represented by the dot at 5 on the first axis and 1 on the second. Customer B has WTP values of 1 and 5 and is represented accordingly.

Assume there are equal numbers of type As and Bs (normalized to 1 each). If the two goods are sold separately, it is apparent that the best prices are $p^1 = 5$, $p^2 = 5$, yielding CS $= 0$, $\pi = 10$, SS $= 10$.

If the two goods are bundled and sold for a single price, the best price is $p^{1\&2} = 6$, yielding CS $= 0$, $\pi = 12$, SS $= 12$. Thus the producer is better off. Consumer surplus remains zero, but consumers acquire both goods.

Bundling of this kind is most effective when customers have complementary preferences, with one group having strong preferences for one good, and another group having strong preferences for another good, as in the foregoing example. In fact with two goods it is possible, depending on the nature of consumer preferences, that bundling can increase by nearly 50 percent the producer surplus compared to the unbundled case. (See exercise 4.) Now consider the situation where preferences are not complementary.

Example 8.4 (No bundle). Suppose everything is the same as in example 8.3 except that the preferences are those of figure 8.5(b). Although the WTP values are different, they are aligned along the 45-degree diagonal through the origin. It should be apparent

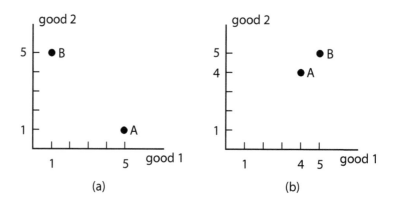

FIGURE 8.5 Bundling. (a) Sold individually, each good will be priced at 5. The bundle can be sold for 6. (b) If preferences are aligned, bundling is not effective.

that there is no advantage to bundling, and the producer should price for A, obtaining the producer surplus $\pi = 16$.

Averaging

People frequently purchase more items when bundled than they would buy if sold separately, even at the same unit cost. For example, magazine subscriptions are bundles (of written articles) that people purchase in great volume. A recently discovered principle of averaging at least partially explains this phenomenon, and the principle may become a useful tool for pricing information products. The concept can be explained in terms of an example.

Imagine a producer of technical articles written for a fairly broad audience, and imagine that these are offered for sale, for example on the World Wide Web, at a fixed price per article. The marginal cost is zero. The perceived value of such an article to a potential customer will typically vary from article to article. To be concrete, suppose that a customer's willingness-to-pay for such an article varies uniformly between $0 and $10. That is, if you are such a customer, then when you see a collection of these articles advertised, you are likely to assign different willingness-to-pay values to different articles: $1 to one of them, $9 to the next, $4 to the next, and so on, with these values being essentially uniformly distributed between $0 and $10. Figure 8.6 (a) shows the probability density of various WTP values. The figure shows the continuous version where any value between $0 and $10 can occur, but a discrete version works just as well.

This WTP pattern is the same for all customers, but the values assigned by different customers are independent. Your neighbor may be interested in different articles than you, but he or she also has the probability structure shown in the figure.

As shown in section 7.6, the corresponding demand curve is a straight line. In this case, the price p per article must satisfy $0 \leq p \leq 10$. If price p is charged, then on average a fraction $1 - p/10$ of potential customers will purchase the article. For example at $5, half of the customers will buy, at $8, 20 percent will buy. Demand falls linearly as price is increased. This demand curve is shown in figure 8.6(b).

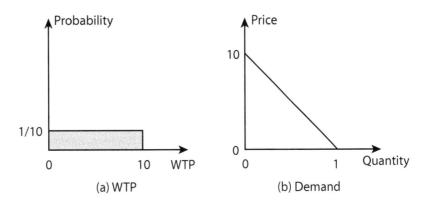

(a) WTP   (b) Demand

FIGURE 8.6 Variable willingness-to-pay. The WTP of various articles varies uniformly between $0 and $10 for each person.

The quantity is expressed in terms of the percentage of the total customer population; thus at $0, everyone will buy ($q = 1$).

From chapter 7, the monopolist solution[2] (which is applicable here because there is a single source for the articles) is to set the price at the midpoint; namely, at $5. At this price, half of the customers buy each article. The overall profit per consumer therefore averages $2.50.

Now suppose that articles are bundled into a large lot, perhaps in the form of a yearly subscription. The WTP of a customer varies over the individual articles just as before. However, when viewed as a collection, the average WTP is $5 per article. If a customer purchases the large bundle, he or she will value it on average at $5 per article, and hence to acquire the bundle that customer is willing to pay $5 times the number of articles in the bundle. In fact, every customer will place nearly the same $5 per article value on the bundle.

Because all consumers value the bundle identically, the producer can set the price at (or near) $5 and all customers will buy the bundle. The profit per consumer is now $\pi = \$5$ per article, which is twice what it was without bundling. Bundling has effectively made all customers appear equal on a per article basis. (See exercise 8 for further analysis of this method.)

Note that this method parallels Shannon's basic idea of bundling message symbols into metasymbols to achieve lower average code length per symbol.

Blocking Entry*

Bundling can be a powerful weapon of competition, blocking potential competitors attempting to enter the market with similar products. An excellent application of this is the area of software. For example, a large software firm that sells several application packages separately may benefit by bundling them into a "suite" at a greatly reduced price over the total price of individual programs. A competitor that produces one of these products could be a potential threat in head-to-head competition in the market for that single product, but the existence of the bundle can thwart the competitor's challenge.

To formalize this argument, suppose there are two products, 1 and 2. Consumers are characterized by a pair (w^1, w^2) of WTP values. Such a pair defines a point on a diagram that has WTP values for product 1 on one axis and for product 2 on the second axis. Assume that there are many potential customers and that their respective WTP values vary from $0 to $1, more or less randomly from customer to customer. In other words, the points on the two-dimensional diagram are scattered uniformly in a unit square, as shown in figure 8.7(a). Each point in the shaded region is a possible customer point, and they are evenly distributed.

Given this distribution of WTP values, a monopolist using separate prices for the two products will select the midpoint WTP values; namely[3] $p^1 = \$.50$, $p^2 = \$.50$. Half of the consumers will buy the first product and half will buy the second, although

---

[2] The solution can be worked out easily. The profit is $\pi = p(1 - p/10)$, which by elementary calculus is found to have a maximum at $p = 5$.

[3] The monopolist maximizes price p times quantity $1 - p$, which is $p(1 - p)$. This gives $p = \$.50$.

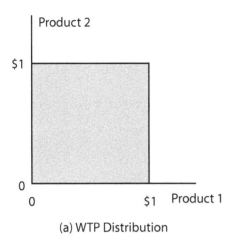

(a) WTP Distribution   (b) Sales regions

FIGURE 8.7 Pairs of points. (a) The WTP points are evenly distributed in the unit square. (b) A monopolist sets both prices at $.50, and sells product 1 to everyone to the right of the vertical at $.50 and product 2 to everyone above the horizontal at $.50. Customers in the darkly shaded region purchase both products.

the two halves are not totally distinct. This is made clear in figure 8.7(b). The total average profit (per possible consumer) is $\pi = \$.25 + \$.25 = \$.50$.

Suppose that a potential competitor (a challenger) plans to enter the market with a version of one of these products, say product 1. If the challenger sets a price of $\bar{p}^1 = \$.50 - \varepsilon$, for some small $\varepsilon > 0$, this challenger will (in theory) capture the entire market for product 1, leaving the original producer (the incumbent) with only product 2. The incumbent's profit would be cut in half to $\pi = \$.25$. The incumbent could of course cut the price of product 1 below that of the challenger, but this cycle of price cutting could continue until the price nearly reaches the duopoly price of $.33.

Consider instead the scheme where the incumbent sells the bundle of the two products at a price of $1, and does not sell the products separately. The challenger sells product 1 for $.50. Consumers buy either nothing, product 1, or the bundle; the decision among these being made according to the maximum of $0.00, WTP$_1$ − $.50, or WTP$_1$ + WTP$_2$ − $1. If only product 1 is chosen, then WTP$_1$ − $.50 > WTP$_1$ + WTP$_2$ − $1 which means WTP$_2$ < $.50. Hence the challenger is able to sell only to those customers having WTP$_1$ ≥ $.50 and WTP$_2$ < $.50. The incumbent will sell to everyone else whose WTP values sum to over $1, and whose WTP$_2$ > $.50. The result is shown in figure 8.8. The incumbent gets the darkly shaded area and the challenger gets the lightly shaded area. The incumbent's profit is now $\pi = \$.25 + \$.125 = \$.375$, which is greater than the value of $\pi = \$.25 + \$.33/3 = \$.36$ obtained without bundling in the face of a potential duopoly. Notice also that more products are sold in this bundled and partially competitive solution than in the monopoly solution.

A complete analysis allows both the incumbent and the challenger to adjust prices, and it turns out to be optimal for both to lower their prices somewhat. However, the incumbent's net producer surplus remains very close to the figure given above. (See exercise 7.)

Product 2

$1

Incumbent's sales

Challenger's sales

0

0                                $1 Product 1

FIGURE 8.8 Blocking. By bundling, the incumbent blocks a challenger's access to much of the market.

8.4 Sharing

Another characteristic of some information products is that they are easily shared. Sharing may involve copying, but it can also be physical sharing of a single copy. Library borrowing, video rentals, used-book sales, book club offers, software site licenses, and interlibrary loans are all examples of sharing. Historically, creators of information products have greatly feared the institution of such sharing, but usually those fears have been unwarranted. In fact, in the past, producers frequently benefited enormously from sharing. Libraries were initially feared by publishers, but libraries motivated millions of people to read, and book sales skyrocketed. Motion picture studios fought the introduction of videotape machines that could duplicate movies, but introduction of videotapes and DVDs produced large increases in motion picture revenues. The subject is complex, and it constantly changes as new technology facilitates opportunities to share information products.

A simple model can be used to investigate the economic consequences of sharing.

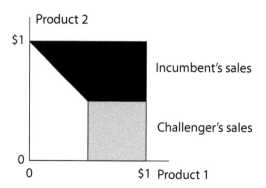

To fix ideas, consider the possibility of sharing a book among a number of potential readers indexed by $y = 1, 2, \ldots, n$. Each reader is characterized by a willingness-to-pay for the opportunity to read the book. Individual y has WTP $w(y)$, and, without loss of generality, the individuals are ordered so the $w(y)$ values range from highest to lowest as y goes from 1 to n. The marginal cost of book production is constant at m, and there is a fixed cost F.

If all books are sold at a single price p, this p will equal the WTP of the last customer to buy the book. More exactly, p will satisfy $w(y) \geq p > w(y + 1)$, where all consumers, $1, 2, \ldots, y$ buy the book. For example, if the consumers have WTP values [9, 8, 7, 6, 5, 4] and the price is $p = 7$, then consumers 1, 2, and 3 purchase the book.

The producer, acting as a monopolist, will find the number y of books to sell by solving

$$\max_{y} \; w(y)y - my - F. \tag{8.1}$$

Denote the optimal value of y by y_b. This problem and its solution is illustrated in figure 8.9.

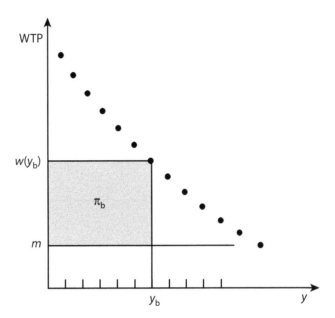

FIGURE 8.9 Monopoly solution. The optimal number of books sold is y_b when there is no sharing.

Now suppose that the consumers form book clubs, each consisting of k members. A club purchases at most one copy of the book and circulates it among its members. All members of the club contribute the same amount toward purchase of the book. There may be transaction costs as well (for example mailing costs if the members are some distance from each other), so suppose that each member pays transaction cost t in addition to a proportionate share of the book price.

A club purchases a book only if the net (after transaction cost) price p is no greater than the lowest WTP of a member of the club. For example, suppose again that the consumers have WTP values [9, 8, 7, 6, 5, 4]. If two-person clubs form as (9, 8) (7, 6) (5, 4) and there are no transaction costs, then at price $p = 12$ the first two clubs will purchase the book, since each member must pay 6 and this is the WTP of the last individual. If instead the clubs form as (9, 8) (7, 5) (6, 4) only the first club will purchase the book at $p = 12$.

Club formation is considered **efficient** if the WTP of every individual in a club that purchases the book is greater than the WTP of any individual in a club that did not purchase the book. If formation were not efficient, some member of a club that did not purchase the book would be willing to pay a member of a club that did purchase it to switch places. In the example suppose the clubs form as (9, 8) (7, 5) (6, 4) and $F = 10$. Then two clubs purchase but individual 4 (with WTP $= 6$) would be inclined to switch with individual 5 (with WTP $= 5$). We assume club formation is efficient.

According to the preceding framework, if y_c people read, then y_c/k books are sold. The consumer with the lowest WTP in clubs that purchase the book is individual y_c. That last individual has a net willingness-to-pay of $w(y_c) - t$ after accounting for the transaction cost. The proportionate book price per member can be no higher than this, which means that the book price itself can be no higher than $k[w(y_c) - t]$. The situation is shown in figure 8.10, which shows groups of three.

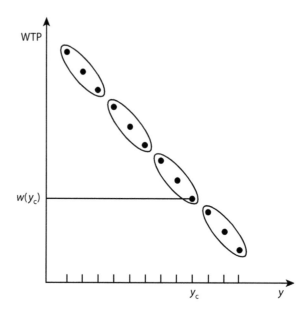

FIGURE 8.10 Clubs. Clubs consisting of three people are shown by the outlined dots. One book will be sold to each club with members 1 through y_c.

The producer's problem in the club situation is

$$\max_{y_c} (\text{number of books sold})(\text{price}) - \text{cost}$$

$$= \max_{y_c} (y_c/k)[k(w(y_c) - t)] - m(y_c/k) - F$$

$$= \max_{y_c} w(y_c)y_c - (t + m/k)y_c - F. \tag{8.2}$$

This problem has the same form as (8.1) except that the marginal cost is changed from m to $m' = m/k + t$. Here y_c is the number of people served, and the number of books sold is y_c/k.

If $m' < m$ (for instance if $t = 0$, $m > 0$, and $k > 1$), the following conclusions follow concerning the shared solution relative to the original solution:

(a) Producer surplus is increased.

(b) More people are served.

(c) Consumers pay less per reading.

(d) Consumer surplus is increased.

Conclusion (a) follows because with $m' < m$ the producer could use the old solution y_b, and that itself would increase profit. Conclusion (b) is easy to see since when marginal cost is lowered, a monopolist will increase supply. (For a formal proof, see exercise 9.) Conclusions (c) and (d) follow from (b).

This analysis is highly simplified, but it indicates that, after the market adjusts, sharing is not necessarily harmful to producers, and in fact producers as well as consumers may benefit from it.

8.5 EXERCISES

1. (Price discrimination) A firm with constant marginal cost sells in two markets with demand functions

$$p_1 = A_1 - a_1 q_1, \quad p_2 = A_2 - a_2 q_2.$$

   (a) Suppose the firm sets the same price in both markets, and selects that price to maximize the combined profit. How much will it sell?
   (b) Suppose the firm sets separate prices in the two markets, each set to maximize profit in its corresponding market. How much in total will the firm sell?
   (c) Show that the difference in profit in the two cases is

$$\Delta\pi = \frac{(A_1 - A_2)^2}{4(a_1 + a_2)}.$$

2. (Some versions) Suppose there are three consumer types A, B, C equal in number. A certain firm sells two products 1 and 2 and has zero marginal cost. The willingness-to-pay values are $w_A = (6, 2)$, $w_B = (4, 3)$, $w_C = (9, 3)$, and consumers purchase at most one version. What prices should the firm set for the two versions? What are the resulting CS, π, and SS?

3. (An attractive bundle) A firm sells two products, each of which has zero marginal cost. There are four customers (or four customer types in equal amounts) with willingness-to-pay combinations (40, 70), (70, 30), (10, 100), and (20, 40).

   (a) What price p_1 should the firm set for the first product to obtain maximum profit for that product?
   (b) What price p_2 should the firm set for the second product?
   (c) Suppose that the firm sells the two products together as a bundle with a fixed price. What price p_B should the firm set for the bundle?
   (d) Compare the total profits for the bundled and unbundled cases.

4. (High-profit bundling) Show that bundling two items can sometimes increase producer surplus by up to 50 percent as compared to no bundling.

5. (Extreme bundling) Suppose that a vendor sells articles for which all potential customers have willingness-to-pay of either $0 or $10, each with probability 1/2. The customers' valuations of a particular article are independent. The vendor's marginal cost of an article is zero. What is the best price per bundle and associated profit if the articles are sold in bundles of 2?

6. (Alternate product competition) Firm A produces product A. Firm B produces an alternative product B. Firm A sets the price of product A at $p_A = 3$ and firm B sets the price of product B at $p_B = 6$.

   The willingness-to-pay values for the two products are distributed evenly on the rectangle shown in figure 8.11, with a total number of people equal to $7 \times 12 = 84$. (The exact positions are not known. Each person's position is random, but uniform over the rectangle, and independent of everyone else's.) The WTP values for the two products correspond to a person's coordinates in the rectangle. (A person at the upper-right-hand corner has $\text{WTP}_A = 7$ and $\text{WTP}_B = 12$.)

   A person might buy A or B, but never both.

   (a) Show the region in the square corresponding to people who purchase A, and another region for those who purchase B.
   (b) Compute the areas of these two regions.
   (c) Assuming zero marginal cost, find the expected producer surpluses π_A and π_B.

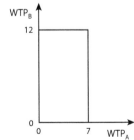

FIGURE 8.11 Figure for exercise 6.

7. (Block bundle) Refer to the blocking entry example. Ignoring the existence of a challenger, what is the optimal price if the two items in the example are sold as a bundle (and not separately)?

8. (How close to average?*) We wonder how many items must be bundled to obtain the major benefits of averaging. Is it only a few, or a thousand? The averaging bundling method in section 8.3 is based on the **law of large numbers** from probability theory; namely

$$\frac{x_1 + x_2 + \cdots + x_n}{n} \longrightarrow \bar{x},$$

where the x_i's are the individual WTP values for items and $\bar{x}$ is the expected value of each x_i. The **central limit theorem** says that

$$\frac{x_1 + x_2 + \cdots + x_n}{n} \approx \bar{x} + N(0, \sigma/\sqrt{n}),$$

where $N(0, \sigma)$ denotes a normal (Gaussian) random variable with mean zero and standard deviation $\sigma \equiv \sqrt{[E(x - \bar{x})^2]}$. If each x_i is a uniform random variable from 0 to $10, then $\sigma = \$2.88$. Suppose n items are bundled, and let p denote the price for the bundle. The per-article profit for the bundle is approximately

$$\pi(n) = p \int_p^\infty e^{-\frac{n}{2}(x-5)^2/\sigma^2} dx.$$

The producer will choose p to maximize $\pi(n)$. Make a table (using a spreadsheet optimizer) of the optimal $\pi(n)$ for $n = 2, 4, 8, 16, 32, \ldots$ until the result is close to $5. How many items are needed to form a highly effective bundle?

9. (Club sharing) Consider the problems

$$\max_y v(y) - m_a y \qquad\qquad (8.3a)$$

$$\max_y v(y) - m_b y, \qquad\qquad (8.3b)$$

where $m_b < m_a$. Show that the corresponding solutions y_a and y_b satisfy $y_b > y_a$. Hint: First write the inequality that states that y_a is better than y_b for problem (8.3a), and then write the equation that states that y_b is better than y_a for (8.3b). Combine.

8.6 Bibliography

A nice nontechnical (nonmathematical) discussion of methods for enhancing the producer surplus of information goods is [1]. A mathematical treatment of the type in the early part of this chapter is [2]. For general theory of economics and pricing rules see [3] or the excellent treatment of nonlinear pricing [4]. The innovative analysis of averaging in bundling was presented in [5]. The role of bundling in entry blocking was presented in [6]. The theory of information sharing (as in a book club) discussed in this chapter is based on [7].

References

[1] Shapiro, Carl, and Hal R. Varian. *Information Rules*. Boston: Harvard Business School Press, Boston 1999.

[2] Varian, Hal R. "Versioning Information Goods." Working paper, University of California, Berkeley, March 13, 1997.

[3] Luenberger, David G. *Microeconomic Theory*. New York: McGraw-Hill, 1995.

[4] Wilson, Robert R. *Nonlinear Pricing*. Oxford: Oxford University Press, 1993.

[5] Bakos, Yannis, and Erik Brynjolfsson. "Bundling Information Goods: Pricing, Profits, and Efficiency." *Management Science* 45 (1999): 1613–30.

[6] Nalebuff, Barry. "Bundling." Yale ICF Working Paper No. 99-14, November 22, 1999.

[7] Varian, Hal R. "Buying, Sharing, and Renting Information Goods." Working paper, University of California, Berkeley, August 5, 2000.

9 VALUE

Information has value—or more correctly, some information in some situations has value to some people. An obvious indication of this value is that people routinely buy or sell information through newspapers, stock reports, telephone calls, Internet traffic, written correspondence, music performances, and many other forms of communication. However, it is usually difficult to objectively assign a value to a specific piece of information from knowledge of its general characteristics. The value of most information is subjective, frequently valued differently by different people. In this regard information is similar to most other commodities; its price is established by a complex process of supply and demand.

It *is* possible to assign an objective value to information when the information influences economic action. The value then is equal to the difference in economic reward of an informed action over an uninformed one. For example, if you are about to purchase an item for $100 and a friend tells you that you can get it next door for $75, that information is worth $25, provided that you change your action accordingly. Situations such as this can be analyzed quantitatively. In fact, a kind of calculus of value is available for assigning value to information associated with economic decisions.

This chapter deals with the value of information in those terms. There must be an action potentially influenced by the information, and the consequences of action must be measured in monetary terms or some other scale of value. Hence music, in the absence of a potentially altered action, although pleasing and valuable in a broad sense, does not have value of the type discussed in this chapter. A weather forecast that may influence a farmer's decision to plant crops *is* of the type considered, for that information may have tangible value. However, if the farmer would plant his crop no matter what the weather forecast reported, the forecast has no objective value. It is simply a curiosity, or entertainment. On the other hand, if the farmer would not plant a crop if the forecast is unfavorable and would if the forecast is favorable and there is some chance that either will be reported, then the forecast has value.

9.1 Conditional Information

One aspect of the situations discussed in this chapter is that useful information is usually received through a channel: a test procedure, a newspaper report, a signal from Mars, a talk with a friend or expert, a survey result, or a laboratory sample. Each of these may provide information about something that could influence a decision and its consequent action. In terms of chapter 5, it is **mutual information**, but in the context of decision making the value assigned to information is measured monetarily, not in terms of bits. However, the notion of a channel and its probabilistic structure is the same as in chapter 5.

The role of conditional information can be illustrated concretely with a simple example of an oil-drilling project, an example that is used throughout the chapter.

Example 9.1 (Red–Black Oil site). The Red–Black Oil company (so named because the company may go in the red or the black) has located a somewhat promising site to drill for oil. It will cost $120 (under some nominal scaling) to drill for oil. It is estimated that there is a probability of 1/3 that oil will be discovered and a probability of 2/3 that it will not. If oil is discovered, the ultimate payoff (over many years) is worth $600. If no oil is discovered, the payoff is zero.

The situation is shown graphically in figure 9.1 by a simple tree of possibilities. The value of such a venture is, for simplicity, taken to be the expected value of the payoff minus the cost. In this case the expected value of the payoff is $200 and the cost is $120; hence the value is $80. This is a baseline. There is no channel information yet.

The action possibilities in this example are simply to drill or stop. Currently, it might be considered a marginally attractive project.

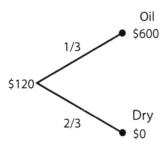

FIGURE 9.1 Red–Black Oil well. It costs $120 to drill and there is a 1/3 chance that oil will be discovered. If there is oil, the reward will be $600; otherwise zero. Hence the value of the well is currently $80.

Example 9.2 (A possible test). A standard geological test is available that, if positive, indicates a strong possibility that oil is present. The test represents an information channel from the actual state of oil at the site to a test result.

Let A denote the random variable with two possible outcomes Oil and Dry depending on whether there is oil at the site. The two possibilities have probabilities 1/3 and 2/3, respectively. Let B be the random variable defining the test result. It has two possible outcomes, Positive and Negative.

The reliability of the test is defined by the conditional probabilities of various test results as a function of the actual situation. These conditional probabilities are listed in the table below. For example, the probability of a positive test result, given that there is oil, is .75.

$p(B|A)$	Positive	Negative
Oil	.75	.25
Dry	.25	.75

Figure 9.2 shows two alternative graphical representations of the channel information. Because the conditional probabilities in the chart are symmetric, this particular channel is a binary symmetric channel (BSC), as defined in chapter 5.

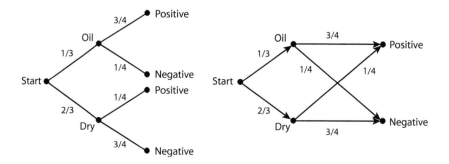

FIGURE 9.2 Test channel. *Left*: tree representation. *Right*: traditional channel representation. The nature of the test is defined by the set of conditional probabilities.

The decision of whether or not to drill is made after seeing the results of the test. An intelligent decision should, however, be based on the revised probabilities of whether or not there is oil. These are found by flipping the channel.

Flipping the Channel (or Tree)

As discussed in section 5.3, the flipping operation is accomplished with Bayes' rule. The probabilities of the output B are

$$p(b_j) = \sum_i p(b_j|a_i)p(a_i)$$

for $j = 1, 2, \ldots, m$, which expresses $p(b_j)$ as the sum of the probabilities of all ways that b_j can occur.

Next, Bayes' rule is used to calculate the **backward probabilities** as

$$p(a_i|b_j) = p(b_j|a_i)p(a_i)/p(b_j).$$

Example 9.3 (Flipping Red–Black). The probabilities of the two test outcomes at Red–Black Oil are

$$p(\text{Pos}) = p(\text{Pos}|\text{Oil})p(\text{Oil}) + p(\text{Pos}|\text{Dry})p(\text{Dry})$$
$$= (3/4)(1/3) + (1/4)(2/3) = 5/12$$
$$p(\text{Neg}) = 1 - p(\text{Pos}) = 7/12.$$

The backward probabilities are calculated with Bayes' rule.

$$p(\text{Oil}|\text{Pos}) = p(\text{Pos}|\text{Oil})p(\text{Oil})/p(\text{Pos})$$
$$= (3/4)(1/3)/(5/12) = 3/5$$
$$p(\text{Oil}|\text{Neg}) = p(\text{Neg}|\text{Oil})p(\text{Oil})/p(\text{Neg})$$
$$= (1/4)(1/3)/(7/12) = 1/7$$
$$p(\text{Dry}|\text{Pos}) = p(\text{Pos}|\text{Dry})p(\text{Dry})/p(\text{Pos})$$
$$= (1/4)(2/3)/(5/12) = 2/5$$
$$p(\text{Dry}|\text{Neg}) = p(\text{Neg}|\text{Dry})p(\text{Dry})/p(\text{Neg})$$
$$= (3/4)(2/3)/(7/12) = 6/7$$

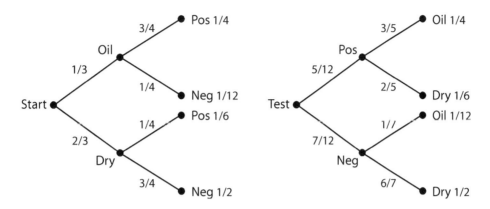

FIGURE 9.3 The oil tree and its flipped version. Note that the center two end nodes represent different combinations in the two trees. However, the end probabilities match for identical combinations such as (Oil, Neg) and (Neg, Oil).

The original tree and its flipped version are shown in figure 9.3. The end nodes represent joint events, and their probabilities must match. For example, the top node in both trees is the joint event (Oil, Positive). The joint event (Oil, Negative) is second from the top in the left tree and third from the top in the right tree.

9.2 Informativity and Generalized Entropy*

An interesting sidelight (which can be omitted at first reading) relates intuitive notions of the relative usefulness of channel information to entropy concepts.

Suppose there is a random variable X and information about X in the form of a random variable Y. Y can be regarded as channel information about X, and the shorthand for this is $X \Rightarrow Y$. Now suppose that there is an alternate source of information Z about X, expressed as $X \Rightarrow Z$. For example, X may be a condition of someone's heart with Y and Z being alternate tests designed to reveal information about that condition. It is natural to consider which of the two tests is better.

We say that Y is **more informative** than Z (about X) if all the probabilistic information about X given by Z is contained in Y. This somewhat vague statement is made precise by its consequence. Y is **more informative** than Z about X if the variables X, Y, Z have the channel structure

$$X \Rightarrow Y \Rightarrow Z.$$

Hence Z is merely a probabilistic transformation, of Y. For instance, Z may be a corrupted version of Y.

The condition of Y being more informative than Z implies that when making inferences about X or making decisions whose outcomes are governed by X, one would always prefer to know Y rather than Z. For example, to estimate the temperature outside your window, a thermometer would be more informative than noting whether or not it is snowing.

Generalized Entropy

If Y is more informative than Z about X, it seems plausible that the mutual information $I(X;Y)$ (see section 5.4) is greater than or equal to the mutual information $I(X;Z)$. Indeed, this is proved as follows.

$H(X|Y,Z) = H(X|Y)$ since Z adds nothing to Y about X.

$H(X|Y,Z) \leq H(X|Z)$ since conditioning never increases entropy.

$H(X|Y) \leq H(X|Z)$ combining the two above.

Hence

$$I(X;Y) \equiv H(X) - H(X|Y) \geq H(X) - H(X|Z) \equiv I(X;Z),$$

which shows that $I(X;Y) \geq I(X|Z)$.

This result can be generalized and then used as an alternative characterization of informativity. A **generalized entropy** function $G(X)$ is a concave function[1] of the probabilities of X. An example is $G(X) = \sum_i p_i \log(1/p_i)$, equal to the normal entropy $H(X)$. Another example is $G(X) = 1 - \max\{p_1, p_2, \ldots, p_n\}$.

Associated with a generalized entropy function is a notion of **generalized conditional entropy** defined as

$$G(X|Y) = \mathrm{E}[G(X)|Y].$$

This is the expected value (before Y is observed) of the generalized entropy of X that will hold after Y is observed. If $G = H$, the usual entropy, then $G(X|Y)$ is the usual conditional entropy $H(X|Y)$. Likewise, the **generalized mutual information** is

$$I_G(X;Y) = G(X) - G(X|Y).$$

It is not hard to prove that for any generalized entropy function G, there holds $I_G(X;Y) \geq 0$ and $I_G(X;Y) \geq I_G(X;Z)$ when Y is more informative than Z about X. (See exercise 2.)

Example 9.4 (Funny die). Let X be result of the toss of a die producing a number between 1 and 6. Define the generalized entropy function as $G(X) = \sum_{i=1}^{6} p_i(1 - p_i)$. The generalized entropy of X is then $G(X) = (1/6)(5/6) \times 6 = 5/6$.

Let Y be the information about X that tells whether X is in the Lower half (1, 2, or 3) or the Upper half (4, 5, or 6). Given that Y says Lower, the probabilities of 1, 2, 3 are each $1/3$, and hence the generalized entropy conditional on Lower is $(1/3)(2/3) \times 3 = 2/3$. The same holds symmetrically when Y says Upper. These each happen with probability $1/2$, so the average conditional generalized entropy is $G(X|Y) = 2/3$. Hence $I_G(X;Y) = 5/6 - 2/3 = 1/6 = .167$.

Suppose Z has the same two signals as Y except that there is a $1/4$ chance that any reported signal is incorrect. Clearly Y is more informative about X than is Z. If Z reports Lower, the probabilities of the low numbers are each $(1/3)(3/4) = 1/4$ (rather than $1/3$) and those of the Upper numbers are $(1/3)(1/4) = 1/12$. These are

---

[1] Suppose $x = (x_1, x_2, \ldots, x_n)$ and $y = (y_1, y_2, \ldots, y_n)$ are vectors. A function $f(x)$ is concave if for all such vectors and all α with $0 \leq \alpha \leq 1$ there holds $f(\alpha x + (1 - \alpha)y) \geq \alpha f(x) + (1 - \alpha)f(y)$.

reversed if Upper is reported. Hence, the average conditional generalized entropy is $(1/4)(3/4) \times 3 + (1/12)(11/12) \times 3 = 19/24 = .79$. Thus $I_G(X;Z) = 5/6 - 19/24 = .042$, which is less than the earlier $I_G(X;Y) = .167$, consistent with Y being more informative than Z.

A deeper result is the following, which closely relates generalized entropy to informativity.

Theorem 9.1. Y is more informative than Z about X if and only if $I_G(X;Y) \geq I_G(X;Z)$ for all concave generalized entropy functions G.

This is an elegant result, establishing a full and symmetric relation between generalized information and informativity.

9.3 Decisions

Channel information can influence decisions because such information changes the probabilities that are assigned to relevant events. Decisions can be matched to information. It is this opportunity for **decision matching** that is responsible for the value of the channel information. In fact, channel information has value only if a decision will vary depending on the specific outcome obtained. The variation of decision with channel outcome is termed a **strategy**.

Example 9.5 (Red–Black drilling decision). From figure 9.3 one can compute the best decisions and their associated net payoffs. For example, if the test is Positive, then the probability of Oil is $3/5$. Assuming that the decision is to Drill, the expected payoff is $(3/5)\$600 - \$120 = \$240$. If the test is Negative, the expected payoff for drilling is $(1/7)\$600 - \$120 = -\$34.29$. Considering the other possibilities, one may construct the following table of net payoffs as a function of the decision Drill or Stop.

Net Payoff	Drill	Stop
Postive	$240	$0
Negative	−$34	$0

From this it is clear that the decision should be to Drill if the test is Positive and Stop if the test is Negative.

The value of the project using the derived strategy can be calculated, and this calculation is carried out in figure 9.4. The resulting expected net profit is $100. Hence, the value of the project has increased from $80 to $100. The difference of $20 is the value of the test information (before the test is made).

9.4 The Structure of Value

We have emphasized that conditional (channel) information can have value if it has the potential to influence a decision. This section summarizes how the value of information is measured, generalizing the ideas illustrated in the earlier examples.

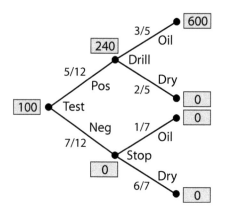

FIGURE 9.4 Red–Black valuation. The numbers at the end are the final payoffs at the corresponding nodes assuming the optimal decision strategy is followed. The value of $240 at top center is found from (3/5)$600−$120=$240, since $120 must be paid upon drilling.

Suppose that the underlying source of uncertainty is the random variable A with possible values $(a_1, a_2, \ldots, a_n)$. There is a decision set $\mathcal{D}$ from which a decision d can be selected. Finally, there are net payoffs $v(a_i, d)$ that depend on the event A and the decision d.

For each event a_i there is a maximum payoff defined by

$$v^*(a_i) = \max_{d \in \mathcal{D}} v(a_i, d).$$

In Red–Black Oil, $v^*(\text{Oil}) = \$600 - \$120 = \$480$ and $v^*(\text{Dry}) = \$0$.

It can be useful, in itself, to determine these optimal values of events, for those values bound the overall possible value. If your car is acting up and the mechanic tells you that it is either the clutch or the brakes, and repair of the clutch would cost $1,500 and the brakes $1,000, you may decide to trade the car in for a new one without bothering to get more information. Knowledge of the bounds is as far as you need go.

Value of A

The baseline situation is the value defined by the random variable A, its probabilities, its payoffs, and its decision set. This value is

$$V^*(A) = \max_{d \in \mathcal{D}} \sum_{i=1}^{n} p(a_i) v(a_i, d).$$

In this definition, a single $d \in \mathcal{D}$ is selected to maximize the expected value of the payoff. In the Red–Black Oil problem, this is determined by the best overall decision, which is to drill. The associated value is $V^*(A) = (1/3)\$480 - (2/3)\$120 = \$80$.

The Value of Perfect Information

Additional concepts of value depend on the nature of the information available. The best information is perfect information—that is, perfect knowledge of the specific event in A, known before the decision must be made. The associated value can be written as $V^*(A|a_i)$ for a particular a_i. Clearly,

$$V^*(A|a_i) = v^*(a_i)$$

because if a_i is known to be the event, the decision can be optimized for it.

The **value *with* perfect information** is $V^*(A|A)$, which is the expected value of $V^*(A|a_i)$ with the expectation taken with respect to the various a_i's. Thus $V^*(A|A) = \sum_i V^*(A|a_i)p(a_i)$.

For Red–Black Oil, the value *with* perfect information is $(1/3)(\$600 - \$120) + (2/3)\$0 = \$160$, because one can drill if there is oil and stop if there is not. These occur with probabilities 1/3 and 2/3, respectively.

The **value *of* perfect information** is the maximum amount a decision maker would pay for perfect information. It is

$$VI(A|A) = V^*(A|A) - V^*(A).$$

A nice way to speak about this is in terms of a **clairvoyant**. The value of perfect information is the amount one would pay for the services of clairvoyant who, before a decision is required, could reveal exactly which event in A occurs.

The Value of Channel Information

If information about A is given by B, the new overall value *with B* is denoted $V^*(A|B)$. It is defined as

$$V^*(A|B) = \sum_{j=1}^{mp} p(b_j) \left[\max_{d \in \mathcal{D}} \sum_{i=1}^{n} p(a_i|b_j)v(a_i, d) \right].$$

The interpretation is straightforward. Once B is observed as some b_j, the new probabilities for the events in A are given by the conditional probabilities $p(a_i|b_j)$. With knowledge of b_j one maximizes the expected payoff, using a single decision for that b_j. This is done for each b_j and the results averaged according to the probabilities of the b_j's.

This procedure was earlier carried out in example 9.5 for Red–Black Oil when B was the outcome of a geologic test. The value with the test was found to be \$100.

The value *of* information B about A is the difference

$$VI(A|B) = V^*(A|B) - V^*(A).$$

For Red–Black Oil, this difference is $\$100 - \$80 = \$20$.

Nonnegative Value of Information

The value resulting from a particular information outcome may be either positive or negative. In Red–Black Oil, a positive test result leads to a new valuation of \$240 instead of the original \$80. However, a negative test result leads to a new valuation of \$0, a big decrease. Nevertheless, the new expected value, before the test is carried out, is \$100, which is an increase. This is true in general,[2] as shown below.

Nonnegative value of Information. The value of information is always nonnegative.

Proof:

$$V^*(A|B) = \sum_{j=1}^{m} p(b_j) \left[\max_{d \in \mathcal{D}} \sum_{i=1}^{n} p(a_i|b_j)v(a_i,d) \right]$$

$$\geq \max_{d \in \mathcal{D}} \left[\sum_{j=1}^{m} p(b_j) \sum_{i=1}^{n} p(a_i|b_j)v(a_i,d) \right]$$

$$= \max_{d \in \mathcal{D}} \left[\sum_{j=1}^{m} \sum_{i=1}^{n} p(a_i|b_j)p(b_j)v(a_i,d) \right]$$

$$= \max_{d \in \mathcal{D}} \left[\sum_{j=1}^{m} \sum_{i=1}^{n} p(a_i,b_j)v(a_i,d) \right]$$

$$= \max_{d \in \mathcal{D}} \left[\sum_{i=1}^{n} p(a_i)v(a_i,d) \right] = V^*(A).$$

Hence $V^*(A|B) \geq V^*(A)$. ■

Hence, just as conditional information never increases overall entropy (and thus mutual information is always nonnegative), the value of conditional information is always nonnegative.

---

[2]It is true in general for single decision-maker situations. Information can have negative value to a group in some situations.

9.5 Utility Functions*

When faced with a prospect whose payoff depends on a random event, many people evaluate the prospect at less than its expected value because of the associated risk. For example, a 50–50 chance at $1,000 or $0 might be valued at less than $500. Aversion to risk can be incorporated into decision making by defining a **utility function** $U(v)$ for money v. The function U is increasing; that is, $v_2 > v_1$ implies $U(v_2) > U(v_1)$: more money is preferred to less. A popular form is the exponential utility function $U(v) = -e^{-av}$, where $a > 0$ is a **risk aversion coefficient**. Notice that for this function, $-1 < U(v) < 0$ for $v > 0$, but of course a constant can be added to make $U(v)$ positive.

Once a utility function is selected, decisions are made so as to maximize expected utility. The best decision for the situation described by the random variable A and decision set $\mathcal{D}$ is found by solving

$$\overline{U}^*(A) = \max_{d \in \mathcal{D}} \sum_{i=1}^{n} p(a_i) U(v(a_i, d)).$$

A utility value $\overline{U}^*$ can be converted into equivalent monetary units V^*. The proper V^* is the value that satisfies $U(V^*) = \overline{U}^*$. This value is called the **certainty equivalent** since it is the risk-free amount that has the same utility as the expected utility of the risky prospect. For an exponential utility function the certainty equivalent V^* of utility level $\overline{U}^*$ satisfies $U(V^*) = -e^{-aV^*} = \overline{U}^*$. Hence $V^* = -\ln(-\overline{U}^*)/a$, (which is positive if $-1 < \overline{U}^* < 0$).

Suppose now that information in the form of a random variable B is available before the decision is made. The value of that information is the maximum amount that one would pay (as a fee to a testing company, perhaps) for the information while maintaining the same level of $\overline{U}^*$. The information must be purchased in advance, independent of the outcome. If F is the fee for information, then the new payoff functions are $v(a_i, d) - F$, for each i.

The exponential utility function has the analytical advantage that $U(v - F) = U(v)e^{aF}$, which simplifies calculations. (See exercise 7.)

Example 9.6 (Red–Black Oil with utility). For the situation of example 9.1 suppose the decision maker uses the utility function $U(v) = -e^{-.001v}$. First, let us find the value without channel information. It is

$$\overline{U}^*(A) = \max\left\{-\frac{1}{3}e^{-.001(480)} - \frac{2}{3}e^{-.001(-120)}, \; -\frac{1}{3}e^{-.001(0)} - \frac{2}{3}e^{-.001(0)}\right\}$$

$$= -.9579257 \text{ (since the first choice is best)}$$

$$V^*(A) = -\ln(.9579257)/.001 = \$42.99,$$

which because of aversion to risk is less than the $80 value when expected value rather than the utility function is used as the criterion for evaluation.

Now let us calculate the value with the test information of examples 9.2 and 9.3. We let b_1 and b_2 denote Positive and Negative test results, respectively. The calculations are identical to those above, except that the probability of oil is changed according to the test result, as shown in the second tree of figure 9.3.

$$\overline{U}^*(A|b_1) = \max\left\{-\frac{3}{5}e^{-.001(480)} - \frac{2}{5}e^{-.001(-120)}, -\frac{3}{5}e^{-.001(0)} - \frac{2}{5}e^{.001(0)}\right\}$$

$$= -.8222686$$

$$\overline{U}^*(A|b_2) = \max\left\{-\frac{1}{7}e^{-.001(480)} - \frac{6}{7}e^{-.001(-120)}, -\frac{1}{7}e^{.001(0)} - \frac{6}{7}e^{-.001(0)}\right\}$$

$$= -1.00$$

$$\overline{U}^*(A|B) = -\frac{5}{12}.8222686 - \frac{7}{12}1.00 = -.9259453$$

$$V^*(A|B) = -\ln(.9259458)/.001 = \$76.94.$$

The value *with* information is therefore $V^*(A|B) = \$76.94$, which compares with $100 found earlier.

The value *of* the information is the value of F that reduces $V^*(A - F|B)$ to $V^*(A)$. In this case (because of the special properties of the exponential utility function) this is the difference $76.94 − $42.98 = $33.96. Note that the value of information is substantially greater than the $20 found earlier using expected value as the measure of value—even though the values $V^*(A)$ and $V^*(A|B)$ are each less than before.

9.6 Informativity and Decision Making*

It might be intuitively clear that if information B is more informative than C about A, then B is more valuable as well (or at least as valuable as C). This is true, and is proved by the following steps, which are similar to those used to prove the generalized entropy relation in section 9.2.

$$V^*(A|B, C) = V^*(A|B) \quad \text{since } C \text{ adds nothing to } B$$

$$V^*(A|B, C) \geq V^*(A|C) \quad \text{additional information never reduces value}$$

$$V^*(A|B) \geq V^*(A|C) \quad \text{combining the above.}$$

In a certain sense the converse of this result is also true. Loosely it states that if all decision makers (facing a variety of situations with payoffs determined by A) are better off knowing B rather than C, then B is more informative than C about A.

Hence, there are three characterizations of B being more informative than C about A: (1) there is a channel representation $A \Rightarrow B \Rightarrow C$, (2) generalized mutual information satisfies $I_G(A; B) \geq I_G(A; C)$, and (3) the value of B is greater than the value of C for decisions whose outcome is determined by A.

9.7 EXERCISES

1. (Mutual oil information) Find the mutual information $I(A;B)$ of the Red–Black Oil situation.

2. (Generalized mutual information*) Let $G(X) = G(p(x_1), p(x_2))$, where $p(x_1)$ and $p(x_2)$ are the probabilities of the two possible values x_1, x_2 of X. Suppose G is concave. The generalized entropy of X conditional on Y is

$$G(X|Y) = E[G(X)|Y]$$
$$= p(y_1)[G(p(x_1|y_1), p(x_2|y_1))] + p(y_2)[G(p(x_1|y_2), p(x_2|y_2))].$$

Show that
(a) $G(X|Y) \le G(X)$.
(b) $I_G(X;Y) \ge 0$.

3. (Odd entropy) Work example 9.4 using the generalized entropy function $G = 1 - \max\{p_1, p_2, p_3, p_4, p_5, p_6\}$. Find $I_G(X;Y)$ and $I(X;Z)$. Which is greater and why?

4. (The diligent professor) Professor Earnesto is designing a final exam for his class. He believes that 40 percent of the students are excellent and deserve As while the other, average, students deserve Bs. However, Earnesto does not know which students are which. He is designing an exam that will help distinguish them, but currently he has two versions. For version 1, he believes that excellent students have a 70 percent chance of scoring high on the exam while only 30 percent of the average students will score high. For version 2, he believes that 80 percent of the excellent students will score high, while 50 percent of the average students will score high. He intends to give As to those who score high and Bs to the others. Earnesto wants to maximize the match of grades: As to excellent students and Bs to average students. Which exam should he use, and in expected value terms, what percentage of grades will match?

5. (Virus test) A certain exotic virus cannot be detected with certainty, but a simple saliva bleach test is available that is somewhat reliable. If the virus is present, the test gives a positive result 60 percent of the time and a negative result 40 percent of the time. If the virus is not present, the test gives a (false) positive result 20 percent of the time. The test is only administered to people who have symptoms characteristic of the virus infection, and it is known that 30 percent of the people with these symptoms actually have the virus.
(a) What is the probability that someone with the symptoms will have a positive test result?
(b) What is the mutual information $I(A;B)$ where A is the state of the virus, and B is the test result?
(c) Suppose that if you have the virus and are treated, you recover 100 percent of the time. However, it is not a comfortable treatment and you will have to terminate your vacation. The treatment has a negative value to you whether or not you have the virus. On the other hand, not treating the virus will mean a difficult time, which has a much greater negative value. You might construct the payoff structure shown in the table below.

Value	Treat	Wait
virus	−10	−50
clear	−10	0

Find the value $V^*(A)$ of the original situation without the test.

(d) Find the value $V^*(A|B)$, which is the value with the test information.

(e) Find the value of information.

6. (Information value) Show that the value of information is nonnegative when decisions are made with a utility function.

7. (Exponential utility) Show that if utility is $U(v) = -e^{-av}$, then the value of information is $VI(A|B) = V^*(A|B) - V^*(A)$.

9.8 Bibliography

A number of researchers were inspired by Shannon's theory of information to expand the concept to general notions of economic value. One of the earliest papers is [1]. The concept was greatly expanded and made practical in the field of decision analysis [2]. See also [3]. The concept of a source of information Y being more informative than Z and characterizing this situation as Z being a probabilistic transformation of Y (through a channel) was presented by Blackwell [4]. The relation of this concept to generalized entropy is due to DeGroot [5], and the relation to decision making is due to Bohnenblust et al [6]. Accessible presentations of all three concepts are [7] and [8].

References

[1] Marschak, Jacob. "Remarks on the Economics of Information." Cowles Foundation Paper 146. Reprinted in *Contributions to Scientific Research in Management*, Los Angeles: University of California, 1960.

[2] Howard, Ronald A., and James E. Matheson, eds. *Readings on the Principles and Applications of Decision Analysis*. Menlo Park, Calif.: Strategic Decisions Group, 1983.

[3] Keeney, Ralph L., and Howard Raiffa. *Decisions with Multiple Objectives: Preferences and Value Tradeoffs*. Cambridge: Cambridge University Press, 1993.

[4] Blackwell, David. "Equivalent Comparison of Experiments." *Annals of Mathematical Statistics* 24 (1953): 265–72.

[5] DeGroot, M. "Uncertainty, Information, and Sequential Experiments." *Annals of Mathematical Statistics* 33 (1962): 813–20.

[6] Bohnenblust, H., L. Shapley, and S. Sherman, "Reconnaissance in Game Theory." Research Memorandum RM-208, RAND Corporation, 1949.

[7] Kihlstrom, R. E. "A Bayesian Exposition of Blackwell's Theorem on the Comparison of Experiments." In: M. Boyer and R. E. Kihlstrom, eds., *Bayesian Models in Economic Theory*. Amsterdam: North-Holland, 1984.

[8] Weber, Thomas, "The Value of Information." Working paper, Department of Management Science and Engineering, Stanford University, 2005.

INTERACTION

10

Interpersonal interaction is a fundamental characteristic of human endeavor, and information profoundly shapes that interaction. Interaction is guided by both micro information (bank balances, temperatures, stock prices, sales forecasts, or grade-point averages) and macro information (CD sales, Internet availability, or software design). This chapter briefly surveys some particular aspects of both the micro and macro levels of interpersonal interaction.

Game theory provides one framework for the study of interaction. This theory began, indeed, with the study of simple games, such as matching pennies or elementary card games, but soon advanced to the analysis of military strategy and simple economic situations. The theory has continued to advance, to the point where today it is used to study quite complex situations, especially those in an economic setting. In a general formal game, players (or parties) have a set of possible actions, and the reward to each player depends on the joint action of all players and possibly on one or more random events. For example, the winner in a game of chess is decided by the joint actions of the two players. A company's profit may depend on what other firms do and on random economic events.

As game theory evolved, the nature of the information possessed by the players was found to be a critical consideration in the formulation of strategy. In some cases, the advantage of a player is determined almost entirely by special knowledge. Playing a game of cards with a marked deck is a good example.

The information structure of a game can be classified as either **symmetric** or **asymmetric**, depending on whether all players have the same information or not. Games with symmetric information structure are easier by far to analyze than those that are asymmetric because when different players have different information, their actions can reveal information. This point is emphasized and illustrated in this chapter.

One difference between multiperson interaction and single-person decision making is that information may have negative value in a multiperson setting. Consider, for example, two farmers: farmer A's crop needs sun and will perish in rain; farmer B's crop needs rain and will perish in sun. Suppose there is a 50–50 chance of sun or rain and the profit to each farmer will be $100 or $0, depending on the weather. The two farmers could execute a contract that would share their profits. They would then each be guaranteed $50. If they are risk-averse, $50 for sure is preferable to a 50–50 chance of $100 or $0. Suppose now that before the contract is signed, a soothsayer offers to announce the future weather. How much is that information worth to the farmers? Each farmer will attach negative value to it, for certainly after it is announced, the contract will not be executed (since one farmer will be certain of $100). If the soothsayer is engaged, then before his announcement, each farmer again faces the 50–50 chance of $100 or $0 with no chance of sharing profit. Hence both farmers would not want the soothsayer to announce the future weather before they sign the sharing agreement.

Study at the macro level focuses on interactions associated with information products or services rather than on specific information content. Some of the most important issues are so-called network effects, whereby information services become more valuable to each customer as more people join the network. Here the relevant interaction is indirect, in that each new customer joins for his or her own benefit but by doing so benefits others as well. It will be seen that the principles of economics studied in earlier chapters can be augmented to provide an analytic characterization of this phenomenon.

10.1 Common Knowledge

If I know something you don't know, you might be able to deduce what I know from my actions. For example, if you see me running to the bus stop, you might deduce that the bus is due to arrive soon, even though you did not previously know the schedule. Things get more complex, however, in a game situation. If I bet high in a game of poker, you might deduce that I have a good hand; but then again I might be bluffing in an attempt to deceive you. While playing bridge with you as my partner, on the other hand, I try to bid in such a way as to transmit as much information to you as possible, while simultaneously optimizing our likely score. In either case, my actions are designed to both advance my position in the game and manage my private information. Games of this sort, asymmetric games where information is not equally known to all players and where actions may reveal information, can be exceedingly difficult to analyze.

To avoid the complexity of the interactions between information and action, one focuses on those (symmetric) games in which all players have the same information. This is formalized by the concept of common knowledge.

A piece of information is said to be **mutual information**[1] if all players know it. You and I both may know (because of an inadvertent exposure by the dealer) that the bottom card of a deck of cards is the ace of clubs. That knowledge is mutual information. However, more is required for information to be common knowledge.

Information is said to be **common knowledge** if all players know the information, each player knows that all other players know it, each player knows that all others

---

[1] This notion of mutual information is unrelated to the standard information-theoretic concept of chapter 5 and used in chapter 9.

know that all others know it, and so forth, ad infinitum. If I did not notice that you saw the ace of clubs on the bottom of the deck, then, even though we both know the card and so it is mutual information, that knowledge is not common knowledge.

To see why we need the infinite sequence of "I know that you know that I know that you know . . . ," imagine that annual salaries are recently established at a company. If they are listed on a bulletin board for all to read and everyone knows that all employees see the list, then these salaries are common knowledge among employees. On the other hand, if each employee receives a sealed envelope with his or her salary stated inside, then the salaries are not common knowledge. However, it is possible that some employees visit the department secretary who shows them a master list of salaries. In fact it may happen that each employee independently observes this master list. Then everyone knows all the salaries. Salaries are now mutual information, but they are not common knowledge because each employee does not know who else knows. There may, however, be a second list (compiled by the secretary) with the names of those who saw the first list, and some employees might see that second list. They would know who knew the salaries, but they would not know who saw that second list. A third list might contain the names of those who saw the second list, and so forth. Only if everyone saw the entire infinity of lists would salaries be common knowledge.

The following example presents a wonderful puzzle that illustrates an aspect of common knowledge.

Example 10.1 (Inner council). On a small Pacific island a tribe of natives practices an important ritual to determine members of the tribal council. The group of eligible men sit in a circle around a fire. The chief goes around the circle and places his thumb on each man's forehead, but he secretly uses either his right or left thumb to do so. His right thumb is covered with ash that leaves a mark on the foreheads it touches. The men cannot tell which thumb was used and hence cannot tell whether they have the mark on their own forehead although they can see whether or not others do. The special mark signifies that that man has been chosen to be elevated to the council.

When the chief completes his path around the circle, he announces that at least one man has the special mark, and that any man who knows he has the mark should stand up to be received into the council. Of course, a man who does not have the mark must not stand, for that would be presumptuous and forever ruin his reputation in the community.

No one stands after the chief's announcement, so the chief beats a drum and asks again. Still no one stands. He beats the drum and asks again. This continues, until finally, all at once, a group of men stand, and they are precisely the men who have the special mark. How did the men know to stand?

The answer is found by induction. Suppose k men have the mark. If $k = 1$, the man with the mark will surely deduce that he has it because he sees no one else with a mark, and the chief announced that at least one man was marked. Hence, he will stand up immediately.

If $k = 2$, then a man with a mark, seeing one marked forehead, will reason that if he does not have a mark, the man who does will stand up immediately. Noticing that the marked man did not stand up at the first round, he concludes that there must be two marked men and he must be one of them. So he (and the other marked man) stand up at the second round. If $k = 3$, this same logic will cause the three men with marks to stand on the third round, and so forth. In general, if k men are marked, they will all stand at the k-th round.

Initially, the pattern of marks was not common knowledge. But actions (or lack of action) revealed enough information so that the distribution of marks eventually became common knowledge.

Games where all knowledge is common knowledge can be analyzed without considering that actions might reveal secret information. For example, in the popular game of Monopoly, everyone knows the odds of the dice, the mixed composition of the Chance and Community Chest cards, and the rules of the game. There is common knowledge with respect to these elements. Good strategic play is not concerned with hiding private information but rather with management of property. Poker, on the other hand, is definitely not a game with common knowledge, and management of private information is a critical element of smart play.

Example 10.2 (Betting). Suppose two players A and B are arranging a bet between them on the outcome of the toss of an unevenly weighted coin: A will bet on heads and B on tails. It is agreed that A will wager $100 on heads, but they will negotiate the odds to be used. Players evaluate a particular bet on the basis of the expected value of the outcome. The actual probability of heads of this special coin is $1/3$, but this probability is not common knowledge. Fair odds for the bet would be 1:2 since the probabilities are $1/3$ and $2/3$.

If player A knows the probability but does not know whether B knows it, A will propose odds more favorable than 1:2 and use as much guile as possible to obtain good odds. B will also draw on his or her negotiating skill to get odds that favor B. It is impossible to predict what odds will eventually be set.

On the other hand, if the probability of heads is common knowledge, the two players most likely will quickly agree on odds of 1:2.

10.2 Agree to Disagree?

People often disagree about the probability of an event. For example, you and I may assign different probabilities to the chance that it will rain tomorrow, or to the chance that the price of XYZ stock will increase by five points within a week. We can discuss these matters, and in the spirit of friendship agree to disagree, each of us respecting the beliefs of the other—or can we?

Robert Aumann showed that in certain rather general cases, if the respective probabilities of an event held by different people are common knowledge, then they must be equal. In other words, under the assumption of common knowledge it is impossible to agree to disagree.

Formally, we assume that there are a finite number of possible states of the world, such as *rain*, *clouds*, or *sun* tomorrow. Individuals A and B initially agree on the probabilities of each of these states. These are termed the **prior probabilities**. In the case of weather, these priors may be based on the 100-year statistics for weather for the day of the year in question. Focusing on a particular event E (such as rain tomorrow), A and B will initially agree on the probability of E. Next A and B separately obtain additional information. For weather, A may consult his barometer, and B may look at the sky. Based on their separate information, A and B update their probabilities concerning the event E to new values termed their **posterior probabilities**, which are denoted q_A and q_B respectively. In general these may be different. However, if both q_A and q_B are common knowledge, they must be equal. That is, if A knows q_B and B knows q_A and each knows that the other knows and so forth, then $q_A = q_B$.

This striking result is illustrated by the following example.

Example 10.3 (Hidden money). Four large cards are placed on a table. They have distinctive shapes (square or round) and colors (white or gray) as shown in figure 10.1. Under the first and fourth cards is $1,000 in cash. Two players A and B (say Alice and Bob) are brought into the room. The setup of the cards and money is known to both, and is in fact common knowledge.

One card is secretly selected at random by a third party (perhaps by flipping two coins). The probability of any one card being chosen is $1/4$. These prior probabilities are common knowledge. Neither A nor B knows which card is chosen.

Next, A is told whether the selected card is gray or white, and B is told whether it is square or round. Neither A nor B hears what the other is told but they know the possibilities. From the information they have, A and B each work out their respective posterior probability that the chosen card covers $1,000.

Suppose in particular that the card chosen is card number 1. A will know that the chosen card is gray, and will set her posterior probability at $q_A = 1/2$. B will know that the card is square and will set his posterior probability at $q_B = 1/3$.

A deduces that B knows that the card is square, and hence A will compute $q_{B|A}$ (the conditional probability of q_B given A's information) as $q_{B|A} = 1/3$. B will know that no matter whether A was told gray or white, A must assign a posterior of $q_A = 1/2$. Hence $q_{A|B} = 1/2$. Therefore $q_{A|B} = q_A$ and $q_{B|A} = q_B$. Thus the posteriors q_A and q_B are **mutual information** in the sense that both A and B know their values, and they are different.

The reason the posteriors are different is that they are not common knowledge. A easily deduces the value that B assigns to q_A is $1/2$; that is, $q_{[A|B]|A} = 1/2$. However, B does not know what value A assigns to q_B. That is, B does not know $q_{B|A}$. As far as B knows, the chosen card may be number 3. This is consistent with B's information

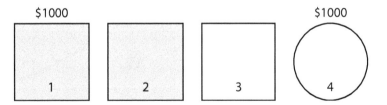

FIGURE 10.1 Agreement experiment. A card is secretly selected by a neutral party. Player A is told the color of a card and player B is told the shape. If the actual card is number 1, the players will each know what probability the other assigns to the chance that the card covers $1,000. But these probabilities are not equal since they are not common knowledge. If each probability were common knowledge, they would be equal; that is, the players would agree, and in this case they would agree that the probability is $1/2$.

about the shape. If it is card number 3, then A would have been told that the color was white, in which case she would deduce that B was told the shape was square or round. In that situation A would conclude that $q_B = 1/3$ (if card 3 was chosen) or $q_B = 1$ (if card 4 was chosen), each with probability $1/2$. Thus B does not know that A knows what B knows, and q_B is not common knowledge. These inferences are summarized in the following table:

A	$q_A = 1/2$	$q_{B\|A} = 1/3$	$q_{[A\|B]\|A} = 1/2$
B	$q_B = 1/3$	$q_{A\|B} = 1/2$	$q_{[B\|A]\|B} = 1/3 \text{ or } 1$

If A tells B that she knows that $q_B = 1/3$, then B will revise his posterior to $q_B = 1/2$. A will know that B will do this and so q_A and q_B will be common knowledge, and they will be equal.

Aumann's result shows that the assumption of common knowledge about probabilities simplifies analysis of game situations, avoiding issues that arise when probabilities are different.

Proof of the Result*

Suppose there are n possible states of the world with prior probabilities p_i for $i = 1, 2, \ldots, n$ that are common knowledge. The set of all these states is denoted Ω. An **event** is a subset $E \subset \Omega$. In the hidden money example, Ω is the set of four cards, and E is the subset of the two cards 1 and 4.

Each player has an information set that is a partition of Ω into disjoint subsets. In the hidden money example, A's partition is the two sets $\{1, 2\}$, $\{3, 4\}$ for gray and white, and B's partition is $\{1, 2, 3\}$, $\{4\}$ for square and round. These are illustrated in figure 10.2. A and B are told which set in their respective partition contains the chosen point i.

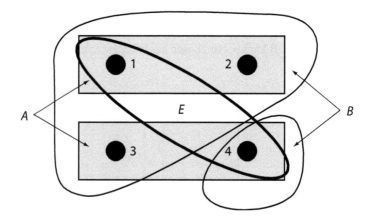

FIGURE 10.2 Partitions. Player A's partition consists of the two subsets of $\Omega = \{1, 2, 3, 4\}$ defined by the two rectangular shaped sets. B's partition consists of the two subsets defined by the rounded sets. The event E consists of those cards covering $1,000, the two points in the elliptical region.

Let P be the smallest subset of Ω such that $i \in P$ and $P = \cup_j P_A^j$ for some members P_A^j of A's partition, and also $P = \cup_k P_B^k$ for some members P_B^k of B's partition. In the example shown in figure 10.2, P is equal to Ω for any i, but in general P may be a strict subset of Ω.

Let q_A be A's posterior probability of E. For q_A to be common knowledge, q_A must be the same for every P_A^j in the set of partitions defining P. This is the condition that results from following the chain of "A knows what B knows, B knows that A knows that. . . ." (This condition is proved explicitly in Aumann's paper.)

The condition that q_A be constant in all P_A^j means that $q_A = \text{Prob}\{E \cap P_A^j\}/\text{Prob}\{P_A^j\}$ for each P_A^j. Hence $\text{Prob}\{E \cap P_A^j\} = q_A \, \text{Prob}\{P_A^j\}$. Summing these over the (disjoint) P_A^j's leads to $\text{Prob}\{E \cap P\} = q_A \sum_j \text{Prob}\{P_A^j\} = q_A \, \text{Prob}\{P\}$. Likewise, $\text{Prob}\{E \cap P\} = q_B \, \text{Prob}\{P\}$, and so $q_A = q_B$.

10.3 Information and Decisions

There are two main components of information in interactive situations: the states of nature and the actions of individuals. In a given situation, a player may have partial information about either of these.

Actions

A classic example of the problem associated with ignorance of actions is the **prisoner's dilemma**. The situation is that two fellows are arrested for having stolen property in their possession. They are questioned in separate rooms, but they both know that there is not enough evidence to convict them of robbery unless one of them confesses. If both stay quiet they will be charged with possession of stolen goods and serve short sentences. On the other hand, if both confess, they will both get long prison sentences. However, if only one confesses and turns state's evidence, he will go free and the other will get an even more severe sentence than if they both confessed.

The dilemma arises because each prisoner is ignorant of the other's action. If they could cooperate, they could obtain the best solution for both. The prisoner's dilemma has been extensively studied because there is no clear obvious solution—because of the lack of action information.

States of the World

The term "state of the world" is a bit overblown. In a game situation it refers to values of random variables and past actions that are relevant to the current decisions of the game. In a game of checkers, for example, the state involves only the current configuration of pieces on the board, not the stock market average or the weather. In a game of poker, the state is the current distribution of cards: in players' hands and in the deck. In general, players have only partial information about the state, and each player has different information, leading to the possibility that players' actions reveal information. Players must account for that in developing their strategies.

A special situation, frequently analyzed, is where the only uncertainty is the private information held by a single player, say player 1, and the other player (or other players) are required to act first, followed by player 1. In this case the actions of these other players do not reveal information, for they had none to reveal. There are many interesting examples of this kind of situation.

Example 10.4 (Adverse selection). Consider an insurance company that offers insurance against car theft. The company knows that the aggregate probability of car theft is 1 percent. Ignoring administrative costs, the fair odds price of the insurance is 1 percent of the insured value. Assuming that the insurance industry is highly competitive, the price will be driven down to the fair odds price.

However, not everyone faces the same probability of theft, and it is reasonable to assume that each individual knows the probability he or she faces but the insurance company does not.

If the company offers insurance at a price 1 percent of the value of coverage, it is likely that the high-risk individuals will purchase more insurance than low-risk individuals, and this will, of course, change the average probability of theft among those who purchase insurance, raising the probability above 1 percent. Consequently, the insurance company will lose money. This is the phenomenon of **adverse selection**. People self-select on the basis of their private information in a way that is adverse to the company.

However, the insurance company acts first in this situation by setting the terms of the contract. The company forecasts the likely responses of various individuals to tentative contracts, and then designs the contract to maximize the overall expected profit associated with the likely responses.

10.4 A Formal Analysis*

To generalize the above discussion, assume that each player seeks to maximize the expected value of a certain objective function which might represent profit, personal preference, or political advantage. The function depends on the actions of all players and on some random events. For simplicity, assume that there are only two players A and B with possible action sets A and B and objective functions U_A and U_B, respectively. A selects $a \in A$ and B selects $b \in B$. There is also a random event e. Player A's payoff is, then,

$$E_A[U_A(a, b, e)],$$

where E_A denotes expectation relative to the probabilities that A has for event e and action b. Likewise, B's payoff is of the form

$$E_B[U_B(a, b, e)].$$

A significant feature of the game is the nature of the information about the random variable e that the two players possess. That may differ among the players. Another significant feature of the game is the order of the actions. If B acts first, for example, then the expectation E_A is computed with b as given, while E_B must account for A's reaction to b. General problems of this form are difficult to solve.

These situations are less difficult when, as mentioned before, (1) player B first selects $b \in B$ and this action is observed by A before selecting $a \in A$; and in addition, (2) B knows nothing about e that A does not already know.

Example 10.5 (Duopoly). Suppose two companies A and B compete in a market. Both have constant marginal cost m and select production levels q_A and q_B, respectively. The overall demand function is linear but with an unknown additive constant. That is,

$$p = e - q,$$

where p is the price, q is the total quantity $q = q_A + q_B$, and e is a random constant whose value is revealed only after the production decisions are made. The expected profits for A and B are

$$\pi_A = E_A[p\, q_A - m\, q_A] = E_A[(e - m - q_A - q_B)q_A]$$
$$\pi_B = E_B[p\, q_B - m\, q_B] = E_B[(e - m - q_A - q_B)q_B].$$

If B selects q_B first and then A selects q_A, the expectation that A forms will be influenced by the observed q_B chosen by B, and B must account for that. Generally the problem is highly complex.

On the other hand, if the information about e is common knowledge, the expectations of e viewed by both A and B are identical and equal to, say, $\bar{e}$. Hence, A can solve the problem

$$\max_{q_A}[(\bar{e} - m - q_B)q_A - q_A^2],$$

leading to

$$q_A = \tfrac{1}{2}(\bar{e} - m - q_B).$$

Knowing how A will respond, B will select q_B to solve

$$\max_{q_B}[\bar{e} - m - \tfrac{1}{2}(\bar{e} - m - q_B) - q_B]q_B,$$

leading to $q_B = \tfrac{1}{2}[\bar{e} - m]$ and $q_A = \tfrac{1}{4}[\bar{e} - m]$.

The Use of Types

To formalize the relatively nice situations described earlier where player B acts first and B knows nothing relevant that A does not know, it is convenient to assume that there are a finite number of possible states, denoted $t_1, t_2, \ldots, t_n$. Player A knows the state exactly. Player B assigns probabilities p_i to the occurrence of each t_i. Player B (who does not know the state) must act first, and player A responds. The finite state possibilities $t_1, t_2, \ldots, t_n$ are termed **types** because they are associated identically with A's knowledge. One says that A is of type t_i, if t_i is the state.

This framework includes the models used earlier in the study of bundling, versions, adverse selection, and other situations where customers were characterized by their types. The other party, B (often a firm), designs price schedules for the various types.

A's problem is simple to solve, since A has perfect information about the type and about B's (previous) action b. If A is of type t_i, A's problem is

$$\max_{a \in A} U_A(a, b, t_i),$$

where b is known. The result of this optimization is a_i, which can be expressed using the $\max^{-1}$ notation[2]

$$a_i = \max_{a \in A}^{-1} U_A(a, b, t_i).$$

B maximizes the expected value of U_B by realizing that A will react according to type as above. Thus, B's problem is

$$\max_{a_1, a_2, \dots a_n, b} \sum_{i=1}^{n} p_i U_B(a_i, b, t_i)$$

subject to

$$a_i = \max_{a \in A}^{-1} U_A(a, b, t_i), \quad \text{for } i = 1, 2, \dots, n.$$

This problem says that B must consider all possible types, and take the expectation with respect to them. B must also account, type by type, for A's reaction a_i to B's action b. These reactions a_i are entered into the objective function. The constraints insure that the reactions a_i that B uses in the objective function are those that type i would use.

Example 10.6 (Print shop pricing). A print shop has two types of customers that occur in equal proportions. Type 1 has objective function $U_1 = c_1 q - \frac{1}{2} d_1 q^2 - pq$ for a quantity q of a print run at price p per unit. Likewise, type 2 has objective $U_2 = c_2 q - \frac{1}{2} d_2 q^2 - pq$.

The print shop has a cost function of $F + mq$. Ideally, the shop would work out the maximum profit price for each customer, but the shop does not know the type of any particular customer. Instead the shop decides to post a unit price schedule (so that the total cost is pq) of the form $p = K - Lq$ that will apply to all jobs, and will maximize average profit. (See figure 10.3.) This schedule charges no one a setup fee and gives a discount to high-volume customers. For simplicity (and to avoid some technical difficulties) the shop fixes L and optimizes K.

Each customer will optimally respond to the price schedule, according to that customer's type. In particular, type 1 customers will solve

$$\max_{q} c_1 q - \frac{1}{2} d_1 q^2 - (K - Lq)q,$$

leading to

$$q_1 = \frac{c_1 - K}{d_1 - 2L}.$$

A similar expression holds for type 2.

[2]In general, if x^* is the value of x that achieves $\max f(x)$, one writes $x^* = \max^{-1} f(x)$.

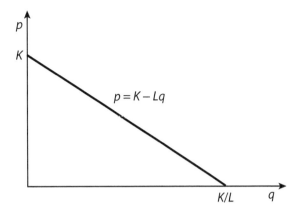

FIGURE 10.3 Price for printing. The print shop price decreases with volume.

The print shop's problem is therefore

$$\max_{K} \; \frac{1}{2}(p_1 - m)q_1 + \frac{1}{2}(p_2 - m)q_2 \tag{10.1}$$

subject to

$$q_1 = \frac{c_1 - K}{d_1 - 2L} \qquad q_2 = \frac{c_2 - K}{d_2 - 2L}$$

$$p_1 = K - Lq_1 \qquad\quad p_2 = K - Lq_2.$$

As a specific case, suppose the parameters are

$$c_1 = 24, \quad d_1 = 2$$
$$c_2 = 14, \quad d_2 = .5$$
$$m = 4$$
$$L = .2.$$

If a single price were used, the maximum profit would be \$61.25, achieved at a unit price of \$9. One the other hand, solving (10.1) yields a profit of \$91.875, achieved at $K = \$12.00$—a 30 percent increase in profit.

10.5 Metcalfe's Law

Interaction of information also takes place at a macro level, as in demand for products, computer virus propagation, compatibility standards, and signal interference. One of the most important interactions is the positive feedback associated with participation in an information network. The value of a network to a participant tends to grow as others join. For example, the telegraph and telephone networks became increasingly valuable as additional lines were constructed and more subscribers joined the network. More recently, the growth of the Internet has made it more useful to everyone.

During his Ph.D. studies at Harvard University, Robert M. Metcalfe left to work at Xerox's Palo Alto Research Center (PARC). There he read about a new concept for

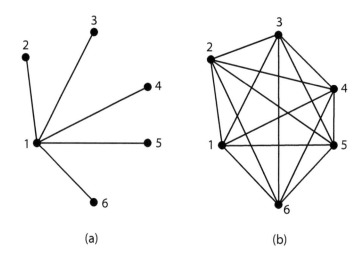

FIGURE 10.4 Metcalfe's law. In a network of N nodes, each node can be connected to $N-1$ other nodes, as shown in (a) for node 1. Altogether there are $\frac{1}{2}N(N-1)$ possible node connections, as shown in (b).

communication embodied in the ALOHA network linking the Hawaiian Islands.[3] In this network, messages were divided into small packets each sent separately, although a certain percentage of these packets collided and were lost. (Chapter 22 studies this system.) Inspired by the general idea, Metcalfe returned to Harvard to apply queuing theory to packet flow so as to vastly improve the performance of the ALOHA network method. This work was his Ph.D. thesis, and the alternative system he developed was termed the **Ethernet**. Metcalfe later founded the 3Com Corporation, which produces networking hardware. To promote his vision of a universal network, Metcalfe expressed the value of a network in terms of a simple rule. Usually, the rule as stated as follows.[4]

Metcalfe's Law. The value of a network to each of its N users is approximately proportional to N^2.

The reasoning behind this "law" is shown in figure 10.4. Each of the N users can potentially connect to $N-1$ others. There are thus $N(N-1)$ possible ordered pairs of users. This number must be divided by 2 because communication between users i and j is accounted for in communication between users j and i; therefore there are $\frac{1}{2}N(N-1)$ possible interconnections. If it is assumed that each of these connections is roughly of equal value, the value of the overall network is proportional to $N(N-1)$, or approximately to N^2 for large N.

Looser forms of the law can be formulated that assume, for instance, that each person's possible interactions are limited or that some interconnections are more valuable than others. Some of these alternative formulations imply that value grows

[3] Developed by Prof. Norman Abramson, author of *Information Theory and Coding*, one of the first popular texts on information theory. (See the reference in chapter 2.)

[4] The original form of the rule may have been the following: "The power of the network increases exponentially by the number of computers connected to it. Therefore, every computer added to the network both uses it as a resource and adds resources in a spiral of increasing value and choice."

less fast than N^2, but still faster than N. However, Metcalfe's law is not intended as a precise statement of value, but rather as a simple expression of the fact that each addition to the network tends to enhance the value to those already participating. It points out the interaction that one individual's decision has on others' value. A careful analysis of value must be based on economic principles, and that is addressed in the next section.

10.6 Network Economics*

Economists say that networks exhibit **externalities** because the value to an individual of a unit of network usage (such as a month of telephone service) depends on how many other people purchase units. This is different from, say, the value of an apple to a consumer, which does not depend on how many people eat apples.

If a network's externalities are positive,[5] with additional users enhancing everyone's value, it is most efficient to have a single standard, or single system, serving as many people as possible. However, there are, as with most information products, questions of how the network service should be priced to maximize benefit. A simplified model illustrates the methods used to analyze such issues.

Let the willingness-to-pay for the n-th unit (say the n-th connection or the n-th customer) be $p(n, \bar{n})$ when people expect there to be $\bar{n}$ customers. For a fixed value of $\bar{n}$ the function $p(n, \bar{n})$ is like an ordinary WTP function, and measures the willingness-to-pay for the n-th unit assuming that the other $n - 1$ have already been purchased. This function decreases with n and serves as a demand function. At price $p(n, \bar{n})$ consumers as a group will purchase n units.

This WTP function is predicated on a certain expected number of units sold, $\bar{n}$. Imagine, for example, that the network's administration announces that market research indicates that 100,000 people will sign up, and as a result everyone assigns the value $\bar{n} = 100,000$ for their expectation. Then the demand curve will be $p(n, 100,000)$.

A WTP curve (or demand curve) can be constructed for each value of $\bar{n}$, resulting in a family of curves, as shown in figure 10.5. The figure assumes that there is a maximum possible network size n_{max}. It also assumes that each demand curve is downward sloping and, for simplicity, linear. It also assumes that customers prefer a larger network over a smaller one, which means that the demand curves shift upward with increasing values of $\bar{n}$.

Fulfilled Expectations

As the network evolves, people revise their expectations about total network size on the basis of actual sign-ups, and eventually produce a fairly accurate estimate $\bar{n}$ of n. In fact, in equilibrium the estimate $\bar{n}$ will be exact, with $\bar{n} = n$. The expectations of individuals are then said to be **fulfilled**. What they expect is what actually happens.

---

[5]An example of a negative externality is congestion—on a road or a communication network. There, more users make the network less desirable to each individual.

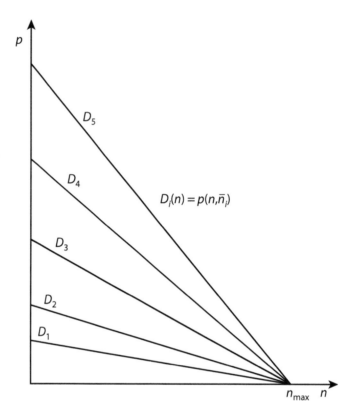

FIGURE 10.5 Network demand curves. For each level of expected usage, there is an associated demand curve.

This equilibrium condition implies that the demand curve $p(n, \bar{n})$ only applies where $n = \bar{n}$. In other words, at equilibrium, $n = \bar{n}$, and hence the demand point must satisfy $p = p(n, n)$.

The curve in figure 10.6 traces out the points $p(n, n)$ on the family of demand lines. In order to rule out explosive demand it is assumed that $p(n, n)$ goes to zero as n goes to infinity. The resulting $p(n, n)$ curve therefore rises initially but eventually falls.

Competitive Solution

Suppose that the marginal cost of supplying a unit of service is the constant m. If m is low enough so that it cuts the $p(n, n)$ curve, as it does in the figure, it is possible to have marginal cost pricing,[6] which is characteristic of perfect competition. For the situation shown in the figure, there are three possible solutions: (a) $n = 0$, which applies if fixed costs cannot be recovered from sales, (b) the first intersection of the marginal cost line with the $p(n, n)$ curve, which corresponds to n_3 in the figure,

[6]See chapter 7.

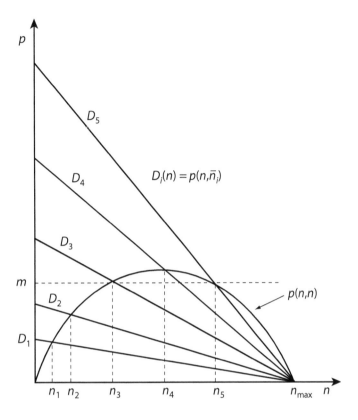

FIGURE 10.6 Fulfilled-expectations curve. Expectations are fulfilled when $\bar{n} = n$. This condition defines a single point on each demand curve, and the family of these points defines the curve $p(n,n)$ of points consistent with fulfilled expectations.

and (c) the second intersection, corresponding to n_5. The first intersection point is unstable because a small increase in n leads to both increased consumer surplus and increased profit, so the network will tend to expand. The second intersection point is stable, for it represents the standard case of a decreasing demand curve and constant marginal cost.

For a network to operate at its stable level under competition it is necessary that the volume of service increase to the equilibrium point. For this reason, network administrators may elect to set price below marginal cost and then slowly raise the price as the network expands by moving up the $p(n, n)$ curve. Indeed, this strategy of setting initial price low to capture market share seems to be favored by companies hoping to make their network, system, or software the industry standard.

Example 10.7 (Quadratic network). Suppose $p(n, \bar{n}) = [A - an]\bar{n}$. Each demand curve (with fixed $\bar{n}$) is linear, and the curves shift upward as $\bar{n}$ increases. For this arrangement, $n_{\max} = A/a$, and $p(n, n) = An - an^2$, which is a parabola.

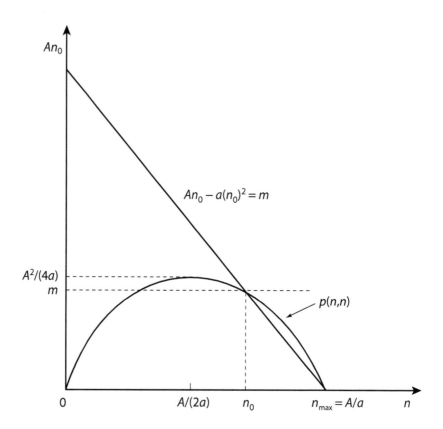

FIGURE 10.7 Network example. It is possible to find explicit values for the case $p(n,\bar{n}) = [A - an]\bar{n}$.

The level n_0 corresponding to price equal to marginal cost satisfies $p(n_0, n_0) = m$, or more specifically,

$$An_0 - an_0^2 = m.$$

The solution is illustrated in figure 10.7.

As a specific example, suppose $A = 150$, $a = 1$, and $m = 5{,}000$. Then $n_{\max} = 150$ and $n_0 = 100$.

The competitive solution corresponding to marginal cost pricing is not optimal from a social surplus viewpoint because of the positive externality. When individuals consider joining the network, they count only their own benefit without reckoning the benefit that their joining would bestow on others. The marginal benefit to society is greater than the marginal benefit to any one individual. This implies that the socially optimal size of the network is larger than the size resulting from competition. This is an argument for the government to subsidize important networks, although the form of subsidy is not clear.

Monopoly control of the network would lead to a network smaller than that of competition and hence inferior in terms of social welfare.

10.7 EXERCISES

1. (Hidden money) For the hidden money game, suppose that the card selected at random is #2. Are the probabilities q_A and q_R mutual information? Are they common knowledge? Repeat for the cases where the selected card is #3 and #4.

2. (Prisoner's dilemma) The prisoner's dilemma is represented concretely by the diagram in figure 10.8. Each prisoner selects either Q (for quiet) or C (for confess). Prisoner A's decision determines a row and prisoner B's decision determines a column. Together their two decisions determine a box in the square, which lists the pair of penalties (a, b), for prisoners A and B, respectively.

   If they could confer and reach agreement, they likely would both elect to be Quiet. However, when they are separated, A fears that B may switch to Confess, which would have disastrous consequences for A. B has the same fears about A's action. The police interrogators will perhaps seek to make these feared outcomes credible by suggesting that the other prisoner is about to confess. If prisoner A must act first and this act is reported correctly to B, what penalties will the two obtain?

3. (Big profit) A consumer has utility function for the quantity of a commodity given by $U(q) = q - \frac{1}{2}q^2 - pq$, where p is the unit price.
   (a) If a monopolist with zero marginal cost charges a fixed unit price, what is the maximum producer surplus?
   (b) Suppose that the monopolist charges according to the price schedule $p = K - Lq$. Show that by proper choice of K and L, the producer can get arbitrarily close to double the surplus of part (a). Hint: Solve this problem graphically, noting that $p(q) = K - Lq$, but the quantity q satisfies $1 - q = K - 2Lq$. A further hint is that K should be nearly 1.

4. (Random duopoly) Consider a market that is shared by two firms A and B. The demand curve is $p = e - q$, where q is the total quantity produced and e is a constant that is either e_1 or e_2. The marginal cost of both firms is m.
   (a) Suppose that B goes first and selects q_B. Then after observing q_B, A follows and selects q_A. Both firms know the demand curve. What are q_A and q_B?

FIGURE 10.8 Prisoner's dilemma. Prisoner A selects a row, and B selects a column. Then A gets a penalty equal to the first entry in the corresponding pair and B gets the second.

(b) Now suppose that the value of e is known to A, but B knows only that e has the possible values e_1 and e_2 with probabilities p_1 and $p_2 = 1 - p_1$, respectively. Define $\bar{e} = p_1 e_1 + p_2 e_2$. Again B must select q_B first and A selects q_A after observing q_B. Assuming A and B each act to maximize the expected value of profit, find q_B and q_A.

5. (The quadratic network*) In the quadratic network, $p(n,n) = An - an^2$. At a point n, the consumer and producer surpluses are

$$CS = [An - p(n,n)]n/2,$$
$$\pi = [p(n,n) - m]n.$$

(a) Show that the social surplus is

$$SS(n) = -\frac{1}{2}an^3 + An^2 - mn.$$

(b) Show that the derivative of the social surplus with respect to n is

$$SS'(n) = \frac{3}{2}[An - an^2] + \frac{1}{2}An - m.$$

(c) Evaluate the sign of the derivative at these three points:
    I. At the marginal cost solution n_0.
    II. At the maximum point $n_{\max} = A/a$.
    III. At $n = 0$.

Accounting for the fact that $SS'(n)$ is quadratic, determine the value of n that maximizes social surplus.

10.8 Bibliography

The classic paper [1] by Aumann is surprising in that it is only three pages long! (plus references). A comprehensive review of subsequent developments and examples of the theory of common knowledge is in [2]. A somewhat advanced introduction to information structures and game theory is [3]. Several applications of game theory are presented in [4]. Metcaffe's [5] application of queuing theory to networks is contained in his dissertation. The general model of networks presented here is adopted from [6].

References

[1] Aumann, Robert J. "Agreeing to Disagree." *Annals of Statistics* 4 (1976): 1236–39.

[2] "Common Knowledge." In *Stanford Encyclopedia of Philosophy*, http://plato.stanford.edu.

[3] Rasmusen, Eric. *Games and Information: An Introduction to Game Theory.* New York: Basil Blackwell, 1989.

[4] Luenberger, David G. *Microeconomic Theory.* New York: McGraw-Hill, 1995.

[5] Metcalfe, Robert M. "Packet Communication." Ph.D. diss., Harvard University, Project MAC-TR-114, December 1973.

[6] Economides, Nicholas. "The Economics of Networks." *International Journal of Industrial Organization* 16 (1996): 673–99.

SUMMARY OF PART II

A fundamental difficulty associated with information products is that competitive markets do not always reward the creators of the products. There are two reasons for this. First, **competition** tends to drive the price of a product down to its **marginal cost**, and this is low (often almost zero) for information products. Hence, producers cannot recover their fixed cost of creation and production. The second reason is that information products are often easy to copy, and this can destroy the market for the original product. These issues and some resolutions are analyzed by using standard economic concepts of **producer**, **consumer**, and **social surplus**, which provide means for tracing how value accrues to different market participants.

Copyright and patent law provide a degree of protection that allows producers of information products to obtain some profit. A single (**monopolistic**) producer can obtain a profit (without accounting for fixed costs) of about one-half of what is theoretically available. Producers can obtain even more by employing more advanced pricing schemes. **Discrimination** by age, locale, or sex, charging different amounts to different groups, is one method. **Bundling** of distinct products into packages sold as a unit is another. Bundling is used, for example, to price magazine subscriptions, season theater tickets, and university degree programs. Another scheme is **versioning**, where various quality levels of a single product are offered at different prices, such as high- and low-resolution photos, professional and academic versions of software, and orchestra versus balcony symphony tickets. Many of these methods rely on indirect discrimination, whereby individuals sort themselves out according to their **willingness-to-pay** values, which producers cannot directly observe.

Soon after Shannon published his theory of communication, researchers considered whether a notion of information value, similar to entropy, could be formulated. The question has its natural setting in the evaluation of decisions that have measurable economic consequences. In this context it is natural to define (as in chapter 5) the notion of a **channel** that transforms (through measurement, tests, or questionnaires) uncertainty at an unobservable source A to an observed outcome B. The transformation is defined by a system of conditional probabilities. Value can be imputed to the variable B by determining the amount by which the expected value of a decision whose consequences depend on A is improved by observation of B. The techniques for carrying out these calculations form a useful methodology for actual decision making.

A **generalized entropy** of a variable X can be defined by a concave function $G(X)$ of the probabilities of X. Many of the notions and results of classical information theory can be extended to generalized entropy.

A fundamental complexity associated with information and economics arises whenever different individuals possess different information about an event and there is an element of competition among the individuals. The complexity arises because actions tend to reveal information as well as directly govern rewards. This complexity occurs in card games, bidding situations, employer–employee relations, insurance contracts, and indeed many human interactions. The complexity largely disappears in the case of **common knowledge,** where everyone knows the same thing, and everyone knows everybody else knows it, and so forth. The formulation of optimal actions is then not confounded by differences in information.

Another simplifying case is when player A has special information, termed the player's **type**, but player B acts first. Then for any decision contemplated by player B, B can anticipate the response that A will have as a function of A's type. Hence B can account for that response when selecting a decision. Bundling and versioning pricing schemes are based on this model.

Economic considerations are also important in information networks. Metcalfe's law states that the value to each participant of an information network increases with the number of participants. This loose statement can be made more precise by considering network service as an economic product.

PART III

ENCRYPTION

Security through Mathematics

11

CIPHERS

Secret communication is probably as old as human history. We can imagine that cave men whispered secrets, and that some symbols written on cave walls were intended only for close friends. Actual evidence of written ciphers goes back over four thousand years to a hieroglyphic substitution written by Egyptian scribes on the walls of the tomb of a nobleman. Such ciphers were intended to enhance the significance of the message or to serve as puzzles, but nevertheless they contained the elements of cryptography.

One of the most famous cipher incidences is recorded in the Old Testament in the book of Daniel. At a banquet held by King Belshazzar for a thousand of his lords, the "fingers of a human hand appeared, writing on the plaster wall of the palace." No one could interpet the riddle of the Aramaic words *Mene, Mene, Tekel, Upharsin*. Finally, Daniel was summoned and he easily read "the writing on the wall," and for this he was made one of the three leaders of the government. Although Daniel's interpretation was not strictly a decipherment, he is widely credited as perhaps being the first cryptanalyst.

Ciphers and cryptography often played a decisive role in history. Mary Queen of Scots was held in the Tower of London on suspicion of treason. Because of her stature she could not be executed unless there was definitive proof of her treachery. She communicated with her outside page by means of a complex cipher. This cipher was eventually broken, and her correspondence revealed her role in a plot to kill Queen Elizabeth and overtake the throne. This clinched the judgment, leading to Mary's beheading.

Ciphers are crucial in military campaigns. The first known use of military encryption is associated with the Spartans, who used a transposition device known as a **scytale** that scrambled the letters of a message. Julius Caesar used a simple substitution code for both military and domestic communication.

At the end of the eighteenth century, Thomas Jefferson created a **wheel cipher** that was far advanced for its time. Unfortunately, his idea was filed away and only rediscovered among his papers in 1922. Because of its strength, the system was subsequently used by various agencies of the government and the military. Thomas Jefferson is accordingly called the father of American cryptography.

One of the most infamous ciphers is the dreaded Enigma cipher used by the Germans in World War II. It was implemented by a complex **Enigma machine** that scrambled and substituted text using a series of complex wheels and circuits. It was considered unbreakable, and for this reason the Germans relied on it. Its analysis and eventual breaking by a British agency was one of the most important military achievements of the war and is credited with shortening the war by at least two years.

Today encryption is a vital part of everyday life. Sensitive phone messages are scrambled, Internet communication is encrypted, and smart credit card and digital cash transactions are secured by encryption techniques far superior to those of early ciphers, even superior to the mysterious Enigma. The fascinating development of encryption is explored in the next few chapters.

11.1 Definitions

A generic cipher system is shown schematically in figure 11.1. The **plaintext** is the original message. It is **encrypted** to produce the corresponding **ciphertext**. This ciphertext may appear to be a jumble of letters, or it may be a series of entirely different symbols, such as ♠∇♯ ⋈ § ♠◇ ○ ♡ □∞ ◇♮$. Once the ciphertext is received by the intended party, it is **decrypted** to reproduce the original plaintext. Of course the sender and receiver must both agree on the encryption process.

As a rule, a particular cipher method is but one of a family of similar ciphers, each separate member of the family being distinguished by a **key**. The key governs the encryption process and also the decryption. In practice, the strength of a cipher system is related to the number of possible keys.

11.2 Example Ciphers

It was as long ago as 500 BC that the Spartan government encoded messages with a scytale (pronounced SITalee), which was a cylinder of fixed radius. The sender spiraled a strip of parchment around the cylinder and wrote across it, each letter being placed on adjacent turns of the parchment. When the strip was unwound, the order of the letters was mixed up. The message was decrypted by generals in the field who possessed a duplicate scytale with the same radius. See figure 11.2.

For example, if the circumference of the scytale were equivalent to four letters of text, a strip with the ciphertext ROEOERNMICTINESNFMCG could be decrypted

FIGURE 11.1 Cipher process. Plaintext is encrypted into ciphertext, and this is sent to the receiver, who decrypts it to recover the original plaintext.

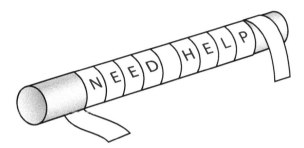

FIGURE 11.2 The scytale. A message is written across the spiraled parchment and then unwound. The scrambled message is decrypted by use of a duplicate scytale. This figure shows the start of a message that would progress across other rows as well.

by spiraling the message on a cylinder of circumference equal to four letters (to give REINFORCEMENTS . . .).

The scytale is an example of a **transposition cipher** in which the ciphertext consists of the same letters as the plaintext but physically transposed in some systematic fashion. The key of the scytale is its radius.

Transposition Ciphers

Practical transposition ciphers are similar to those produced by a scytale, but generated on the interleaving principle discussed in chapter 6 in the context of error-correcting codes. The plaintext message is written letter by letter in a matrix row by row, but converted to ciphertext by reading the letters out column by column. For good measure, the columns can be permuted.

Suppose a five by five array is used. The message THE INVASION WILL BEGIN TODAY is read in by rows as shown below.

	2	4	3	1	5
T	T	H	E	I	N
H	V	A	S	I	O
E	N	W	I	L	L
I	B	E	G	I	N
N	T	O	D	A	Y

The message is read out by columns to obtain the ciphertext. A keyword can be used to mix the order of the columns. In this case we have selected the keyword MONEY to define the order. Remembering a word is easier than remembering a specific order. The keyword is translated into digits by following the alphabetical order of the letters in the keyword. Since E is the lowest letter in the MONEY, it becomes 1, M is the second lowest, so it is 2. Following this procedure with each letter, the keyword translates to 24315. This is the order to be used when writing out the columns. The resulting ciphertext is IILIA TVNBT ESIGT HAWEO NOLNY, which can be spaced differently so as not to reveal the column size, to say, IIL IAT VNB TES IGT HAW EON OLN YDL with two letters added to fill out the last apparent three-letter word.

A general **transposition cipher of order p** reorders a block of p plaintext symbols according to a given permutation. Interleaving is a simple way to construct such a permutation, but it does not include all possible permutations. For example, the symbols could be read into a triangular array row by row and read out column by column. There are in fact $p!$ possible permutations of order p and hence $p!$ transposition ciphers of order p.[1] Said another way, there are $p!$ possible keys (a key being a permutation) associated with transposition codes of order p.

Substitution Ciphers

The most common simple ciphers are **substitution ciphers** where each letter of plaintext is transformed into another letter or symbol, but the order of the letters is not changed.

The **Caesar cipher** is one of the simplest and most well known of the substitution ciphers. In this cipher each letter of the alphabet is shifted by, say, three letters. Thus: a becomes D, b becomes E, and so forth. At the end of the alphabet, the shifting is wound around back to the beginning. The complete set of substitutions is therefore

plaintext	a	b	c	d	e	f	g	h	i	j	k	l	m	n	o	p	q	r	s	t	u	v	w	x	y	z
ciphertext	D	E	F	G	H	I	J	K	L	M	N	O	P	Q	R	S	T	U	V	W	X	Y	Z	A	B	C

Of course shift lengths other than three can be used. The length of the shift is the key, and knowing its value enables one to decrypt the plaintext message.

A modern-day version of this shift cipher is embodied by the Unix operator ROT 13, which shifts all letters by 13. Decryption is achieved by applying the same operator again, since two shifts by 13 produces a shift of 26.

Alternative Alphabets

A substitution cipher may employ a different alphabet for the ciphertext than that of the plaintext. One example is the **pigpen cipher**, said to have been used by Confederate soldiers in the Civil War and still a favorite of school-age children. It is confusing to someone who does not know the secret, but easily reconstructed by those who do. The substitution is made by drawing two simple figures: one being the same set of four lines used in a tic-tac-toe game, and the other being a large X. Letters of the alphabet are entered in pairs into the spaces created by these figures, as shown in figure 11.3.

A letter is encoded as the outline of the area in which it is contained. For example, the letter A is encrypted as ⌐. If the letter is the second of the pair

---

[1] There are p ways to select the new location of the first letter, $p - 1$ ways to select the new location of the second, and so forth.

A, B	C, D	E, F
G, H	I, J	K, L
M, N	O, P	Q, R

FIGURE 11.3 Pigpen cipher. The placement of letters defines symbols used in the ciphertext.

in the space, a "pig" is included in the form of a dot. Thus, B is encrypted as ⊥. A complete message might be

which is easily decrypted.

11.3 Frequency Analysis

Simple substitution codes, such as the Caesar cipher, are vulnerable to attack based on frequency analysis using the known letter frequencies of English. For example, it is known that the most common letters, in order from most common, are E T A O I N S. When attacking the ciphertext of a substitution code, one first determines the most common letters. For the ciphertext

     ZNKINKIQOYOTZNKSGOR

K, N, O, and Z each occur three times, while the others occur only once or twice. It is natural to assume that one of these most frequent symbols represents the letter e. Trying K = e and then guessing all letters are shifted by six in a Caesar code, causes everything to fall in place, producing the message, "The check is in the mail."

More complex substitution codes are designed to increase the number of possible keys and render frequency analysis less potent.

11.4 Cryptograms

Advanced substitution codes substitute letters or symbols of the ciphertext alphabet according to an arbitrary pattern. That is, a general substitution cipher may represent the letter a by K, b by X, c by F, and so forth, with the correspondence being unique in each direction. Symbols other than letters can be used for the cipher alphabet. In any case, there are 26 different symbols, each corresponding to a plaintext letter. If ordinary letters are used for the cipher alphabet, the code system can be described by a permutation of the 26 letters of the alphabet, as shown in the example below.

Plaintext	A B C D E F G H I J K L M N O P Q R S T U V W X Y Z
Ciphertext	F J W S N O B K M U E I H Z X C Q R D T A V L Y P G

If spacing between words and punctuation is preserved, the ciphertext of such a system is termed a **cryptogram**. A cryptogram is vastly more complex than a Caesar cipher or even a pigpen cipher, for while there are 26 possible Caesar cipher keys, corresponding to the 26 possible shifts of the alphabet, there are $26! \approx 4 \times 10^{26}$ possible cryptogram keys corresponding to the 26! different permutations of the alphabet. This is an enormous number of possibilities.

One approach to breaking a substitution code is by trial and error, trying each possible key until a result makes sense. We can imagine a fast computer applied to the problem of solving a cryptogram by this trial-and-error procedure. The computer would cycle through the possible permutations of 26 letters, checking if the result were reasonable. Assuming that the computer could check 100 million permutations per second (which is optimistic since there would be considerable effort to determine if the result were reasonable), it would take about $2 \times 10^{26}/10^8 = 2 \times 10^{18}$ seconds to check one-half of the permutations (which on average is all that would need to be checked). There are $60 \times 60 \times 24 \times 365 = 31,536,000$ seconds in a year. So it would take $2 \times 10^{18}/(.31536 \times 10^8) \approx 6 \times 10^9 = 6$ billion years to complete the computation.

Despite the complexity of a general substitution code, it preserves much of the character of plaintext language. Letter frequencies are preserved, being merely translated to the substitute letters or symbols. If e is coded as K, then K will likely appear more frequently than any other letter in the cipher, and this will suggest that K is the substitute for e. Word structure is also preserved. For example, double letters in plaintext appear as double letters in the ciphertext.

Edgar Allan Poe heightened public curiosity about cryptograms with publication of his engaging and now classic short story *The Gold Bug*, in which Captain Kidd's treasure is discovered with the help of a special species of golden bug and the breaking of a cryptogram left by Kidd. Poe was fascinated by cryptograms and in his regular column in the Philadelphia newspaper *Alexander's Weekly Messenger* he challenged readers to submit cryptograms and boasted that he would solve them all. He was inundated with submissions, but he readily solved all that were legitimate.

Poe's method was the same that amateur fans of cryptograms use today, a combination of frequency analysis and word structure analysis, although there is evidence that Poe emphasized the latter over the former.

Example 11.1 (An important message). To attack the ciphertext

GFX XCXRU WK QJKWGWJCXD JC GFX FWVV GJ GFX XBKG

we first perform a frequency analysis, realizing that it may not be accurate for such a short message. The following counts are obtained:

X	7	G	6
J	4	F	4
W	4	C	3
K	2	V	2

All the rest have counts of 1.

Frequency analysis suggests X = e and G = t. Then the fact that the word GFX appears three times suggests that it is *the*, the most common three-letter word. This gives F = h. We note that the two-letter word GJ starts with t under our assumption, and it is logical therefore to assume the word is *to*, which gives J = o. This means that the two-letter word JC starts with an o and hence it likely that C = n.

The two-letter word WK contains no t, o, h, or n. Hence, a likely choice is is, which gives W = i and K = s. At this point we have the message

```
 t  h e    e n e        i s        o s i t i o n e
 G  F X    X C X R U    W K    Q J K W G W J C X D
───────────────────────────────────────────────────
 o  n    t h e    h i        t o    t h e    e  s t
 J  C    G F X    F W V V    G J    G F X    X B K G
```

From here it is easy to fill in the missing letters to obtain the message: the enemy is positioned on the hill to the east.

This approach has been duplicated in a computer program for solving cryptograms, which includes a dictionary of the 1,000 most common words in English, partitioned into words of different lengths and different structure. For example THAT, HIGH, and AREA are in the same group because the first and fourth letters agree in each of these words. The method systematically tries letter assignments in an attempt to maximize the number of words that match those in the dictionary.

Cryptograms of about 30 letters in length appear in puzzle books as challenges, and most can be easily solved by hand in half an hour or so.

11.5 The Vigenère Cipher

The substitution ciphers discussed so far are termed **monoalphabetic** since there is a single alphabet (and single substitution order) used to construct the cipher. By the sixteenth century, the weakness of this type of cipher was recognized, and more complex ciphers that varied the substitution process from letter to letter were proposed. Such ciphers are termed **polyalphabetic** since more than one alphabet substitution is used. The most practical and popular of these was invented by Blaise de Vigenère in about 1562.

The **Vigenère cipher** uses a keyword to vary the substitution formula with each new letter. Each substitution is determined by a simple shift of the alphabet as in a Caesar cipher, but the length of the shift is determined by the key. To assist in the process of shifting, one may use the Vigenère table of table 11.1.

As an example, suppose the keyword is chosen to be CODE. To encrypt a message, it is written letter by letter with the keyword lined up above it and repeated over and over so that it spans the entire message. The keyword letter that is written above the plaintext letter is the shift as determined by table 11.1. An example is shown below.

Keyword	C O D E C O D E C O D E C O D E C O D E C O D E C O D E C O D E C O D E
Message	T H E P R E S I D E N T I S I L L W I T H A H I G H F E V E R T O D A Y
Ciphertext	V V H T T S V M F S Q X K G L P N K L X J O K M I V I I X S U X Q R D C

TABLE 11.1
Vigenère Table. Each row is a shift of the one above it.

	A	B	C	D	E	F	G	H	I	J	K	L	M	N	O	P	Q	R	S	T	U	V	W	X	Y	Z
A	a	b	c	d	e	f	g	h	i	j	k	l	m	n	o	p	q	r	s	t	u	v	w	x	y	z
B	b	c	d	e	f	g	h	i	j	k	l	m	n	o	p	q	r	s	t	u	v	w	x	y	z	a
C	c	d	e	f	g	h	i	j	k	l	m	n	o	p	q	r	s	t	u	v	w	x	y	z	a	b
D	d	e	f	g	h	i	j	k	l	m	n	o	p	q	r	s	t	u	v	w	x	y	z	a	b	c
E	e	f	g	h	i	j	k	l	m	n	o	p	q	r	s	t	u	v	w	x	y	z	a	b	c	d
F	f	g	h	i	j	k	l	m	n	o	p	q	r	s	t	u	v	w	x	y	z	a	b	c	d	e
G	g	h	i	j	k	l	m	n	o	p	q	r	s	t	u	v	w	x	y	z	a	b	c	d	e	f
H	h	i	j	k	l	m	n	o	p	q	r	s	t	u	v	w	x	y	z	a	b	c	d	e	f	g
I	i	j	k	l	m	n	o	p	q	r	s	t	u	v	w	x	y	z	a	b	c	d	e	f	g	h
J	j	k	l	m	n	o	p	q	r	s	t	u	v	w	x	y	z	a	b	c	d	e	f	g	h	i
K	k	l	m	n	o	p	q	r	s	t	u	v	w	x	y	z	a	b	c	d	e	f	g	h	i	j
L	l	m	n	o	p	q	r	s	t	u	v	w	x	y	z	a	b	c	d	e	f	g	h	i	j	k
M	m	n	o	p	q	r	s	t	u	v	w	x	y	z	a	b	c	d	e	f	g	h	i	j	k	l
N	n	o	p	q	r	s	t	u	v	w	x	y	z	a	b	c	d	e	f	g	h	i	j	k	l	m
O	o	p	q	r	s	t	u	v	w	x	y	z	a	b	c	d	e	f	g	h	i	j	k	l	m	n
P	p	q	r	s	t	u	v	w	x	y	z	a	b	c	d	e	f	g	h	i	j	k	l	m	n	o
Q	q	r	s	t	u	v	w	x	y	z	a	b	c	d	e	f	g	h	i	j	k	l	m	n	o	p
R	r	s	t	u	v	w	x	y	z	a	b	c	d	e	f	g	h	i	j	k	l	m	n	o	p	q
S	s	t	u	v	w	x	y	z	a	b	c	d	e	f	g	h	i	j	k	l	m	n	o	p	q	r
T	t	u	v	w	x	y	z	a	b	c	d	e	f	g	h	i	j	k	l	m	n	o	p	q	r	s
U	u	v	w	x	y	z	a	b	c	d	e	f	g	h	i	j	k	l	m	n	o	p	q	r	s	t
V	v	w	x	y	z	a	b	c	d	e	f	g	h	i	j	k	l	m	n	o	p	q	r	s	t	u
W	w	x	y	z	a	b	c	d	e	f	g	h	i	j	k	l	m	n	o	p	q	r	s	t	u	v
X	x	y	z	a	b	c	d	e	f	g	h	i	j	k	l	m	n	o	p	q	r	s	t	u	v	w
Y	y	z	a	b	c	d	e	f	g	h	i	j	k	l	m	n	o	p	q	r	s	t	u	v	w	x
Z	z	a	b	c	d	e	f	g	h	i	j	k	l	m	n	o	p	q	r	s	t	u	v	w	x	y

The Vigenère cipher destroys ordinary frequency and word structure. A given plaintext letter is likely to be encrypted differently in its several occurrences in the message. Likewise a double letter will not appear as a double letter. The Vigenère cipher is therefore much more difficult to attack than a standard monoalphabetic substitution code. The strength of the cipher, however, depends on the length of the key. If the key is short, the cipher may be broken by using an enhancement of frequency analysis.

Cryptanalysis of the Vigenère Cipher

Vigenère ciphers with relatively short keywords can be attacked by using the numerical values of relative letter frequencies. A table of such frequencies is shown in table 11.2.

Suppose a received message is

CMFUSBIEXKLMDETGNU.

Assume first that the length of the keyword is known. Suppose it is three. Then the collection of every third letter can be analyzed as if these were produced by a simple shift cipher. That is, one looks only at the letters 1, 4, 7, 10, ... For the sample message above these letters are CUIKDG.

A variation of frequency analysis can be applied to this collection of letters using a simple optimization procedure. A certain keyword letter is proposed and the letters in the collection are transformed by this shift. Then each letter in this transformed collection is assigned a value equal to its standard occurrence frequency as given in table 11.2.

The first three key letter possibilities are shown in table 11.3. If the key letter were A, the letters in the collection would be identical to the corresponding cipher letters. The first section of the table lists these letters together with their standard frequencies (as a percent). These values are summed to obtain a total score of 19.582. Next, assuming the key letter were B, the message letters would have been shifted forward by one letter to construct the cipher letters, so now they are shifted backward by one

TABLE 11.2
Letter Frequency Occurrences in English.   The frequencies are given as percentages.

A	8.167	J	0.153	S	6.327
B	1.492	K	0.772	T	9.056
C	2.782	L	4.025	U	2.758
D	4.253	M	2.406	V	.978
E	12.702	N	6.749	W	2.36
F	2.228	O	7.507	X	0.15
G	2.051	P	1.929	Y	1.974
H	6.094	Q	0.095	Z	0.074
I	6.966	R	5.987		

TABLE 11.3
Analysis of a Vigenère Cipher.   Actual frequency values are assigned to each possible shift and the maximum is indicative of the possible key letter.

		Key Letter				
	A		B		C	
	C	2.782	B	1.492	A	8.167
	U	2.758	T	9.056	S	6.327
	I	6.966	H	6.094	G	2.051
	K	.772	J	.153	I	6.966
	D	4.253	C	2.782	B	1.492
	G	2.051	F	2.228	E	12.702
Score		19.582		21.805		37.705

letter to obtain the hypothetical message letters. These are shown in the second section together with their corresponding scores. The total score under the assumption that B is the key letter is 21.0805. Similarly, the score under the assumption that C is the key letter is found to be 37.705. This procedure can be carried out for each possible key letter, producing a score for each one. In this example, it turns out that the key of C gives the highest score, so it is a prime candidate for the actual key letter.

The same process can be carried out for the collection consisting of the letters in positions 2, 5, 8, 11, 14, 17 and for the collection of letters in positions 3, 6, 9, 12, 15, 18. The maximum scores for these sets are obtained by the key letters A and B respectively, implying that the entire keyword is CAB. Indeed, using this as the keyword converts the message to AMESSAGEWILLBESENT, or A MESSAGE WILL BE SENT.

This simple technique may not always be successful on messages as short as this example, but it is highly successful on messages that have a length at least 10 times the key length.

If the length of the keyword is not known, this procedure can be repeated for various lengths until a high total score is achieved, indicating the true key or at least the true length.

The entire procedure can be easily carried out in a spreadsheet program.

The Autokey Cipher

An ingenious variation of the Vigenère cipher that does not require a long key but has some of its advantages, is the **autokey cipher** also devised by Vigenère. In this cipher, the message itself is used as the key. The process is started with a **seed key**, which could be as short as a single letter, but it is better to use a longer one. When as many letters as in the seed key have been encrypted, new key letters are taken from the message itself starting at the beginning. For example, if the key is the single letter C and the message is "We have captured a spy," The actual key would be CWEHAVECAPTUREDASP, producing the ciphertext YALHVZGCPINLVHDSHN.

11.6 The Playfair Cipher

Another way to confound frequency analysis is to encode letters in pairs rather than singly. For example, the pair *th* might be encrypted as 356, and the pair *te* by 12. In its most general form, specification of a symbol for each pair requires a table of size 26^2 by 26^2, which is quite unwieldy. However, such a table can be constructed so that the resulting cipher is completely immune from first-order frequency analysis, and second-order analysis would be effective only on lengthy messages.

The **Playfair cipher** is a simple procedure for encrypting pairs. It was popularized by Lyon Playfair, first Baron Playfair of St. Andrews, but it was actually invented by his good friend Sir Charles Wheatstone, a scientist of unusual breath and creativity. Wheatstone's contributions to telegraphy and his influence on the invention of the telephone are mentioned briefly in chapters 19 and 20.

8	J	E	Q	D	N	5	O
P	U	3	A	R	F	L	W
4	V	C	2	T	M	B	I
K	7	Z	S	G	X	H	Y

FIGURE 11.4 A Playfair matrix. The matrix defines a pair-wise substitution cipher that is difficult to break.

The cipher is defined by an array such as shown in figure 11.4. Any size array is suitable as long as it has at least two columns and two rows and contains at least 26 elements (or 25 if i and j are considered identical). The figure shows a four by eight array constructed by scattering the alphabet among the cells and filling the remainder with integers.

A message to be encrypted is first partitioned into pairs of adjacent letters. If this would lead to a double letter, an X is inserted between the two. For example, the message LET US MEET AT NOON is rewritten as

LE TU SM EX ET AT NO ON.

An X is inserted between the two Es, but not between the two Os, which are in different pairs. If there is a final single letter, an X is appended.

A pair of letters is encoded according to the following rules:

1. If the letters are in the same row, encode the letters by using those to the immediate right in the same row. If at the right end, use the first letter in the row (that is, consider that the rows wrap from the right end to the left). For example, the pair TI is encrypted as M4.

2. If the letters are in the same column, encode the letters by using those immediately below in the same row. If at the bottom, use the top letter of the column. For example, the pair RG is encrypted as TD.

3. If the letters appear in different rows and columns, encode each as the letter in the same row but in the column of the other letter. For example, the pair LE is encrypted as 35.

The message LET US MEET AT NOON becomes, when grouped in fours, 35VR X2NZ DCR2 5885.

The Playfair cipher is easy to implement but extremely difficult to break. For this reason Wheatstone and Playfair described the system to the under secretary of the Foreign Office, suggesting that it was ideal for field work. The under secretary complained that the system was too complex. Wheatstone said that he could readily teach it to three out of four elementary school boys in 15 minutes, but the under secretary responded, "That is very possible, but you could never teach it to attachés."

11.7 Homophonic Codes

Standard frequency analysis is powerless against a **homophonic code** that assigns more than one symbol to each letter in such a way that the frequency of the code alphabet is uniform. If 100 symbols are used (such as the two-digit numbers from 00 to 99), the number of symbols assigned to a letter can be chosen to closely match the relative frequency of that letter. For example, since the letter A occurs approximately 8 percent of the time, 8 symbols are assigned to it. Likewise, E which occurs approximately 12 percent of the time is assigned 12 symbols. Table 11.4 shows such a code. During encryption the particular symbol to be used to represent a letter is chosen

TABLE 11.4
A Homophonic Code. If the symbols assigned to a letter are selected randomly when encrypting that letter, each symbol will occur with approximately equal probability, rendering ordinary frequency analysis virtually useless.

A	04, 25, 30, 43, 45, 47, 68, 86
B	51
C	67, 72, 93
D	22, 41, 55, 84
E	02, 12, 36, 48, 50, 53, 59, 66, 70, 77, 82, 89
F	06, 71
G	23, 29
H	11, 17, 52, 74, 78, 96
I	16, 20, 27, 46, 49, 62, 99
J	87
K	69
L	09, 32, 54, 73
M	44, 85
N	00, 14, 21, 33, 56, 90
O	01, 34, 37, 57, 61, 80, 91
P	07, 94
Q	63
R	05, 19, 28, 38, 58, 60
S	08, 24, 39, 65, 95, 81
T	10, 18, 26, 35, 42, 75, 76, 79, 83, 88
U	15, 40, 64
V	13
W	31, 97
X	98
Y	03
Z	92

randomly from those assigned to the letter, randomness being applied on each occurrence. The great mathematician Gauss is reported to have discovered the homophonic code and believed it to be unbreakable. (It is not.)

11.8 Jefferson's Wheel Cipher

Thomas Jefferson probably invented his **wheel cipher** during the 1790s. He described his invention in his personal papers, but apparently never put it to practical use. According to his description, the wheel is made from a wooden cylinder about 2 inches in diameter and 6 inches long, with a 1/4 inch hole bored through the center. This cylinder is sliced up into 36 disks, each about 1/6 of an inch thick. Around the circumference of each disk are inked the 26 letters of the alphabet in random order, each disk with a different order. The disks are then threaded through their center holes onto a shaft. The series of disks then define 26 rows, each of 36 random letters. A message of up to 36 characters is encoded by rotating the disks one at a

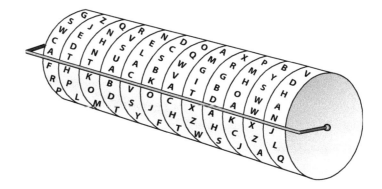

FIGURE 11.5 Simplified Jefferson's wheel cipher. The disks can be threaded on in any order. They are then rotated so as to spell out a message, and the ciphertext is taken as the text of any other row.

time so that one of the rows (and it does not matter which row) matches the text of the message. Once in position, Jefferson suggests locking the arrangement with a screw. The ciphertext is then taken to be the letters read across any other row. See figure 11.5 for a simplified version with only 12 disks.

This ciphertext is decrypted by the intended recipient, who has an identical wheel, by rotating the disks so as to match the ciphertext across one of the rows. The recipient then locks the disks in that position and turns the whole apparatus until a message that makes sense is observed across a row.

The disks of the wheel are numbered, and they can be threaded onto the shaft in numerous orders. The threading order is the key of the cipher system. Jefferson realized that there are 36! orders and computed (correctly) that this is about 3.72×10^{41} possible keys.[2]

Jefferson's wheel cipher is extraordinarily difficult to break. However, he apparently felt that it was too complex, and mysteriously, he selected instead a Vigenère cipher as the official cipher for the Lewis and Clark expedition. Had his wheel cipher been adopted by the U.S. government, it would certainly have been the most advanced encryption device available well into the 20th century.

11.9 The Enigma Machine

In 1918 the German inventor Arthur Scherbius filed a patent for a mechanical encryption machine that became known as the **Enigma machine**. It was not the first encryption machine, for seeds of the Enigma concept were contained in the early scytale, in Thomas Jefferson's cipher wheel, and in several other cipher machines. However, the Enigma remains the most advanced machine of that type actually manufactured, and was to play a vital (even pivotal) role in World War II.

The machine looks like an overgrown typewriter, for in fact the message is typed in on typewriter keys. (See figure 11.6.) The machine contains several other features, but the most important is the series of three disks (termed **rotors**) that implement complex

[2]The exact answer is 371,993,326,789,901,217,476,999,448,150,835,200,000,000.

FIGURE 11.6 The Enigma machine. The machine has typewriter keys for entry, three rotors to define a letter substitution, a plugboard to implement a further substitution, and finally, lightbulbs to indicate the result.

substitutions. The rotors, similar to Jefferson's disks, have circumferences divided into 26 segments. The Enigma rotors are made of nonconducting material such as hard rubber or ceramic. On the face of each rotor are 26 electrical contacts evenly spaced near the outer rim to match the 26 divisions. These contacts are electrically connected in pairs, one on the front face of the rotor, the other on the back face, but these pairs are in a random pattern. The first contact on the front might be connected to the sixteenth on the back. A rotor therefore defines a substitution cipher. If the machine were made with only a single rotor of this type, it would be arranged so that striking a keyboard key would apply voltage to the corresponding contact on the front face of the disk, and that current would pass to the corresponding contact on the back face, which would activate a lightbulb connected to that contact. Twenty-six lightbulbs are arranged in the same configuration as keys, so that each bulb represents letters by their position.

The three rotors are connected in series. Each rotor has a different arrangement of inner connections and thus represents a different set of permutations. When a key is struck, an electrical circuit passes through the first rotor, into and through the second, then into and through the third.

Now the true enigmatic nature comes to play. After a key is struck and a bulb lit, the first rotor rotates one space, thereby changing the permutation of the first rotor for the next letter. The second rotor rotates a single space after the first rotor makes a complete cycle of 26 spaces, like an automobile odometer; the third rotor steps one space only after the second makes a complete cycle.

The rotors are numbered 1, 2, and 3 and are themselves interchangeable. Thus the rotors might be placed in the machine in order 3, 1, 2. In later models a total of five rotors were available, from which three were chosen for a given setting.

There is a fourth disk termed a **reflector** that does not rotate and has 26 contacts only on one face. These contacts are connected in pairs so that current entering one contact emerges from another, sending the current back through the other three rotors along a path different from that taken in the forward direction. When it completes its circuit, the current lights one of 26 lamps, which indicates the encrypted version of the typed letter. The reflector adds no new complexity but greatly simplifies decryption because now an electrical path from a typewriter key (say Z) to a bulb (say K) is the same as the path from key K to bulb Z. Hence a message can be decoded by keying in the ciphertext, with the bulbs now giving the plaintext. The basic Enigma structure is illustrated in figure 11.7.

There is yet another complication. A **plugboard** allows for the interchange of two letters before they enter the rotors. This is accomplished by plugging the ends of a cable into holes corresponding to two letters. In later models, a total of up to 13 pairs of letters could be swapped in this way. In practice only 10 were used.

Finally, another complication is that associated with each rotor is a **ring** with a notch that determines when in a rotor's cycle it advances the next rotor. This would be like an odometer that advances the next digit on, say mile 4, instead of mile 0. The rings can be set at any of the 26 possibilities.

In practice, the German military provided a list of basic settings for each day, consisting of three ordered rotor numbers, ring settings, initial rotor positions, and plugboard arrangement. Then each message during the day varied only the initial rotor positions, which were sent as the first part of the message.

The number of possible Enigma keys is enormous. The number can be computed in steps, the largest contribution being from the plugboard.[3]

$$\text{Rotor choice} = 5!/(2!\ 3!) = 10 \text{ (selecting 3 from 5)}$$

$$\text{Rotor order} = 3! = 6$$

$$\text{Rotor positions:} = 26^3 = 17{,}576$$

$$\text{Ring positions:} = 26^2 = 676 \text{ (since only the first two matter)}$$

$$\text{Plugboard combinations} = 26!/[6!\ 10!\ 2^{10}] = 150{,}738{,}274{,}937{,}250$$

---

[3] The total number of combinations when n cables are used is $26!/[(26-2n)!(2n)!]$ (the number of ways to select $2n$ holes from 26), times $(2n)!$ the number of ways of inserting $2n$ cable ends), divided by 2^n (because connecting A to B is the same as connecting B to A), divided by $n!$ (the number of ways the cables can be ordered).

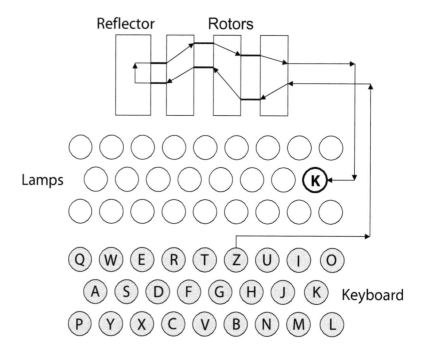

FIGURE 11.7 Basic Enigma structure. Pressing key Z completes an electrical circuit that passes through the three rotors, the reflector, and back through the rotors to light lamp K that indicates the encrypted version of Z. The path is reversible so that pressing K will light lamp Z. The actual Enigma includes additional complications of a plugboard and rings.

The total number of keys is therefore the astronomical number 107, 458, 687, 327, 250, 619, 360, 00 $\approx 1.0 \times 10^{23}$.

Actually the number of keys, though huge, is less than the 26! $\approx 4 \cdot 10^{26}$ keys of a single substitution code of 26 letters and much less than that of Jefferson's wheel cipher. Most of the Enigma keys are due to the plugboard substitutions. If these substitutions constituted the entire encryption, the Enigma would be susceptible to elementary frequency analysis and hence could be easily broken. The strength of the Enigma is that it scrambles the message in so complex a manner that basic letter frequency and word structure are effectively obliterated.

The Enigma served as the primary encryption system for the entire German military beginning in about 1928. During the war years, the Germans purchased over 30,000 Enigmas. The first successful attack on the Enigma was accomplished by the Polish cryptologist Marian Rejewski in about 1932. He had obtained design specifications for the early version of the machine used at that time. This used only three rotors and six plugboard possibilities. His method exploited a protocol of German messages that required each message to begin by repeating a three-letter rotor key to be used for that message. This minor redundancy was enough to decode the message by the use of a complex machine termed a **bombe** that Rejewski had built for the purpose.

When the Enigma was enhanced in 1938 by the addition of two more Enigma rotors, from which to select three, and the number of plugboard cables was increased from six to 10, this additional complexity was enough to render Rejewski's method impotent.

During the Second World War the later models of the Enigma provided essentially complete security for the German military. It was used to issue orders to submarines, coordinate land and air battles, respond rapidly to special situations, and in general direct the war in a way that was potentially devastating. Hundreds or thousands of messages were sent each day, all with perfect secrecy.

During the war, the British established a cryptography center at Bletchley Park to attempt to decipher German messages. The brilliant Alan Turing was recruited for the effort. Turing was a young mathematician who had already answered one of the greatest mathematical–philosophical questions of the time by showing, with the invention of an imaginary computer termed a **Turing machine**, that there are mathematical propositions that mathematics itself cannot resolve. At Bletchley Park, in a feat of tremendous genius and hard work, Turing devised a method for decoding the German transmissions. His method relied on an insightful analysis of the structure of the coded messages produced by the Enigma, the use of guesses or knowledge of the plaintext fragments (such as the German word WETTER, which appeared regularly in daily weather reports), and the development of a huge machine, again termed a bombe, that tried the large number of combinations that remained after incorporation of the first two method of attack. Turing's method was operational during 1940–42, and the British were able to read a high percentage of German communications. Of course, it was critical that the Germans not know that their "unbreakable system" had been broken. A challenge faced by the British then was to decide whether and how to intervene in military operations without raising the suspicion that the Enigma had been breached.

It is generally agreed that the war was shortened by about two years because of the breaking of the Enigma system using Alan Turing's method.

When the war ended, Turing worked at the National Physical Laboratory in London, and in 1948 he became the deputy director of the Computing Laboratory at Manchester, where the first electronically programmable computer was built. But later, Turing was not celebrated as a hero. Instead, this man who solved one of the most outstanding mathematical–philosophical problems of the age, played a decisive role in the war effort, and helped launch modern computing was arrested and had his secret clearance suspended because he admitted to having a homosexual relationship (which was illegal at the time). In 1954 at the age of 42 Alan Turing died from potassium cyanide poisoning widely believed to have been purposely self-administered.

11.10 The One-Time Pad

The strength of the Vigenère cipher increases with the length of the key: frequency characteristics and word structure are essentially eliminated. In fact, if the key is as long as the message, and itself completely random, the associated ciphertext will be random as well, with absolutely no structure that can form the basis for cryptanalysis. This special version of the Vigenère cipher is termed a **one-time pad**, the name coming from the practice of writing the random key letters on a pad of paper and using this pad to encrypt messages. Each page of a pad contains a different random sequence, and each key letter is used once only. When an entire pad has been used, it is discarded. A one-time pad provides perfect secrecy according to the precise definition given in the next chapter. It is mathematically impossible to decrypt it without the key.

Because of the ultimate security provided by the one-time pad, it has often been used in sensitive situations. In the military, the pads of random key letters took the form of a codebook, with a separate page to be used for each date. Capture of an enemy codebook was a great military prize.

The security of the one-time pad is sometimes approximated by using an actual published book such as a novel, with the letters of the novel used sequentially to define the Vigenère shift of successive message letters. The key letters are not strictly random in this case, and this may compromise the code's security.

Although the one-time pad is ideal in theory, there are number of practical difficulties that discourage it from being used widely. How can the random letters be generated? How can the long sequences be distributed to the various communicating parties who are a great distance apart? What if the pad falls into enemy hands? If the same pad is used by all communicating parties on any one day, doesn't that introduce redundancy that the enemy might use to advantage? These questions highlight the fundamental issue associated with classical encryption methods. Security rests with the key, and the question of distributing keys is itself an issue of secret communication. It is this question that motivated the development of the entirely new approach to encryption described in chapter 13.

11.11 EXERCISES

1. (An easy cipher) Decode the following:

    ZNOY OY GT KGYE IOVNKX ZU YURBK

2. (Autokey cipher) The following message was coded with the autokey system using a seed only one letter long. What is the message?

    PCN LMI ZNVQ WAL WWIH?

3. (Transposition cipher) Consider a transposition cipher that uses a five by five matrix and permutes the columns before reading them out.
    (a) How many keys are possible in such a cipher?
    (b) Decrypt the following.

    HTASL LEEOT ASWME TSSAP EMGCE.

4. (Beaufort cipher) Let $k = (k_1, k_2, \ldots, k_n)$ be a keyword of length n, and let $p = (p_1, p_2, \ldots, p_n)$ be a plaintext message of length n. The Beaufort encryption of the message is the ciphertext $c = (c_1, c_2, \ldots, c_n) = (k_1 - p_1, k_2 - p_2, \ldots, k_n - p_n)$, where $k_i - p_i$ denotes a backward shift of k_i by the shift corresponding to the letter p_i. For example, if $k = (H, I)$ and $p = (B, Y)$, then $c = (G, K)$.
    (a) Encode the message HELLO with the keyword JUMPS.
    (b) Show that encryption and decryption of a Beaufort cipher are identical processes. That is, to decipher it is only necessary to encrypt the ciphertext with the same keyword.

5. (Vigenère and transposition) Suppose a Vigenère cipher is constructed with a key that is three letters long. The result of this first encryption is further encrypted with a Vigenère cipher, with another three-letter key.
    (a) How many possible keys are embodied in the final ciphertext?
    (b) Suppose that after the first Vigenère encipherment, the resulting text is transformed by a transposition code that reads the text into a three by three matrix row by row and reads it out column by column after the columns are permuted. Effectively how many possible keys are embodied in the resulting double-encrypted ciphertext?
    (c) Suppose that after the Vigenère encryption and the transposition encryption, the text is subjected to another Vigenère cipher with a key of length 3. How many possible keys are embodied in this final result?
    (d) Suppose that after the first Vigenère cipher, the transposition cipher, and the second Vigenère cipher, the transposition is reversed (perhaps because the transposition key is discovered). The result is then equivalent to a compound Vigenère cipher. What is the length of the keyword in this compound Vigenère cipher?
    (e) How many possible keys are embodied in this compound Vigenère cipher?

6. (Affine ciphers) Let a be an integer between 1 and 26, and let x be the integer corresponding to one of the 26 letters of the alphabet. The corresponding linear cipher transforms the message x into the ciphertext y by

$$y = ax \quad \mod 26.$$

For example, if $a = 3$ and the message is "d," the ciphertext is $y = 3 \times 4 = 12$. If the message is "k," the ciphertext is $y = 3 \times 11 \mod 26 = 33 \mod 26 = 7$.

To be acceptable, the value of a must be **invertible** mod 26, such that the correspondence from x to y can be uniquely inverted. A case that does not have this property is $a = 4$, for then both $x = 3$ (for "c") and $x = 16$ (for "p") lead to $y = 12$. An a is invertible if it has no common factor, other than 1, with 26. Hence, 4 is not invertible since both 4 and 26 share the factor 2. (See section 13.5.)

(a) List the acceptable values of a.

(b) Decipher the following linear ciphertext:
$$13 \quad 9 \quad 9 \quad 10 \quad 7 \quad 10 \quad 20 \quad 11 \quad 20 \quad 9.$$

(c) An **affine cipher** is of the form $y = ax + b$ mod 26, where both a and b are integers between 1 and 26, with a being one of the values in part (a). How many keys are there in affine ciphers?

(d) An affine cipher can be combined with a Vigenère cipher by fixing a but using k different values of b and cycling through these b values, letter by letter. How many keys are there in this compound cipher?

7. (Hill cipher) Hill devised a cipher that extends both transposition and linear ciphers. It has the form

$$y = xA \bmod 26$$

where A is an $n \times n$ matrix of integers. The n-dimensional message (row) vector x is transformed into the ciphertext vector y. For example, with

$$A = \begin{bmatrix} 11 & 8 \\ 3 & 7 \end{bmatrix}$$

the message $x = (8, 9)$ can be transformed by the Hill transformation as

$$y = (8, 9) \begin{bmatrix} 11 & 8 \\ 3 & 7 \end{bmatrix} = (88 + 27, 64 + 63) \bmod 26 = (11, 23).$$

To decipher the result, the inverse of the matrix A (mod 26) is applied. This inverse will exist if the determinant of A has no common factor, except 1 and 26, with 26. Often it is arranged that the determinant of A mod 26 is in fact 1. For example, the determinant of the matrix A given above is $11 \times 7 - 8 \times 3 = 53 = 2 \times 26 + 1 \rightarrow 1$ in mod 26 terms.

(a) Find the inverse of the A matrix given above. (Recall that the inverse of a two by two matrix is

$$\begin{bmatrix} a_{11} & a_{12} \\ a_{21} & a_{22} \end{bmatrix}^{-1} = \frac{1}{D} \begin{bmatrix} a_{22} & -a_{12} \\ -a_{21} & a_{11} \end{bmatrix}$$

where D is the determinant of A.)

(b) Decipher the ciphertext $(1, 2)$.

11.12 Bibliography

There are a number of excellent presentations of classical cryptography, including the elementary and entertaining [1], [2], and [3] and the more advanced [7]; as well as the texts referenced for chapter 12. Comprehensive histories of cryptography and its role in significant life circumstances are the large and wonderful books [4] and [5]. A computer program to solve cryptograms was outlined in [6]. See [7] for a good discussion of the affine, Hill, and Beaufort ciphers.

References

[1] Gardner, Martin. *Codes, Ciphers, and Secret Writing*. Mineola, N.Y.: Dover, 1984.

[2] Pickover, Clifford A. *Cryptorunes: Codes and Secretc writing*. Rohnert Park, Calif.: Pomegranate Communications, 2000.

[3] Beutelspacher, Albrecht. *Cryptology*. Trans. J Chris Fisher. Washington, D.C.: Mathematical Association of America, 1996.

[4] Kahn, David. *The Code Breakers*. New York: Scribner, 1996.

[5] Singh, Simon. *The Code Book*. New York: Doubleday, 1999.

[6] Hart, George W. "To Decode Short Cryptograms." *Communications of the ACM* 27, no. 9 (1994): 102–8.

[7] Mollin, Richard A. *An Introduction to Cryptography*. Boca Raton: Chapman & Hall/CRC, 2001.

12

CRYPTOGRAPHY THEORY

Perhaps it is not surprising that the basic theory of classical cryptography was established by Claude Shannon. Indeed, just two years after publication of his *Mathematical Theory of Communication* he published a definitive analysis of classical cryptography. In fact, Shannon developed his theory of cryptography simultaneously with his theory of communication because the two are intimately linked, each using similar concepts and addressing similar issues. Encryption is a form of encoding, and hence the same tools used to analyze codes designed for accurate transmission can also be used to analyze ciphers designed to render transmission and recovery difficult. Indeed, the general objective of a system of encryption is to render a message completely random, so that its entropy is maximal.

Shannon's theory makes two important and fundamental contributions. The first is a precise definition of perfect security in an encryption system. His theory also gives conditions that imply such security. The second contribution is a quantitative relation between the redundancy of a language as measured by entropy and the degree of security offered by a specific encryption system. The greater the redundancy, the easier it is to break a code.

This chapter presents the essentials of Shannon's theory and relates it to some familiar ciphers and to some practical considerations.

12.1 Perfect Security

In a cryptosystem let M denote a plaintext message, C the ciphertext representing the encrypted message, and K the key. The key is simply an intermediate element that guides the translation of the message M into the encrypted version C.

In cryptography theory it is assumed that the form of the encryption system is known by all parties: the sender, the receiver, and an outsider seeking to break the code. The encryption system is completely defined once the key is known, and at the

186

outset at least, this key is known only to the sender and intended receiver, not the outsider.

It is assumed that there are a finite number of possible messages and keys with known probabilities. Usually it is assumed that the key and message are probabilistically independent. The initial (*a priori*) probability of a message M is $p(M)$.

The outsider sees only C, and from this seeks to deduce M. The outsider will therefore compute the conditional probability $p(M|C)$.

Perfect security. A system is perfectly secure if

$$p(M|C) = p(M) \tag{12.1}$$

for all possible messages M and all sets of *a priori* probabilities of messages.

According to this definition, the probability of any message is not changed by observation of the ciphertext.

Example 12.1 (Single Caesar cipher). Consider the Caesar cipher used to transmit a single letter. The set of messages is the set of 26 alphabet letters, the set of keys is the 26 possible shifts, and the set of possible ciphertexts is again the set of 26 letters. Suppose the probabilities of the message letters correspond to their natural English frequencies, and assume that the keys are chosen with equal probabilities (each equal to 1/26). Then the probability $p(M|C)$ of any given one-letter message M given a cipher letter C is again $p(M)$. Hence, this system has perfect secrecy.[1] This is not true if the probabilities of the key letters are unequal.

There is a useful test for perfectly secure systems, stated below.

Lemma 12.1. A system is perfectly secure if and only if

$$p(C|M) = p(C) \tag{12.2}$$

for every possible message and every possible ciphertext.

Proof: By Bayes' rule[2]

$$p(M|C) = \frac{p(C|M)p(M)}{p(C)}.$$

For perfect security we require $p(M|C) = p(M)$ for all M and all probability distributions $p(M)$. Hence, it must follow that $P(C|M) = p(C)$. Conversely, if $P(C|M) = p(C)$, it is clear that the system is perfectly secure. ∎

---

[1] Here is a short formal proof. By definition

$$p(M|C) = \frac{p(C|M)p(M)}{p(C)}.$$

Clearly $p(C|M) = 1/26$ since the key letters are equally probable. Also $p(C) = \sum_M p(M, K = C - M)$, where $C - M$ denotes the amount of shift between the letters C and M. Since the key is independent of M, this says that $p(C) = \sum_M p(M)/26 = 1/26$. Hence, $p(M|C) = p(M)$.

[2] See section 5.3.

Example 12.2 (Caesar for two letters). Consider a system similar to that of the previous example except that now each message consists of pairs of letters m_1m_2, but the same Caesar shift key applies to both letters. If all two-letter combinations are possible in messages, then all two-letter combinations are possible ciphertexts. However, if the message M is a doubleton such as *aa*, the ciphertext will also be a doubleton. Other ciphertext combinations are not possible. Hence $p(C|M) \neq p(C)$ for that message M. Hence the system is not perfectly secure.

The above two examples illustrate a basic condition for perfect security; namely, the number of keys must be at least as great as the number of possible messages. This basic result is stated formally below.

Theorem 12.1. A necessary condition for a system to be perfectly secure is that the number of possible keys be at least as large as the number of possible messages.

Proof: For simplicity, suppose the system is not homophonic; that is, corresponding to a fixed key and message, there is a unique ciphertext. Suppose there are n possible messages. Let K be a certain fixed key. Then K maps the n messages $M_1, M_2, \ldots, M_n$ into n distinct ciphers $C_1, C_2, \ldots, C_n$ and for convenience assume that M_i is mapped to C_i for each $i = 1, 2, \ldots, n$.

Assuming perfect security, $p(C_1) = p(C_1|M_1) > 0$ since the key K maps M_1 into C_1. Now consider some $j \neq 1$. For perfect security $p(C_1|M_j) = p(C_1) > 0$. Hence there must be a key that maps M_j into C_1. This key is of course different from K. Likewise for every i there is a key that maps M_i into C_1. These keys are all different and hence there must be at least n possible keys. ∎

It must be remembered that the number of keys is the total number of possible combinations. For example, in a Vigenère cipher with key length L, there are 26^L possible keys. If the message is of length N, there are 26^N possible messages. A necessary condition for perfect security is therefore $26^L \geq 26^N$ or equivalently $L \geq N$.

12.2 Entropy Relations

Encryption can be viewed in terms of entropy as well as in terms of probabilities. Indeed, it follows almost immediately that the condition for perfect security can be restated in entropy terms.

Theorem 12.2. A system is perfectly secure if and only if for every set of probabilities

$$H(M|C) = H(M). \tag{12.3}$$

Proof: We have

$$H(M|C) = -\sum_{i,j} p(M_i, C_j) \log p(M_i|C_j) \quad \text{by definition}$$

$$= -\sum_{i,j} p(M_i, C_j) \log p(M_i) \quad \text{by perfect security}$$

$$= -\sum_{i} p(M_i) \log p(M_i) \quad \text{the sum over } C_j\text{'s is 1}$$

$$= H(M).$$

This proves that the system is perfectly secure only if $H(M|C) = H(M)$. Conversely, for this to hold for any set of probabilities, it must be true that $p(M_i|C_j) = p(M_i)$ for all i and j. ∎

The equation $H(M|C) = H(M)$ can be interpreted as saying that no information about the message is conveyed by the ciphertext. The mutual information[3] $I(M; C) = H(M) - H(M|C) - 0$. It is as if the encryption system is a completely noisy channel, from which it is impossible to deduce any information about the input.

Entropy can also be used to quantify the degree of security offered by systems that are not perfectly secure. The analysis is based on the following basic relation.

Theorem 12.3. For any encryption system with K and M independent,

$$H(K|C) = H(K) + H(M) - H(C). \qquad (12.4)$$

Proof:

$$
\begin{aligned}
H(K|C) &= H(K, C) - H(C) \quad \text{by the definition of conditional entropy} \\
&= H(M, K, C) - H(C) \quad \text{since } M \text{ adds nothing to } K \text{ and } C \\
&= H(M, K) - H(C) \quad \text{since } C \text{ adds nothing to } K \text{ and } M \\
&= H(M) + H(K) - H(C) \quad \text{since } M \text{ and } K \text{ are independent.}
\end{aligned}
$$

(Second-step detail: $H(M, K, C) = H(M|K, C) + H(K, C) = H(K, C)$.) ∎

Unicity Point

Suppose that a key chosen from a finite set is used to encrypt messages of variable length. For example, a Caesar cipher has a key chosen from the set of 26 letters, but can be used to encrypt a message of any length. It is of interest to know how long, on average, a message must be in order that the message and the key can be deduced by cryptanalysis of the ciphertext.

For simplicity we consider only systems in which the key length is fixed and does not vary with the length of the message, and in which the message and key uniquely determine the ciphertext.

Denote the message of length n by M^n and the corresponding ciphertext by C^n. Equation (12.4) becomes

$$H(K|C^n) = H(K) + H(M^n) - H(C^n).$$

If the key is chosen randomly from N_K possibilities, then $H(K) = \log N_K$. A good approximation[4] for $H(M^n)$ is $H(M^n) = nH$ where H is the per-symbol entropy of the language. Finally, if there are N_C possible ciphertext symbols and the cipher system is

---

[3] See chapter 5.
[4] This works well for large n, since $H = \lim_{n \to} H(M^n)/n$.

sufficiently complex, then the ciphertext appears random and a good approximation is $H(C^n) = n \log N_C$. Combining these substitutions,

$$H(K|C^n) = \log N_K + n(H - \log N_C). \tag{12.5}$$

The **unicity point** is the value of n that, on average, reduces the number of possible keys to one—or equivalently, reduces the entropy of the key given the ciphertext to zero. This value is found by setting the above equation to zero. Hence

$$n_{\text{unicity}} \approx \frac{\log N_K}{\log N_C - H}. \tag{12.6}$$

Example 12.3 (Caesar cipher). Consider the possibility of breaking a Caesar cipher of English using frequency analysis. Then $N_K = 26$, $N_C = 26$, and $H = 4.2$, which is the first-order entropy of English based on the frequency of letters (see chapter 3). Hence

$$n_{\text{unicity}} \approx \frac{\log 26}{\log 26 - 4.2} = \frac{4.7}{4.7 - 4.2} = \frac{4.7}{.5} = 9.4.$$

This means that about 9 or 10 ciphertext letters should be sufficient to determine the key of a Caesar cipher using frequency analysis.

Example 12.4 (Vigenère cipher). Consider a Vigenère cipher with a key of length L_K. Assuming the length were known, one could attack the cipher by frequency analysis applied to each group of letters separated by L_K letters, as suggested in chapter 11. The analysis is therefore the same as that of a Caesar code on each of the key symbols. Hence, we conclude that a message of about 10 times the length of the key should be sufficient to break a Vigenère cipher, and this is in accord with experience with the method of chapter 11 for these ciphers.

Example 12.5 (Substitution cipher). Consider a full substitution cipher of English, producing cryptograms. In this case the number of possible keys is $26! \approx 4 \times 10^{26}$. We know that frequency analysis alone is not effective for breaking such codes, but use of word structure can be very effective. Accounting for this, it is appropriate to use the estimate discussed in chapter 4, section 4.2, that the entropy of English is about 1.5 bits per letter. This then gives the unicity point as

$$n_{\text{unicity}} \approx \frac{\log 26!}{\log 26 - 1.5} = \frac{88.4}{4.7 - 1.5} = \frac{88.4}{3.2} = 27.6.$$

Hence it is expected that most cryptograms of roughly 30 letters can be broken, and again this is in accord with experience.

The unicity measure assumes unlimited computing time, whereas the methods typically used to break these ciphers use quite limited computing capacity. It is thus perhaps surprising, and gratifying, that the unicity measure agrees quite well with actual practice.

Formula (12.5) can also be used to determine the strength of a cipher when the message length is shorter than the unicity length.

Example 12.6 (A long key). Suppose that a Vigenère cipher has key length 100 and message length 130. Then $N_K = 26^{100}$ giving $H(K) = 100 \log 26 = 470$. Assuming that an attack uses the full structure of English, one may again set $H = 1.5$. Equation (12.5) gives

$$H(K|C^{130}) = 470 - 130(4.7 - 1.5) = 470 - 416 = 54.$$

This means that the remaining entropy in the key is 54, equivalent to there being 2^{54} **spurious keys**. That is, there remain 2^{54} reasonable keys (down from $26^{100} = 2^{470}$) that if applied backward to the ciphertext would produce meaningful English. If the message were as long as 150 letters, the unicity point would be reached.

Redundancy

Shannon defined the **redundancy** R_L of a language by the formula

$$R_L = 1 - \frac{H}{\log N_M},$$

where H is the per-symbol entropy of the language and N_M is the number of symbols in the language's alphabet. If, for example, every possible sequence of letters were possible and equally probable, then $H = \log N_M$ and the redundancy would be zero. If the entropy is less than the maximum possible for the size of the alphabet, the redundancy will have a value satisfying $0 < R_L < 1$.

If for English $H = 1.5$, the corresponding redundancy is $R_L = 1 - 1.5/4.7 = .68$. Hence English is approximately 70 percent redundant.

The unicity point can be expressed neatly in terms of redundancy. For a substitution cipher $N_C = N_M$ and if one takes $H_C = \log N_C$, then (12.6) yields

$$n_{\text{unicity}} = \frac{\log N_K}{\log N_M - H} = \frac{\log N_K}{R_L \log N_M}.$$

Hence low redundancy implies a large unicity point. In the limit, if the language has zero redundancy, the unicity point is infinite.

Note that if the entropy of English is 1.5 bits per letter, this same entropy would be approximately achieved by a language with an alphabet of only three letters (instead of 26) if any combination of letters were a legitimate message in the language. Hence, a redundancy of 70 percent represents a significant departure from the absolute minimum requirements of a language.

Approaching Unicity*

In actuality, equation (12.5),

$$H(K|C^n) = \log N_K + n(H - \log N_C),$$

cannot hold for all n if $H - \log N_C < 0$ since it would give a negative value to $H(K|C^n)$ for large n. Hence the unicity point is based on only a rough approximation of the rate of decrease of $H(K|C^n)$.

A more refined approach is based on examination of each step separately. Let M_n and C_n denote the n-th transmitted message symbol and corresponding cipher text, distinguished from M^n and C^n, which are, as before, the sequences of the first n such symbols. Let K be the key. Then

$$
\begin{aligned}
H(K|C^n) &= H(K, C^n) - H(C^n) \\
&= H(K, C_n|C^{n-1}) + H(C^{n-1}) - H(C^n) \\
&= H(K|C^{n-1}) + H(C_n|K, C^{n-1}) + H(C^{n-1}) - H(C^n) \\
&= H(K|C^{n-1}) + H(M_n|K, C^{n-1}) - H(C_n|C^{n-1}) \\
&= H(K|C^{n-1}) + H(M_n|M^{n-1}) - H(C_n|C^{n-1}).
\end{aligned}
\tag{12.7}
$$

This shows that $H(K|C^n)$ varies from $H(K|C^{n-1})$ by the addition of the single-step entropy of the message minus the single-step entropy of the ciphertext. As the key becomes known, the difference between the entropy of the message and that of the ciphertext falls, so the entropy of the key does not change very much, and learning becomes more gradual.

Example 12.7 (Binomial decryption). Let us apply this recursive formula to the binary case. Take the case where the message alphabet consists of just 0 and 1 with probabilities p and $1 - p$, respectively. Successive symbols are independent. The entropy of the source is then $H(M_n|M^{n-1}) = H(p) \equiv -p \log p - (1 - p) \log (1 - p)$.

The key is fixed at either 0 or 1, and the ciphertext is $C_n = M_n + K \mod 2$. Thus the message 0's and 1's are either all received correctly or they are all received incorrectly. From the perspective of an analyst trying to break the cipher, the key is initially unknown and hence the key values of 0 and 1 are assigned probabilities q_0 and $1 - q_0$, both 1/2.

Later, when the probability of the key being 0 is q, the probability of a ciphertext symbol being 0 is $pq + (1 - p)(1 - q)$. See figure 12.1.

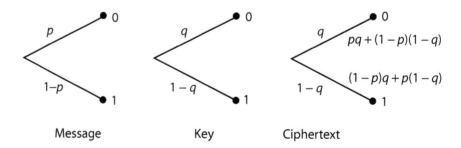

Message   Key   Ciphertext

FIGURE 12.1 Binomial message. A message is either 0 or 1; it is added to either 0 or 1 (mod 2) to produce the ciphertext.

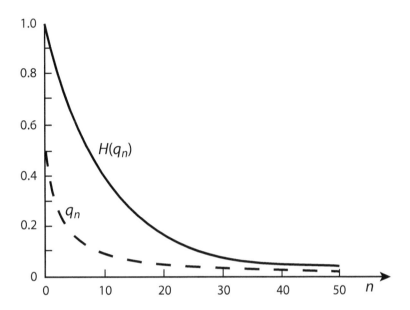

FIGURE 12.2 Decrease in key entropy and convergence of q_n with length of observed ciphertext. Entropy decreases in an exponential fashion as more ciphertext is made available.

Basically, as one watches the stream of ciphertext, it is possible to refine the estimated probability of q until it is found that $q = 0$ or $q = 1$. After $n - 1$ ciphertext symbols have been observed, the corresponding estimate q_{n-1} translates into a value of $H(K|C^{n-1})$ through $H(K|C^{n-1}) = H(q_{n-1})$.

The recursion (12.7) can be expressed in terms of the q_n's as

$$H(q_n) = H(q_{n-1}) + H(p) - H(pq_{n-1} + (1 - p)(1 - q_{n-1})). \tag{12.8}$$

This can be regarded as a recursion for q_n except that there is always a choice of two possible values, each giving the same entropy. This choice would be resolved by the specific ciphertext, but for the recursion we can take the lowest value (which means that q_n will converge to 0). The result of such a recursion using $p = .3$ is displayed in figure 12.2, which shows both q_n and $H(q_n)$ as a function of n. The standard unicity value n is defined as

$$n = \frac{H_K}{H_C - H_M} = \frac{1}{1 - H(.3)} = 8.4.$$

At that point the actual entropy has dropped only to about one-half of its initial value.

12.3 Use of a One-Time Pad*

Suppose a one-time pad is used for encryption, with successive key letters chosen randomly. According to theorem 12.1, perfect security requires that the number of possible keys be at least as great as the number of possible messages. Hence, the alphabet used in the one-time pad must be the same size as or larger than that of the message alphabet. Generally, for English the ordinary alphabet is used for both, or for computer systems the two-symbol alphabet of 0 and 1 is used for both.

If the key symbols are chosen completely randomly, then $p(M|C) = p(M)$, as shown in example 12.1, and the one-time pad is perfectly secure.

It is important that the key symbols in a one-time pad be random, but from a practical point of view it is not entirely clear how to accomplish this while securely distributing the pad to both receiver and sender. If the random numbers are generated by the sender, they must be transmitted to the receiver, and this message itself may be intercepted by an enemy. In practice, the key sequence is produced by both sender and receiver using a predetermined pseudorandom number generator. Hence only the design of the generator need be known by the communicating parties. Two such generator methods are discussed here.

Shift Registers

Shift registers (see chapter 6) can generate **pseudorandom sequences**, and this method is often applied to binary encryption systems where the message, key, and ciphertext alphabets consist of 0 and 1. An individual message symbol m_i is encrypted by adding the random key symbol k_i to produce the cipher symbol $c_i = m_i + k_i$ with addition carried out modulo 2.

A general linear shift register of order n is shown in figure 12.3. It is determined by the n coefficients $a_1, a_2, \ldots, a_n$ and the initial conditions, which need not be those of the figure. Usually, it is understood that in an n-th order shift register $a_n \neq 0$, for otherwise the performance would be that of an $(n-1)$-th order register, simply delayed one step.

Three important properties of linear shift registers are easily stated.

1. The maximum possible period length of an n-th order linear shift register is $2^n - 1$.

    A period ends when the register combination is identical to one that occurred earlier. Subsequent combinations will repeat those that were derived earlier from

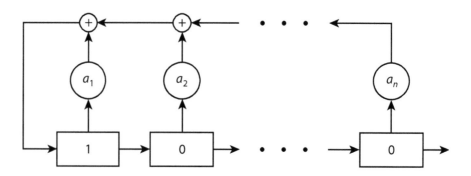

FIGURE 12.3 General linear shift register. The general linear shift register of length n has n feedback coefficients (which are zero or one) and n initial conditions.

that combination. The n registers can hold only 2^n distinct combinations, corresponding to 2^n binary numbers. However, if the combination of all zeros is obtained, the shift register will generate 0 in all subsequent steps, and hence the period length will be 1. Hence, the best case is for $2^n - 1$ combinations to be generated.

2. If a linear shift register of order n has a period of length $2^n - 1$, then the output of a complete cycle will consist of exactly 2^{n-1} ones and $2^{n-1} - 1$ zeros. (That is, the number of ones and the number of zeros are almost equal.)

   If the period is $2^n - 1$, then the shift registers hold all combinations of length-n binary numbers except the combination of all zeros. In the entire set of binary numbers exactly half end in 0 and half in 1. The shift register does not include one of these; namely the one with all zeros. Hence there are the full 2^{n-1} ones and one less zero.

3. For any n there is a linear shift register with period $2^n - 1$.

   We shall not prove this statement, but in fact it is relatively easy to construct such shift registers.

Shift registers are simple to implement and if designed to have maximal period, they can produce long sequences of pseudorandom numbers. For example, a shift register of order 25 is fairly simple and yet, if it has maximal period, that period will be $2^{25} - 1 = 33,554,431$, which is much longer than most messages. Shift registers provide convenient keys, for it is necessary only that the sender and receiver agree on the n feedback coefficients and n initial conditions, a total of $2n$ values, rather than each having an entire list of $2^n - 1$ random digits.

A weakness of shift registers in encryption, however, is that their structure can be deduced if a few message symbols are correctly identified. Knowing $2n$ (appropriate) output values (deduced perhaps from knowledge of the $2n$ corresponding message symbols), the $2n$ parameters of an n-th order shift register can be calculated, and the entire cipher decoded.

There are more complex techniques, using for instance nonlinear feedback, that offer greater security; but still, systems with keys generated this way are not immune from dedicated attacks.

Linear Congruences

Another simple and popular method for generating sequences of random numbers is based on linear congruences with modular mathematics. Specifically, a sequence of random numbers is generated from a recursion of the form

$$r_{i+1} = ar_i + b \quad \mod m,$$

where m is termed the **modulus**, a is the **multiplier**, and b is the **increment**. The process is initiated with a **seed** r_1.

Not every choice of parameters is suitable, since the resulting series of numbers may have a short period. For example, taking $m = 10$, $a = 1$, $b = 5$, and $r_1 = 1$ leads to the series 1 6 1 6 1..., which is quite degenerate.

However, some combinations lead to periods of length $m - 1$ and have good randomness properties. For instance $m = 23$, $a = 5$, $b = 7$, and $r_1 = 1$ produces a sequence having a period of 22.

In practice much larger numbers are used. For example, $m = 120,050$, $a = 2,311$, $b = 52,367$ is reported to be a good set of constants.

Because it is easy to implement and the resulting sequences have good randomness properties, the linear congruence method is often used for simulating a series of random events. However, the method is not perfectly secure when used in encryption, for like shift registers, it suffers from the fact that it is possible to deduce the constants of the recursion from knowledge of relatively few terms.

More secure random number generators have been developed, using the mathematical techniques discussed in the next chapter.

12.4 The DES and AES Systems

The Data Encryption Standard (DES) was initially developed at IBM and (with minor modifications) adopted by the U.S. government in 1976. It served as a useful standard method of encryption for 20 years, although it is now superseded by the Advanced Encryption Standard (AES) and the modern public key methods discussed in the next chapter.

DES is based on a series of substitutions and transpositions determined by a key of 64 binary bits, eight of which are parity checks. The effective key size is therefore 56 bits.

A primary weakness of DES is that there are (only) $2^{56} \approx 7.2 \ 10^{16}$ (72 quadrillion) possible keys. There are, therefore, fewer keys than that of a single simple substitution code of 26 letters, of Jefferson's wheel cipher, or of the Enigma machine. Indeed, in the late 1990s a huge DES "cracker" machine was able to break DES-encoded messages in a few hours, showing that it was time to develop a stronger standard. The result is the AES system.

The AES system has three versions with key sizes of 128, 192, and 256 bits. The 128-bit AES system has $2^{128} = 3.4 \ 10^{38}$ possible keys as compared to the $2^{56} = 7.2 \ 10^{16}$ of DES. Hence, there are about 10^{21} times more AES keys than DES keys. To illustrate this magnitude, suppose that it were possible to test all DES keys in 1 second. At that speed, it would take 149 trillion years to check all AES keys. The age of the universe is estimated to be only(!) 13.7 billion years.

12.5 EXERCISES

1. (Secure condition) Show that a necessary and sufficient condition for perfect security is $H(C|M) = H(C)$ for every set of probabilities.

2. (Homophonic keys) Prove theorem 12.1 for homophonic systems.

3. (Coding identity) Show that for any cryptosystem

$$H(K,C) = H(M) + H(K),$$

where M is the message, K is the key, and C is the ciphertext.

4. (Identities*) Show that for a cryptosystem

$$H(M|C) = H(M,C,K) - H(K|M,C) - H(C)$$

and

$$H(K|C) = H(M|C) + H(K|M,C).$$

5. (More identities) Show that for a cryptosystem

$$H(C|M) = H(M,K,C) - H(M) - H(K|C,M) = H(K) - H(K|C,M).$$

6. (Unicity of transposition) Notice that for a transposition cipher, the resulting letters maintain their original frequency distribution. From this, deduce the unicity point of a transposition cipher of period p.

7. (How big?) What is the unicity point of a one-time pad?

8. (Chessboard unicity) A transposition code is devised that permutes the letters in each block of 64 letters according to a key that is a fixed permutation used for each of the blocks. (Perhaps the transposition is constructed by randomly assigning the numbers 1–64 onto the squares of a chessboard, and then reading them out by row.) This transposition cipher will of course not change the frequency distribution of the letters. Find the unicity point for this cipher. Use the data below to the extent that it is helpful. (Entropy values are per-letter values.)
    (a) Zero-th order entropy of English $= \log 26 = 4.7$.
    (b) First-order (frequency) entropy of English $= 4.2$.
    (c) Entropy of English $= 1.2$.
    (d) $\log 26! \approx 88$.
    (e) $\log 64! \approx 300$.

9. (Running cipher) One method for increasing the security of Vigenère-type encryption is to use a running text as key. For example, the key might be the letters, in order, from an obscure novel that is possessed by both the sender and intended receiver of the message. The message itself is encoded by the Vigenère system—encoding the i-th letter of the message by shifting the alphabet by an amount determined by the i-th letter of the key. Both the message and the key are in the same language, which has L letters in its alphabet, and which has a redundancy of r_L. As an approximation, assume that the resulting ciphertext is essentially completely random, with all letters having the same probability and each letter being independent of the others.

(a) What is $H(C^n)$?

(b) What are $H(M^n)$ and $H(K^n)$?

(c) What degree of redundancy r_L (a percentage) of the language is necessary in order that this code system be breakable?

(d) Is the system breakable if the language is standard English?

10. (Homophonic cipher) Suppose that a language using an alphabet of J letters has the property that the i-th letter has a frequency of k_i/N, for each i, where the k_i's and N are integers with $\sum_i k_i = N$. A homophonic cipher is made up of exactly N symbols, with k_i of these assigned to the i-th letter of the alphabet (and with no duplications). When a letter i is to be encrypted, one of the k_i cipher symbols assigned to it is chosen at random. This will make all N code symbols appear with probability $1/N$.

To determine the unicity point of such a code, it is necessary to modify the basic relation for $H(K|C)$ because $H(C|M, K)$ is not zero.

(a) Show that in general

$$H(K|C) = H(C|M, K) + H(M) + H(K) - H(C).$$

(b) Show that for the situation of this exercise

$$H(C|M, K) = \sum_{i=1}^{J} \frac{k_i}{N} \log k_i.$$

(c) If the language has an entropy determined only by the frequency of its letters, can the homophonic cipher be broken?

(d) The number of ways of assigning symbols for this cipher is

$$W = \frac{N!}{k_1! k_2! \cdots k_J!}.$$

If the entropy of the language is H per letter, what is the unicity point of the cipher?

(e) Assuming that the entropy of English is 1.5 bits/letter and that the homophonic code of 100 symbols in the text is used, what is the unicity point? (Note that $\log W = 377$.)

11. (Maximal period shift register) Show that if a linear shift register has maximal period, then a sequence having that period is generated by any nonzero initial condition of the registers.

12. (Two shift registers) Construct two third-order shift registers (and initial conditions) such that the period of one of these is 1 and the other is 7. Always take $a_3 = 1$ so that the registers are fully third-order.

12.6 Bibliography

Shannon's basic paper originating the theory of classical cryptography is [1]. The extension of the concept of unicity to the notion of spurious keys is due to Hellman [2]. The step-by-step analysis of $H(K|C^n)$ for the binary language is apparently new. See [3] for a good discussion of crypto-identities such as those of exercises 3, 4, and 5 and other aspects of the theory. A comprehensive treatment of cryptography is [4].

References

[1] Shannon, Claude E. "Communication Theory of Secrecy Systems." *Bell System Technical Journal* 28 (1949): 656–715.

[2] Hellman, M. E. "An Extension of the Shannon Theory Approach to Cryptography." *IEEE Transactions on Information Theory* 23 (1977): 289–94.

[3] Welsh, Dominic. *Codes and Cryptography*. Oxford: Oxford University Press, 1988.

[4] Schneier, Bruce. *Applied Cryptography*. 2nd ed. New York: Wiley, 1996.

13 ·E·E·E·E·E·

PUBLIC KEY CRYPTOGRAPHY

The weak link in classical cryptology is the transmission of keys between sender and receiver. In communication networks—such as in military operations, international business, or personal finance—a different key should ideally be used by each pair of participants. This massive and frequent exchange of keys is vulnerable to interception and thus to serious security compromise. Indeed, the management of keys dominates the practical implementation of classical cryptography.

In the 1970s Prof. Martin Hellman at Stanford University, working with graduate student Whitfield Diffie, addressed the key management problem directly. They believed, intuitively, that it should be possible to securely transfer a key by digital means. In 1976, in a celebrated paper, they outlined the basics of what is now known as **public key cryptography**, and they provided a specific method for two separated parties to securely construct a key using what is now known as the **Diffie–Hellman key exchange** method. This general approach, which emphasizes computational complexity as a means for providing security, is today the basis for secure ciphers, digital cash, digital signatures, secure Internet transmission, and other digital products that depend on security.

This chapter presents the elements of the public key concept and its implementation. The following chapter discusses many of the theory's interesting applications.

13.1 A Basic Dilemma

Alice and Bob are standard fictional characters in discussions of cryptography. We introduce them now in the context of a simple dilemma or puzzle, the solution of which hints at the techniques of public key cryptography.

Suppose that Alice wishes to send to Bob a handwritten letter through the mail, and she does not want anyone else to read it. She feels that ordinary envelopes, which can be steamed open, do not offer enough security. She has at her disposal a solid

FIGURE 13.1 Alice wants to send a letter to Bob.

metal box that can be secured with a padlock to which she has the only key. If Alice places her letter in the box, locks it with the padlock, and sends it through the mail, Bob will receive it safely. However, since Bob does not have a copy of the key, he is unable to open the box. How can Alice securely send the letter to Bob?

She might first consider sending a copy of the key to Bob, for once he had it, he could use it to open the box. But if she uses the mail for this purpose, the key might be intercepted and duplicated. This is exactly the dilemma of two people wishing to communicate with a classical cipher based on a key. The key must first be sent from one party to the other.

There is a simple but clever solution to this puzzle: Alice places the letter in the box and locks the box with her padlock. The box travels securely through the mail. Bob receives it in fine order but cannot open it. Instead, he further secures the box with an additional padlock to which he has the only key. He then returns the box to Alice through the mail, its security protected now by two locks.

When Alice receives the double-locked box, she unfastens her lock and sends the box back to Bob, again through the mail. When Bob receives the box, he opens the remaining padlock with his key, and reads Alice's letter. Transmission of the letter was secure even though neither Bob nor Alice ever possessed the other's key.

This solution suggests that security may be achieved by clever schemes when brute force or direct methods fail, and this is a theme of public key cryptology. This puzzle is referred to in later sections, in order to relate digital procedures to the physical procedures used by Alice and Bob.

13.2 One-Way Functions

One feature of a padlock is that it can be locked without the key. The key is needed only to unlock it. This is a characteristic of public key cryptology as well. The mathematical analogy of a lock is a **one-way function**. Such a function is easy to apply to an input number, but extremely difficult to unscramble so as to recover the input.

Many functions exhibit a modest degree of one-way-ness. Consider the function f applied to integers x defined by $f(x) = x^2$. The square of x is easily found by ordinary multiplication. Finding the square root with the ordinary method is more difficult. It is easier to go forward than backward.

Although $f(x) = x^2$ gives a general idea of the one-way concept, it does not have all of the characteristics of a lock. Going backward is not difficult enough, and there is no key to unlock it.

Another example is the **phone book function**. If I wanted you to secretly send me the name of a person suspected of a crime, I could instruct you to send me the suspect's phone number. You could easily look up the number in the phone book and send it to me. If someone intercepts the message, he or she would find it difficult to deduce the corresponding name. However, if I had a key—namely a reverse phone book that lists phone numbers sequentially and gives corresponding names—I could easily decrypt your message.

Here are some other functions possessing a degree of one-way-ness:[1]

1. Multiply by n. That is, $f(x) = nx$. Multiplication is easier than division.

2. Raise to power n. That is, $f(x) = x^n$. Raising to a power is easier than taking the n-th root.

3. Put in exponent. That is, $f(x) = n^x$. Exponentiation is easier than taking the logarithm.

4. Put in exponent and drop off digits. For example, let $f(x) = 12^x$ but with only the final four digits retained. Hence $x = 9$ leads to $12^9 \bmod 10^4 = 5,159,780,352 \bmod 10^4 = 0352$. It is very difficult to discover that $x = 9$ from the last four digits.

Of these examples, the last is the most secure. It is difficult to go backward from the last few digits of n^x to find x. It is these kinds of functions that serve as building blocks for the one-way functions used in public key systems.

13.3 Discrete Logarithms

When Diffie and Hellman were searching for an appropriate one-way function, they discussed their need with John Gill, who at the time was a young professor at Stanford. Gill suggested the discrete logarithm problem.

Consider the equation

$$b = a^k \bmod n \tag{13.1}$$

(meaning $a^k = cn + b$ for some integers c and b with $0 \le b < n$). Assume first that a and n are known positive integers. Then, given an integer k, it is not difficult to evaluate b. This can be regarded as $b = f(k)$, the forward direction of a function f. The reverse problem is that of finding k when b, as well as a and n, are given.

If it were not for the modulo n in the equation, the value of k could be found by taking ordinary logarithms. Specifically, $k = (\log b)/(\log a)$. However, when b is defined with the modulo n, it is much more difficult to find k. This problem is termed the **discrete logarithm problem**.

---

[1] The inverses for the first three examples can be found relatively efficiently for very large numbers by the iterative technique called **Newton's method**. Hence the one-way properties of these examples are quite modest.

Item 4 listed in the previous section, with $b = f(k) = a^k \bmod 10,000$, is an example of the discrete logarithm problem. For instance, with $a = 12$, $n = 10,000$, and $b = 0352$, it is difficult to deduce that a solution is $k = 9$ (since $12^9 = 5,159,780,352$).

Primitive Values

A family of discrete logarithm problems is defined by fixing the integers a and n in equation (13.1). Generally, one takes $1 < a < n$ to avoid trivialities. However, even with that obvious restriction, not all combinations of a and n define useful one-way functions.

For example, suppose $a = 5$, $n = 10$. The resulting function can be studied by looking at a table of all resulting k and b combinations, as shown below.

k	1	2	3	4	5	6	7	8	9
$b = 5^k \bmod 10$	5	5	5	5	5	5	5	5	5

This highly degenerate case is clearly not suitable since the relation from b to b is not invertible to a unique k.

Consider instead the values $a = 3, n = 10$. The corresponding table of possibilities is then

k	1	2	3	4	5	6	7	8	9
$b = 3^k \bmod 10$	3	9	7	1	3	9	7	1	3

This combination is better, but it also contains duplicates.

Finally, consider the combination $a = 2, n = 11$ (with a modulus different from before). The corresponding table of resulting k and b combinations is then

k	1	2	3	4	5	6	7	8	9	10
$b = 2^k \bmod 11$	2	4	8	5	10	9	7	3	6	1

In this case b ranges over all integers between 1 and $10 = n - 1$ as k ranges over these same values. The function f from k to b is therefore one-to-one for k in this range.

In general, a is **primitive** relative to n if $a^k \bmod n$ includes all values $1, 2, \ldots, n-1$ as k ranges over $1, 2, \ldots, n-1$. When a is primitive relative to n, it is also said that a is a **generator** (with respect to n) because all the integers between 1 and $n - 1$ are generated by powers $a^k \bmod n$.

13.4 Diffie–Hellman Key Exchange

The **Diffie–Hellman key exchange** system was the first concrete example of the general theory outlined in Diffie and Hellman's original paper. The background in the previous section is sufficient to follow this innovative method of key exchange.[2]

[2] During the period that Diffie and Hellman were working on their general approach to encryption, Ralph Merkle, a graduate student at the University of California, Berkeley, independently outlined an approach to key exchange based on the generation and transmission of thousands of puzzles, one of which would be chosen at random by the receiving party. Merkle soon joined Prof. Hellman's group to work with Hellman

The key exchange method can be described for the players Alice and Bob who wish to agree on a common key for encryption without others being able to discern the key even if they intercept all messages between them. The method consists of three steps.

1. Alice and Bob agree on integers a and n with a primitive[3] with respect to n; and n large. These choices need not be secret.

2. Alice privately selects an integer k_A, $1 < k_A < n$, which is her part of the key. She then computes

$$y_A = a^{k_A} \bmod n \tag{13.2}$$

and openly sends y_A to Bob.

   Likewise, Bob selects an integer k_B, $0 < k_B < n$, which is his part of the key. He computes

$$y_B = a^{k_B} \bmod n \tag{13.3}$$

and openly sends y_B to Alice. Both Alice's and Bob's messages can be sent insecurely.

3. Alice and Bob then agree on the joint key

$$k_{AB} = a^{k_A k_B} \bmod n.$$

Alice computes the key from the formula

$$k_{AB} = (y_B)^{k_A} \bmod n,$$

which she can compute because she (and no one else) knows k_A and she knows the value y_B that Bob sent.

   Likewise, Bob computes

$$k_{AB} = (y_A)^{k_B} \bmod n.$$

Both Alice's and Bob's messages can be sent over an insecure channel because learning y_A (or y_B) does not divulge k_A (or k_B) due to the one-way nature of the function involved.

   Of course the two methods of computing the key are equivalent since

$$a^{k_A k_B} = (a^{k_A})^{k_B} = (a^{k_B})^{k_A}.$$

The key is secure because an outsider who has intercepted y_A and/or y_B cannot compute k_{AB} without knowing either k_A or k_B, and determining one of these requires solving one of the discrete logarithm problems defined by equation (13.2) or (13.3). For a large, but still practical, n (on the order of a few hundred digits) this inverse problem is extremely difficult to solve, requiring perhaps millions of years using today's fastest computers.

---

and Diffie and pursue his Ph.D. degree. Because the Diffie–Hellman key exchange method is close in concept to Merkle's ideas, Hellman has suggested that it should be more properly called **Diffie–Hellman–Merkle key exchange**.

[3] It is not strictly necessary that a be primitive with respect to n, only that a generate a large quantity of numbers of the form $a^k \bmod n$.

Example 13.1 (Small *n*). For illustrative purposes we use a small value of n. Let us take $n = 97$, $a = 5$ (5 is primitive with respect to 97). Suppose Alice and Bob independently select $k_A = 16$, $k_B = 11$, respectively. They compute as follows.

Alice using $k_A = 16$. (It is simplest to break powers into a series of lower powers and reduce each to the modulo equivalent. Alice uses a simple decomposition as follows.)

$$\begin{aligned}
y_A = 5^{16} \bmod 97 &= 5^8 \times 5^8 \bmod 97 \\
&= 390,625 \times 390,625 \bmod 97 \\
&= (4,027 * 97 + 6) \times (4,027 * 97 + 6) \bmod 97 \\
&= 6 \times 6 \bmod 97 \\
&= 36.
\end{aligned}$$

Bob using $k_B = 11$. (Bob uses a more detailed decomposition as follows.)

$$\begin{aligned}
y_B = 5^{11} \bmod 97 &= 5^8 \times 5^2 \times 5 \bmod 97 \\
&= [[(5)^2]^2]^2 \times 5^2 \times 5 \bmod 97 \\
&= [[25]^2]^2 \times 5^2 \times 5 \bmod 97 \\
&= [625]^2 \times 125 \bmod 97 = [43]^2 \times 28 \bmod 97 \\
&= 1849 \times 28 \bmod 97 = 6 \times 28 \bmod 97 = 71.
\end{aligned}$$

Then, knowing k_A and y_B, Alice computes

$$\begin{aligned}
k_{AB} &= (y_B)^{k_A} \bmod 97 \\
&= 71^{16} \bmod 97 \\
&= 62.
\end{aligned}$$

Likewise, Bob calculates

$$\begin{aligned}
k_{AB} &= (y_A)^{k_B} \bmod 97 \\
&= 36^{11} \bmod 97 \\
&= 62.
\end{aligned}$$

Hence they both agree on the key 62.

13.5 Modular Mathematics

Modular mathematics is heavily used in public key encryption systems, but for introductory purposes only a little background is required. The basics are reviewed in this section.

First is notation. The "mod m" notation is used in two ways. The first usage is as an operator that reduces a number to its modulo m value. The value of "$c \bmod m$" is the remainder when c is divided by m. Specifically, if c is expressed as $c = km + b$ where k is an integer and $0 \leq b < m$, then $c \bmod m = b$. For instance, 32 mod 10 = 2.

The second usage of the notation "mod m" is as an equivalence relation, in which case the equivalence symbol $\equiv$ is used. A statement of the form $b \equiv c \bmod m$, means that the equality is true only in mod m terms. Specifically it means that $b - c$ is an integral multiple of m. For example, $32 \equiv 2 \bmod 10$ is such a statement. In a string of such equivalences, the final mod n applies to the entire set. Thus $33 \equiv 19 \equiv 5 \equiv 1 \bmod 2$.

The mod operator follows some simple rules with respect to addition and multiplication that facilitate computation.

Addition:

$$(a + b) \bmod m = [(a \bmod m) + (b \bmod m)] \bmod m.$$

Multiplication:

$$a \cdot b \bmod m = [(a \bmod m) \cdot (b \bmod m)] \bmod m.$$

These rules show that when computing the modular values of sums and products, the mod operator can be applied at any point. For example,

$$(56 + 27)(112 + 13) \equiv (6 + 7)(2 + 3) \equiv 3 \cdot 5 \equiv 5 \bmod 10,$$

which is easier than computing

$$(56 + 27)(112 + 13) = 83 \cdot 125 = 10,375 \equiv 5 \bmod 10.$$

The Euclidean Algorithm

A concept related to both primality and modular mathematics is that of the **greatest common divisor** of two integers. This is the positive integer that divides evenly into both numbers and which is a multiple of every other common divisor. For example, the greatest common divisor of 6 and 10 is 2. The greatest common divisor of the two integers A and B is denoted $\gcd(A, B)$.

If the integers A and B satisfy $\gcd(A, B) = 1$, then A and B are said to be **relatively prime**. For example, if A and B are distinct prime numbers, then they are relatively prime since there is no positive number that divides them both.

The gcd of two integers can be found efficiently by use of the **Euclidean algorithm**, which was described in Euclid's *Elements*, written about 300 BCE, although it believed that the algorithm was discovered 200 years earlier. This wonderful algorithm has the distinction of being the oldest known nontrivial mathematical algorithm.

The algorithm to find $\gcd(A, B)$ is easily described. Assume that $0 < B < A$.

1. If $B = 0$, the algorithm terminates with A as the answer.

2. Replace the pair (A, B) by the pair $(B, A \bmod B)$. Return to step 1.

The algorithm must terminate because the second entry is reduced during step 2.

The process is probably best understood by going through an example.

Example 13.2 (Simple to find). Let us find the greatest common divisor of 70 and 63. We write

$$(70, 63) \rightarrow (63, 70 \bmod 63) = (63, 7) \rightarrow (7, 63 \bmod 7) = (7, 0).$$

Hence $\gcd(70, 63) = 7$.

TABLE 13.1
Computation of gcd(935, 273).

	A	B
$935 = 3 \cdot 273 + 116$	935	273
$273 = 2 \cdot 116 + 41$	273	116
$116 = 2 \cdot 41 + 34$	116	41
$41 = 1 \cdot 34 + 7$	41	34
$34 = 4 \cdot 7 + 6$	34	7
$7 = 1 \cdot 6 + 1$	7	6
$1 = 1 \cdot 1 + 0$	6	1
$1 = 0 + 1$	1	0

Example 13.3 (A longer one). As another example, take $A = 935$, $B = 273$. The appropriate computations are shown in table 13.1. Since the final value of A is 1, we conclude that $\gcd(273, 935) = 1$. The two values are relatively prime.

Modular Inverses

Given a with $0 \leq a < m$, an integer b with $0 \leq b < m$ is the **inverse** of a modulo m if

$$ab \equiv 1 \bmod m.$$

In this case a is said to be **invertible** mod m. It is not always true that such a b exists. However, there is a unique b if a and m are relatively prime. For example, if $a = 7$, $m = 10$, then $b = 3$ is the appropriate inverse because $ab = 7 \cdot 3 = 21 \equiv 1 \bmod 10$.

It is left as an exercise to show that when it exists, the inverse is unique.

When a and m are relatively prime, the inverse of a modulo m can be found from the Euclidean algorithm. However, it is necessary to keep track of the factors used in the algorithm. A systematic procedure is the **generalized Euclidean algorithm** described in the appendix to this chapter.

In the examples of this book, which use relatively small numbers, modular inverses are easily found by trial and error by hand or by using a spreadsheet program.

The main conclusion, however, is the following important theorem.

Theorem 13.1. If a and m are relatively prime, with $0 < a < m$, then there is a unique b, $0 < b < m$, such that $ab \equiv 1 \bmod m$.

Fermat's Theorem

Cryptography theory often takes the modulus m to be a prime number p. There is a simple but important result for that case.[4]

---

[4] This is not the famous "Fermat's last theorem."

Theorem 13.2 (Fermat's theorem). If p is prime, then for any integer x,

$$x^p \equiv x \bmod p. \tag{13.4}$$

Proof: The proof is by induction. The result is trivially true for $x = 0$ and $x = 1$. Assume it is true for some x. Then by direct expansion

$$(x+1)^p = x^p + px^{p-1} + \frac{p(p-1)}{2}x^{p-2} + \cdots + px + 1.$$

The coefficient of the k-th term in the above expansion is the binomial coefficient $p!/[(p-k)!k!]$. All of these coefficients, except for $k = 0$ and $k = p$, contain a factor of p in the numerator. This factor cannot be canceled by any denominator term because p is prime. Therefore, all coefficients except the first and the last are multiples of p. Hence

$$(x+1)^p \equiv x^p + 1 \bmod p.$$

By the induction hypothesis, $x^p \equiv x \bmod p$, and thus the above relation becomes

$$(x+1)^p \equiv x + 1 \bmod p,$$

which shows the result is true for $x + 1$, completing the induction process. ∎

As an example, take $p = 3$, $x = 2$. Then $x^p = 2^3 = 8 \equiv 2 \bmod 3 = x$.

There is an important corollary based on the result concerning modular inverses discussed in the previous subsection.

Corollary. If p is prime and x is not a multiple of p (including 0), then

$$x^{p-1} \equiv 1 \bmod p. \tag{13.5}$$

Proof: The equivalence $x^p \equiv x \bmod p$ can be divided by x if x is not a multiple of p. ∎

13.6 Alternative Puzzle Solution

Perhaps you did not notice that there is an alternative solution to the Alice and Bob locked box dilemma; a solution in which Alice's letter travels only once through the mail instead of three times as in the original solution. This solution begins with Bob, and hence this might be termed the "Bob solution" as contrasted with the earlier "Alice solution."

Bob sends Alice an open lock to which he has the only key. The lock travels safely through the mail, for no one has an incentive to intercept it. Alice receives the lock and snaps it shut on her metal box, securing the letter inside. She then sends the locked box to Bob. He opens it with his key, and reads the letter.

The Bob solution is more efficient, and it provides a good analogy for modern public key systems.

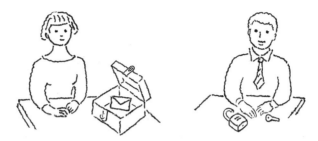

FIGURE 13.2 Alternative solution. Bob sends his open padlock to Alice, who uses it to secure her letter in a box that she sends back to Bob.

13.7 RSA

After Diffie and Hellman published their groundbreaking paper on a new approach to cryptography, a group at MIT—Ronald Rivest, Adi Shamir, and Leonard Adleman— began a quest for an appropriate public key system as envisioned by the Diffie– Hellman paper. After a year of effort by the team, Rivest put it all together one day and wrote it up that night. The paper was published by the three of them. Their method is now termed RSA, after the initials of their last names.

RSA begins with the choice of a modulus, but it is selected to be the product of two primes.

In terms of the Alice and Bob situation, RSA is analogous to the solution titled as Bob's. Bob constructs a lock to which only he has the key, and he sends the open lock to Alice. Alice secures her message with Bob's lock and sends it to him. He opens it with his key.

Bob could send copies of his open lock to everyone, and announce that if anyone wishes to send him a message, they should simply secure with his lock. Only he has the key. In digital form, Bob's lock is a pair of numbers $\langle n, b \rangle$. These are referred to as a **public key**, rather than an open lock, but this public key serves a role analogous to a physical lock.

Here are the steps of the RSA method that you would use to have people send you secure messages, each step capable of being carried out efficiently even for numbers that are a thousand bits or more in length:

1. Select two primes p and q and form $n = pq$ and $\phi = (p - 1)(q - 1)$. (Section 13.9 indicates how large primes can be found.)

2. Select b, $1 < b < \phi$ such that b is relatively prime to ϕ. That is, $\gcd(b, \phi) = 1$. (Values of b can be checked with the Euclidean algorithm.)

3. Find a such that $ab \equiv 1 \bmod \phi$. (This can be done with the generalized Euclidean algorithm as discussed in section 13.5.)

4. Publish n and b. These together form the public key. Instruct people to encrypt messages m (with $0 < m < n$) intended for you by $y = m^b \bmod n$. (Section 13.8 explains how to calculate large modular powers.)

5. Decrypt messages by $m = y^a \bmod n$. (Use the same method as in step 4.)

The important feature of RSA is that outsiders cannot compute ϕ directly from n, and hence they cannot compute a, the mod ϕ inverse of b. The only way someone can find ϕ, and hence a, is by factoring n into its prime components p and q so that they can evaluate $(p-1)(q-1)$. Such factoring is extremely difficult when n is large. In a sense, the underlying one-way function is multiplication of two prime numbers. Such multiplication is easy; factoring is (very) hard.

Example 13.4 (A simple RSA). Here is a step-by-step example using small numbers.

(**Set up public key**) Alice selects the two primes $p = 3$, $q = 11$ and defines $n = pq = 33$. She calculates

$$\phi = (p-1)(q-1) = 2 \times 10 = 20.$$

Next she selects b relatively prime to ϕ. There are several choices: 3, 7, 9, 11, 13, 17, or 19. She settles on $b = 13$. Her public key is the combination $\langle n, b \rangle = \langle 33, 13 \rangle$. She announces this to anyone wishing to send her a message.

(**Find the private key**) Alice finds a such that $ab = 1$ mod ϕ. She can find this a since she knows ϕ. Other people cannot find a unless they can factor n (which although easy for this example, is nearly impossible for very large values of n).

She could find a with the extended Euclidean algorithm, but for this example Alice simply calculates $13a$ mod 20 for all possible values of a until she finds the right one. Here is the table of calculations:

a	1	2	3	4	5	6	7	8	9	10	11	12	13	14	15	16	17
$13a$	13	6	19	12	5	18	11	4	17	10	3	16	9	2	15	8	1

Thus $a = 17$, and Alice's private key is $\langle n, a \rangle = \langle 33, 17 \rangle$.

(**Test message**) Suppose Bob wishes to send Alice the letter d, which he converts to the message $m = 4$, since d is the fourth letter of the alphabet. He encrypts the message m with Alice's public key by computing $y = 4^{13}$ mod 33. (A spreadsheet can calculate this directly using the mod function, or it can be calculated by noting that $4^6 = 4{,}096 \equiv 4$ mod 33; hence $4^{12} \equiv 16$ mod 33; and thus $4^{13} \equiv 16 \times 4 \equiv 64 \equiv 31$ mod 33.) Thus the encrypted version of the message $m = 4$ is $y = 4^{13}$ mod $33 = 31$.

(**Decoding**) Alice receives the encrypted message $y = 31$. To decode it she computes y^{17} mod 33. (This can be done by expressing y^{17} in terms of lower powers. 31^{17} mod $33 = 31(31^4)^4$ mod 33. Using $31^4 = 923{,}521 \equiv 16$ mod 33, she finds 31^{17} mod $33 = 31 \times 16^4$ mod $33 = 2{,}031{,}616$ mod $33 = 4$.) Hence she concludes that $m = 4$, and thus that the message from Bob is the letter d.

Public key systems are sometimes contrasted with classical encryption systems by referring to the classical systems as **symmetric key** systems, since both parties use identical keys. Public key systems are **asymmetric key** systems.

Proof of RSA*

Assume the message m is encrypted to produce $y = m^b \bmod n$ as described. We will evaluate $y^a \bmod n$. By definition

$$y = m^b - sn \quad \text{for some integer } s.$$

Hence

$$y^a \bmod n = [m^b - sn]^a \bmod n$$
$$= m^{ab} \bmod n \quad \text{(omitting terms containing } n\text{)}$$
$$= m^{t(p-1)(q-1)+1} \bmod n, \text{ for some } t \quad \text{(since } ab \equiv 1 \bmod \phi\text{)}$$
$$= m \cdot m^{t(p-1)(q-1)} \bmod n \quad \text{(factoring out } m\text{)}. \tag{13.6}$$

Next we write (for some r)

$$m \cdot m^{t(p-1)(q-1)} \equiv m(rp+1)^{t(q-1)} \bmod p \quad \text{(by Fermat's theorem for } p\text{)}$$
$$\equiv m \bmod p \quad \text{(by eliminating all terms containing } p\text{)}.$$

Likewise,

$$m \cdot m^{t(p-1)(q-1)} \equiv m \bmod q.$$

It follows that[5]

$$m \cdot m^{t(p-1)(q-1)} \equiv m \bmod pq.$$

Finally, from equation (13.6) it follows that (since $pq = n$)

$$y^a \bmod n = m$$

as desired. ∎

13.8 Square and Multiply*

Many public key systems, such as RSA, require the computation of $m^a \bmod n$ for extremely large values of a and n. As found in our numerical examples, it is often useful to break the computation into stages. A systematic method for breaking up the calculation is essential when the numbers are large.

An algorithm for computing powers known as **square and multiply** has a long history. It was described by Legendre in 1798, but in the form for multiplying two large numbers. This latter procedure was apparently known to the Egyptians as early as 1800 BC and hence, like the Euclidean algorithm, it is one of the most ancient of mathematical algorithms.

First it must be noted that squaring and multiplying (mod n) can be carried out using numbers that never exceed n by more than a single digit. As a simple example,

---

[5]If $x \equiv m \bmod p$ and $x \equiv m \bmod q$, then $x = cp + m = dq + m$ for some c and d. Subtracting, we find $cp = dq$. Since p and q are prime, it follows that c is divisible by q. Hence $x = c'qp + m$, and therefore $x \equiv m \bmod pq$.

100×456 is a five-digit number. But 100×456 mod 873 can be computed without exceeding four digits as

$$100 \cdot 456 \equiv 10 \times 4,560 \equiv 10 \times 195 \equiv 1,950 \equiv 204 \text{ mod } 873.$$

Multiplication of more general numbers, not powers of ten, can be carried out the same way by breaking one of the numbers into its multiples of 1, 10, 100, and so forth. For example, $378 \cdot 456 = 100 \cdot 3 \cdot 456 + 10 \cdot 7 \cdot 456 + 8 \cdot 456$.

The method of square and multiply to compute large powers (mod n) also uses the fact that large powers can themselves be broken into small pieces. For example, one can compute m^{16} mod n with only 4 squares as $m^{16} = (((m^2)^2)^2)^2$. Because the computation is reduced mod n at each stage, the size of the numbers remains manageable. For example, if n is 100 digits long, each result is no more than 100 digits long, whereas a typical 100-digit number raised to a typical 100-digit power is close to ten thousand digits long.

To raise a number to a power using the square and multiply method, the exponent must first be expressed as a sum of powers of 2. For instance, m^{13} is expressed as m^{8+4+1}.

Generally, to keep track of these powers when evaluating m^a mod n the exponent a is represented in binary form as

$$a = \sum_{i=0}^{k-1} b_i 2^i,$$

where for $i = 0, 1, 2, \ldots, k-1$, each $b_i = 0$ or 1. For example $13 = 1 \cdot 2^3 + 1 \cdot 2^2 + 0 \cdot 2^1 + 1 \cdot 2^0$, or 1101 in binary. The b_i's guide the transformation of the calculation into a series of squarings and multiplications. The first guide number is the leftmost binary digit, and successive guides are found by moving to the right. If the guide binary digit is 0, the operation is squaring. If the guide binary digit is 1, the operation is square and multiply. The steps for 7^{13} mod 10 are therefore (from 1101): square and

TABLE 13.2
Square and Multiply. The table evaluates 124^{187} mod 313. The process is guided by the binary representation of the exponent 187, which is in the b_k column. At each step the result is reduced modulo 313. The result is 245.

k	b_k	Value
7	1	$1^2 \times 124 = 124$
6	0	$124^2 \quad\quad = 039$
5	1	$039^2 \times 124 = 178$
4	1	$178^2 \times 124 = 040$
3	1	$040^2 \times 124 = 271$
2	0	$271^2 \quad\quad = 199$
1	1	$199^2 \times 124 = 180$
0	1	$180^2 \times 124 = 245$

multiply, square and multiply, square, square and multiply. The process is initiated with the number 1. Thus,

$$7^{13} \bmod 10 = (\{[(1^2 \cdot 7)^2 \cdot 7]^2\}^2 \cdot 7) \bmod 10.$$

Evaluation of what is inside each bracket, starting with the inside parentheses and working to the outer parentheses, gives $7^{13} \bmod 10 \rightarrow 7 \rightarrow 3 \rightarrow 9 \rightarrow 7$.

The calculation of $124^{187} \bmod 313$ is illustrated in table 13.2. It uses $a = 187$ expressed as $a = (1 \cdot 1) + (1 \cdot 2) + (0 \cdot 4) + (1 \cdot 8) + (1 \cdot 16) + (1 \cdot 32) + (0 \cdot 64) + (1 \cdot 128)$. In short, $b = 10111011$ is the binary representation of a.

13.9 Finding Primes*

A public key system such as RSA requires the selection of prime numbers, and these must be large to insure strong security.

A question that arises immediately is whether there is an ample supply of large primes. If they are rare, they may be difficult to find, there may not be enough for everyone, and once known they might be compiled in a list that could be used to factor someone's public key. However, one need not worry, for the celebrated **prime number theorem** states that for any integer N the number of primes no greater than N is approximately $N/\ln N$. Thus the number of 100-digit primes is about $10^{100}/\ln 10^{100} = 10^{100}/100 \ln 10 \approx 4 \times 10^{97}$. This is more than 10 billion times 10^{77}, the number of atoms in the universe. There are plenty of primes to go around.

If a number is selected randomly between 0 and N, the chance that it is a prime is about $1/(\ln N)$. Hence, a randomly selected 100-digit number will be prime with a probability of about $1/\ln 10^{100} \approx 1/230$. If the search is restricted to odd numbers, the probability doubles, to about $1/115$. Hence selecting at random a few thousand odd 100-digit numbers (and checking for primeness) almost guarantees that one will be prime. For instance with ten thousand attempts the chance of *not* finding a prime of 100 digits is $(1 - 1/115)^{10,000} \approx 10^{-38}$.

Checking Primes

Once a candidate number is selected, it must be determined if it is in fact prime. The straightforward way to do this is to attempt to divide the number by all numbers less than the square root of the number. For a 100-digit number this means checking all numbers having 50 digits. Of course, one would not have to try even numbers, and there may be a scheme whereby only division by prime numbers need be tried. But there would be about $10^{50}/\ln 10^{50} \approx 10^{48}$ of those.

In practice, prime checking is carried out with a test procedure that is not perfect, but has a known maximum probability of error. If different versions of this test are applied enough times and the candidate number continues to pass them all, the confidence that the number is in fact prime can be made arbitrarily high.

The most popular of these tests is the **Rabin–Miller method**, which is based on Fermat's theorem. As a simplification of the method, one selects a random number a and evaluates $y = a^{n-1} \bmod n$. If $y \neq 1$, then according to Fermat's theorem, n is not prime. If $y = 1$, there is a chance that n is prime, but additional tests with different values of a must be performed. In the Rabin–Miller procedure, every successful test reduces the probability that the number is not prime by one-fourth. A succession of twenty successful tests reduces the probability of n not being prime to 10^{-12}. For details see exercise 13.11.

13.10 Performance*

Public key cryptology relies on one-way functions. It should be easy to go forward, and extremely difficult to go backward without the key. Proper evaluation of a system, however, requires that the qualitative notion of difficulty be made quantitative.

The difficulty of computational algorithms is usually measured by how the run time of the algorithm varies as a function of the length of the input data. For integer algorithms, such as those used in RSA, the input length is generally taken as the number of digits n in the input string. Difficulty is therefore measured by the run time as a function of n.

Simple algorithms have run times that are proportional to a polynomial in n, and these are said to be **polynomial time** algorithms. The largest power in the polynomial is the most important, and for rough comparisons, it is the only thing considered. For example, the conventional method for multiplying two n-digit numbers is an n^2 algorithm.

An algorithm is considered difficult if its run time as a function of n increases faster than any polynomial. **Exponential-time** algorithms, whose run times are proportional to e^{an} for some $a > 0$, take a great deal of time for large n, and such algorithms are considered too time consuming and their underlying problems are considered to be very hard. A simple method for factoring a number with n digits is to try to divide it by all odd numbers less than its square root. This algorithm is exponential because there are $\frac{1}{2}\sqrt{10^n} = \frac{1}{2}10^{n/2}$ numbers to be tried. Hence factoring (at least with this method) is very hard.

In practical terms, a one-way function used in a public key system should require low-order polynomial algorithms in the forward direction used by the legitimate sender and receiver, and a non-polynomial-time algorithm, such as an exponential-time algorithm, in the backward direction faced by someone attempting to break the cipher.

The forward algorithms associated with RSA involve multiplication, evaluation of modulo values, taking powers with respect to a modulus, and finding primes. These are generally polynomial, most of order n^3 or less. Thus, the encryption (forward) process is fairly simple.

For RSA, the difficulty of the backward process (faced by a cryptanalyst) is dominated by the requirement of factoring. The simple method mentioned above is exponential, but there are faster methods. Currently, the best algorithms for factoring n-digit numbers have run times roughly proportional to $e^{(\log n)^{1/3}}$, which lies between polynomial and exponential time. More progress may be made, for a lower bound on the time to factor a number has not been established. In fact, it has not even been proved that factoring cannot be done in polynomial time.

Progress in factoring has led to the use of longer keys. In 1977 Ronald Rivest estimated that factoring a 125-digit number would require 40 quadrillion years. But in 1994 a challenge number of 129 digits was factored. Currently, it is considered prudent to use keys with a length of 512 bits, and many experts suggest a length of 1,024 bits.

13.11 The Future

There are newer public key systems. **Elliptic curve** methods use an algebraic structure more complex than the integers mod n, and implement analogs of integer-based public key systems on this structure. Elliptic curve methods are attractive because breaking them appears to require full exponential time. Hence, they can be made highly secure with relatively short keys. For example, it is claimed that an elliptic curve key of 200 bits gives the same security as that of an RSA key of 1,024 bits. These methods are well suited to encryption of binary data.

The field of cryptography is still advancing rapidly. There is a continuing race between advances in cryptanalysis and development of new, more secure cryptographic systems. One of the most intriguing innovations on both sides of this race is **quantum cryptography**. A photon has the mysterious property of being in several physical states simultaneously. It may be possible to construct a massively parallel computer using this property that would provide a means for rapidly factoring large integers.

On the other hand, the quantum properties of photons can form the basis of a cryptography system that has the perfect security of a one-time pad. In this system a sender and receiver communicate over a channel that sends photons that can be polarized in either of two states, representing 0's or 1's. It is secure because an eavesdropper's observations must, by the **Heisenberg uncertainty principle**, cause the state to be randomized, and this can be detected by the receiver. Such systems currently exist but transmit over fiber optic lines of only modest length. In these systems quantum cryptography is used to distribute a key for a symmetric key encryption process such as the AES. It is clear that the story of cryptography still has a few surprises in store.

APPENDIX

The Extended Euclidean Algorithm

To find the inverse of $r_1 \bmod r_0$, where $r_1 < r_0$, one records the details of the Euclidean algorithm for $\gcd(r_0, r_1)$, as follows:

$$r_0 = k_1 r_1 + r_2 \qquad\qquad 0 \leq r_2 < r_1$$
$$r_1 = k_2 r_2 + r_3 \qquad\qquad 0 \leq r_3 < r_2$$
$$r_2 = k_3 r_3 + r_4 \qquad\qquad 0 \leq r_4 < r_3$$
$$\vdots \qquad\qquad\qquad \vdots$$
$$r_{n-2} = k_{n-1} r_{n-1} + r_n \qquad 0 \leq r_n < r_{n-1}$$
$$r_{n-1} = k_n r_n.$$

We assume that the algorithm terminates after n steps. If r_0 and r_1 are relatively prime, then $r_n = 1$.

The **extended Euclidean algorithm** introduces a series of values s_j defined by

$$s_0 = 0$$
$$s_1 = 1$$
$$s_j = s_{j-2} - k_{j-1} s_{j-1} \bmod r_0, \quad \text{for } j \geq 2.$$

Then s_n is the desired $\bmod r_0$ inverse of r_1.

Example 13.5 (Inverse of 273 mod 935). From example 13.5 we know that 273 and 935 are relatively prime. The k_j values are read off the steps of the algorithm used to determine that 273 and 935 are relatively prime. Thus,[6]

$$
\begin{aligned}
s_0 &= 0 \\
s_1 &= 1 \\
s_2 &= s_0 - 3s_1 &\equiv& \quad -3 \bmod 935 \\
s_3 &= s_1 - 2s_2 &\equiv& \quad 7 \bmod 935 \\
s_4 &= s_2 - 2s_3 &\equiv& \quad -17 \bmod 935 \\
s_5 &= s_3 - 1s_4 &\equiv& \quad 24 \bmod 935 \\
s_6 &= s_4 - 4s_5 &\equiv& \quad -113 \bmod 935 \\
s_7 &= s_5 - s_6 &\equiv& \quad 137 \bmod 935.
\end{aligned}
$$

Thus we find $s_7 = 137$. Indeed it is easily verified that

$$273 \cdot 137 = 37401 = 40 \cdot 935 + 1 \equiv 1 \bmod 935.$$

---

[6] It is only necessary to determine the s_j's mod 935, except for s_7. At each of the steps the negative values obtained may be converted to positive numbers by adding 935.

13.12 EXERCISES

1. (A table) Make a complete table of $ab \bmod 10$ for a and b between 1 and 10. How many a's have mod 10 inverses?

2. (Uniqueness) Show that if a and m are relatively prime, then the inverse of $a \bmod m$ is unique.

3. (Find an inverse) Find the inverse of 16 mod 101.

4. (Affine code inverse) Consider the affine code that transforms a message x into ciphertext y as
$$y = 19 \cdot x + 6 \bmod 31.$$

   (a) Show that the message can be recovered by an affine transformation of the form
$$x = c \cdot y + d \bmod 31$$
   and determine the constants c and d.

   (b) What is the general method for determining c and d in such a cipher?

5. (Digraphs) Encoding single letters one by one is rather limiting, since there are only 26 possible messages. It is better to use longer basic messages. One way is to send letters in pairs, such as ab or kn. These pairs are termed **digraphs**, and it is clear that there a total of $26^2 = 676$ of them. To be practical, there must be a simple formula for assigning digraphs to numbers. We assign $aa \to 0, ab \to 1, \ldots, az \to 25, ba \to 26$, etc.

   (a) What is the number corresponding to the digraph th?

   (b) What is the general procedure for finding the number corresponding to a digraph?

6. (Break the cipher) Alice has published her RSA public key as $\langle n, b \rangle = \langle 91, 5 \rangle$. Accordingly, Bob sent her the cipher text 71. What letter of the alphabet was Bob's message?

7. (A card experiment) Take a packet of p playing cards where p is prime, for example, seven cards. Have a friend select one of these cards, noting its value. Pretend to mix the cards, but arrange that the chosen card ends on top of the face-down packet.
   Hand the packet to your friend and ask him or her to select any number less than the number of cards in the packet. For example, with seven cards the friend can select any number from one to six. Suppose the number four is selected. Have the friend deal, one at a time, four cards from the top of the packet to the bottom. Then turn the next card face up on top. The process is to be repeated $p - 1$ times, dealing four to the bottom and turning the next card face up on top. After $p - 1$ times, all but one card will be face up, and that one will be the originally selected card. This is true no matter what number between 1 and $p - 1$ the friend uses in the deal and turnover process. Try it yourself.
   Why does this work?
   (In card magic literature, this is termed George Sands' prime number principle.)

8. (Simplified Chinese remainder theorem) The proof of RSA used the fact that if $x = m \bmod p$ and $x = m \bmod q$, then $x = m \bmod pq$. More generally, knowing $x \bmod p$ and $x \bmod q$ determines $x \bmod pq$. For example, consider the table below:

x	1	2	3	4	5	6	7	8	9	10	11	12	13	14	15
$x \bmod 3$	1	2	0	1	2	0	1	2	0	1	2	0	1	2	0
$x \bmod 5$	1	2	3	4	0	1	2	3	4	0	1	2	3	4	0

Knowing the second and third entries in a column determines the top element. A formula for the solution to $x = a$ mod 3, $x = b$ mod 5 is

$$x = [a \cdot 5 \cdot (5^{-1} \bmod 3) + b \cdot 3 \cdot (3^{-1} \bmod 5)] \bmod 15$$
$$= 10a + 6b \bmod 15.$$

Find the solution x mod 45 to the equations $x = 3$ mod 5 and $x = 1$ mod 9.

9. (Square and multiply) Find 41^{105} mod 92.

10. (Generators*) Let p be prime and suppose that you know the factors of $p - 1$. Namely, $p - 1 = q_1 q_2 \cdots q_n$ where each q_i is prime. Show that a is a generator for p if and only if $a^{(p-1)/q_i}$ mod $p \neq 1$ for all $i = 1, 2, \ldots, n$. Use this test to determine if 3 is a generator for $p = 11$.

11. (Rabin–Miller method*) Here is an algorithm for determining if n is prime:
    (a) Represent $n - 1$ as $n - 1 = 2^k m$, where m is odd. (That is, m is what remains when all factors of 2 are taken out of $n - 1$.)
    (b) Select a random number a, $1 < a < n$.
    (c) Compute
    $$b = a^m \bmod n.$$

    If $b = 1$ or -1 ($-1 = n - 1$), declare PRIME and end.
    (d) Repeat $k - 1$ times or until end: Define a new b by
    $$b = b^2 \bmod n.$$

    If $b = -1$, declare PRIME and end.
    If the algorithm ends without declaring PRIME, then n is definitely not prime. If n is prime, the algorithm will declare PRIME. However, for general n, if PRIME is declared, there is a $1/4$ probability that n is not actually prime.
    Apply a single cycle of the method to $n = 5,009$ using $a = 3$.

12. (Cesàro estimate*) Let a and b be integers chosen randomly. Show that the probability that $\gcd(a, b) = 1$ is $6/\pi^2$ using the following steps:
    (a) Assume that the probability of $\gcd(a, b) = 1$ is p. For any integer d it will happen that $\gcd(a, b) = d$ only if a and b are each multiples of d and $\gcd(a/d, b/d) = 1$. Hence find the probability that $\gcd(a, b) = d$ in terms of p.
    (b) Use the fact that sum of all those probabilities must be 1. Hint: $\sum_{i=1}^{\infty} \frac{1}{i^2} = \pi^2/6$.

13.13 Bibliography

The Diffie–Hellman paper [1] that initiated the concept of public key systems is easily accessible. The small book [2] is a readable introduction to public key systems. Intermediate texts are [3], [4], and [5]. Showing how to solve $y = m^b$ mod n for m and using this solution in cryptography is presented in [6]. The original RSA paper is [7]. The Rabin–Miller prime-testing procedure is presented in the intermediate textbooks mentioned, and was originally published as [8]. Another method is that of [9]. A standard method for factoring large numbers is the method of Pollard [10]. For an interesting discussion of the Euclidean and square-and-multiply algorithms, see [11]. The history of cryptography, including public key cryptography, is presented nicely in [12] and [13]. Exercise 10 is based on [5], p. 254.

References

[1] Diffie, W., and M. E. Hellman. "New Directions in Cryptography." *IEEE Transactions on Information Theory* 22 (1976): 644–54.

[2] Beutelspacher, Albrecht. *Cryptology*. Trans. J. Chris Fisher. Washington, D.C.: Mathematical Association of America, 1996.

[3] Stinson, Douglas R. *Cryptography: Theory and Practice*. Boca Raton: CRC Press, 1995.

[4] Mollin, Richard A. *An Introduction to Cryptography*. Boca Raton: Chapman & Hall/CRC Press, 2001.

[5] Schneier, Bruce. *Applied Cryptography*. 2nd ed. New York: Wiley, 1996.

[6] Pohlig, S. C., and M. E. Hellman. "An Improved Algorithm for Computing Logarithms of GF(p) and Its Cryptographic Significance." *IEEE Transactions on Information Theory* 24 (1978): 106–110.

[7] Rivest, Ronald L., A. Shamir, and L. Adleman. "A Method for Obtaining Digital Signatures and Public Key Cryptosystems." *Communications of the ACM* 21 (1978): 120–26.

[8] Rabin, M. O. "Probabilistic Algorithms for Testing Primality." *Journal of Number Theory* 12 (1980): 128–38.

[9] Lehmann, D. J. "On Primality Tests." *SIAM Journal on Computing* 11 (1982): 374–75.

[10] Pollard, J. M. "Theorems on Factorization and Primality Testing." *Proceedings of the Cambridge Philosophical Society* 76 (1974): 521–28.

[11] Knuth, Donald E. *The Art of Computer Programming*, Vol. 2. Reading, Mass.: Addison-Wesley, 1969.

[12] Singh, Simon. *The Code Book*. New York: Doubleday, 1999.

[13] Levy, Steven. *Crypto*. New York: Viking, 2001.

14

SECURITY PROTOCOLS

M odern cryptography has ramifications reaching far beyond the problem of securing message transmission. The principles of cryptography—especially public key cryptography—provide the foundation for a variety of secure digital operations, including digital signatures, practical methods of digital cash, secure transmission over the Internet, anonymous transactions, and proving one's identity. These functions are carried out by systematic procedures termed **protocols**, consisting of specific steps taken by the parties involved. For example, the RSA protocol consists of Alice's selection of two prime numbers p and q to form $n = pq$, her determination of corresponding public and private values b and a, the publication of her public key $\langle n, b \rangle$, and the subsequent steps for encryption and decryption.

Protocols are usually expressed first in generalized terms to convey the essential outline of the procedure. Greater detail is required to specify a practical or commercial system. It is the generalized versions that are discussed in this chapter, with the objective of indicating the range and power of modern cryptography, how it already influences our daily lives and is likely to to do so in the future. There are dozens if not hundreds of protocols in use or in published form. This chapter highlights some of the most common, which in fact are building blocks to more complex protocols.

14.1 Digital Signatures

One of the most basic security protocols enables a person to sign a document digitally, in such a way that other parties can be assured that the signature is authentic. At first this may seem impossible, for couldn't the signature simply be copied? The protocol for digital signatures is, however, one of the simplest, and is a component of several other more complex protocols.

The basic digital signature protocol simply interchanges the roles of public and private keys. Or, said another way, the order of encryption and decryption is reversed.

Here is a digital signature protocol based on the RSA system that allows Alice to sign a message m in such a way that Bob or anyone with knowledge of Alice's public key can be assured that the signature is indeed Alice's:

1. Alice publishes an RSA public key $\langle n, b \rangle$.

2. Alice encrypts her message m with her private key $\langle n, a \rangle$, forming

$$y = m^a \bmod n,$$

   and she presents y as the signed version of m.

3. Bob (or anyone) decrypts y using Alice's public key to recover

$$m = y^b \bmod n.$$

   Bob sees that the decrypted document is the original, and hence knows that only Alice could have encrypted it.

The method is based on the complementary properties of public and private keys. If a message is encrypted using a public key $\langle n, b \rangle$, it can be decrypted with the corresponding private key $\langle n, a \rangle$. Likewise if it is encrypted with a private key $\langle n, a \rangle$, it can be decrypted with the corresponding public key $\langle n, b \rangle$. It is a beautifully elegant system.

The method can be generalized to other asymmetric key systems. Suppose Alice's private and public keys are represented by the parameters k_a and k_b, respectively. And suppose that Bob wishes to send her a message x. For shorthand notation we denote by $E[x, k_a]$ the encryption of x with key k_a by whatever method of encryption is used. In this notation Bob sends $y = E[x, k_a]$. Alice decrypts the result y by the inverse $x = E[y, k_b]$. The digital signature protocol simply reverses the order of encryption and decryption.

Hashing

Although the digital signature protocol described above is simple in concept, it requires a great deal of computation because the entire document m must be encrypted with Alice's private key. Normally, one thinks of a signature as something short appended to a document, rather than an alteration of the entire document. The basic digital signature can be modified so that it, itself, is of modest length.

A **hash** of a document is a mapping of the document into a smaller message—a sort of **message digest**—and that digest can be signed using the digital signature protocol and appended to the full plaintext message. The hash mapping is accomplished by a mathematical formula applied to the numerical version of the original message text. It is represented symbolically by $h = H(m)$. For example, the hash function might calculate $h = H(m) = m \bmod n$, where m is the numerical version of the message and n is a fixed large integer. The result is an integer between 0 and $n - 1$. This particular method, however, is not practical for other than very short messages, and in actual application far more complex functions are used, for reasons to be explained shortly.

Once a hash function is agreed upon, Alice can hash her message and sign the hashed version. Her signature can be verified by checking that the decrypted version of her signed hashed version is identical to the hashed version of the full message. The process is indicated schematically in figure 14.1, which shows how a signature can be appended to a document.

Although almost any hash function can be used as a method to verify Alice's signature in this way, the method may be susceptible to alterations of the document after it has been signed. This would be possible if an alteration could be constructed that resulted in the same hash value. For example, Alice might agree to pay Bob $100,000 for a house. She signs a document to that effect and gives it to Bob. Bob prepares another document that states that she will pay $120,000, and he arranges the document so that its hashed version is identical to the version that Alice signed. Bob destroys the original document, keeping only the signed hashed segment, which he appends to the alternate version. Bob then has a signed document stating that Alice will pay $120,000.

To avoid the potential for such alterations, a hash function should satisfy the following three properties:

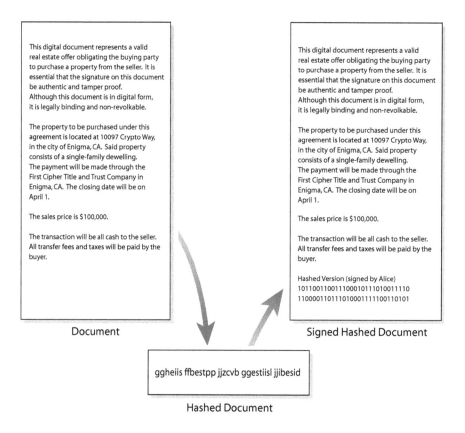

Document

Signed Hashed Document

ggheiis ffbestpp jjzcvb ggestiisl jjibesid

Hashed Document

FIGURE 14.1 A signed document. The original document is hashed to form a short digest (which is represented here in alphabetical form). This digest is signed by Alice; (it is assumed here that this produces a binary stream). The signed digest is appended to the original document.

1. Given a message m, the hash value $h = H(m)$ should be simple to calculate.

2. The function H should be **one way**. That is, given a hashed value h, it should be difficult to find a message m such that $h = H(m)$.

3. The function H should be **collision resistant**. That is, it should be extremely difficult to find two messages m_1 and m_2 such that $H(m_1) = H(m_2)$.

Standardized hash functions satisfying these properties have been developed. They are rather complex, and typically involve several steps much like a classical encryption algorithm such as DES, but good hash functions are now readily available. Still, as a precaution, it is advisable to guard against possible alteration when signing documents prepared by others. One stratagem is to insist on a slight insignificant modification of the document, like the addition of a comma. Such a change will alter the hashed version, and make it extremely unlikely that a specially prepared alternate document will hash equivalently if the minor change is placed in it as well.

14.2 Blinded Signatures

Suppose Alice wishes to have a document notarized by Bob. That is, she wants Bob to sign her message with his private key so that others can decrypt it with Bob's public key, but she does not want Bob to be able to read the document itself. She only wants Bob to sign the message to certify that the document was prepared at a certain date, and she wants this done so that Bob cannot read it even if he keeps a copy. This can be accomplished with a **blinded signature**.

Suppose Bob has a private RSA key $\langle n, e \rangle$ and a corresponding public key $\langle n, f \rangle$. Alice wants Bob to sign her message m with his private key, while being unable to read m. Alice accomplishes this by disguising her message before having it signed and then removing the disguise after it is signed. Here is the protocol:

1. Alice formulates her message m as a number.

2. Alice selects a random **blinding number** k with $0 < k < n$ and gcd $(k, n) = 1$ (the n being part of Bob's key). Then she finds g, which is the mod n inverse of k; that is, $kg \equiv 1 \bmod n$, as discussed in chapter 13. Alice then computes

$$t = mk^f \bmod n,$$

where $\langle n, f \rangle$ is Bob's public key.

3. Alice takes the blinded message t to Bob to sign. Bob cannot extract m from t because the blinding factor k is unknown to him.

4. Bob signs t with his private key to produce

$$y = t^e \bmod n,$$

following the basic digital signature protocol.

5. Alice unblinds y by calculating $z = yg \bmod n$. This gives

$$z = yg \bmod n$$
$$= t^e g \bmod n$$
$$= (mk^f)^e g \bmod n$$
$$= m^e kg \bmod n$$
$$= m^e \bmod n.$$

Hence z is m signed with Bob's private key $\langle n, e \rangle$.

Example 14.1 (Thousand-dollar message). Alice wishes to have her offer to Cynthia signed by her agent Bob, but she does not want Bob to see the amount of her offer. Bob has a published public key $\langle n, f \rangle = (5{,}561, 235)$. Alice's offer is \$1,000.

1. Alice's message is $m = 1{,}000$. She selects the random number $k = 91$ and computes its inverse mod n, which is $g = 550$, since

$$kg = 91 \times 550 \equiv 1 \bmod 5{,}561.$$

Alice blinds her offer by computing

$$t = mk^f \bmod n = 1{,}000 \times 91^{235} \bmod 5{,}561 = 1{,}715.$$

She uses the square-and-multiply method of section 13.8 to compute the large power of 91 mod 5,561.

2. Bob signs the blinded message by using his private key $\langle n, e \rangle$, computing the blinded signed message

$$y = t^e \bmod n = 1715^e \bmod 5{,}561 = 216.$$

Bob also uses the square-and-multiply method to compute the large power of 1,715 mod 5,561, and finds $y = 216$ (but he does not reveal e).

3. Alice realizes that now

$$y \equiv t^e \equiv (mk^f)^e \equiv m^e k \bmod n.$$

She unblinds this by multiplying by the inverse of k, which is $g = 550$. Thus the final message is

$$z = yg \bmod n = 216 \times 550 \bmod 5{,}561 = 2{,}019.$$

4. Cynthia can find Alice's original message by using Bob's public key to compute.

$$m = z^f \bmod n = 2{,}019^{235} \bmod 5{,}561 = 1{,}000.$$

Note that nobody except Bob (not even you) knows Bob's private key $\langle n, e \rangle$ and unless they factor $n = 5{,}561$, they cannot find it. (See exercise 1.)

14.3 Digital Cash

Cash has two features that many consumers like: it is readily transferable and it is anonymous in the sense that one can spend it without identifying oneself. Cryptography offers the possibility of designing digital cash with these same features.

The main outlines of digital cash were developed by David Chaum. Before presenting the protocol for digital cash, let us consider the physical analogy that he suggested. Alice wants to obtain an anonymous $20 cashier's check from her bank.

1. Alice obtains a blank form from the bank and writes a 100-digit identification number on it that she selects randomly.

2. Alice puts the form together with a piece of carbon paper in an envelope and takes it to the bank.

3. The bank debits Alice's account by $20 and stamps the envelope with the bank's special (nonduplicable) $20 stamp. The impression made by this stamp goes through the envelope and through the carbon paper, leaving its image on the form inside, thus certifying the form as representing a valid $20 cashier's check from the bank.

4. Alice removes the certified form from the envelope.

5. Alice gives the $20 check to a merchant in exchange for goods.

6. The merchant sends the check to the bank for payment. The bank verifies that a $20 cashier's check with that identification number has not previously arrived at the bank. Then the bank credits the merchant's account for $20 and adds the 100-digit identification number to the master list of identification numbers. The bank has no way of knowing that the check was issued to Alice. From Alice's viewpoint, the check is (almost) as good as cash.

Figure 14.2 illustrates the process.

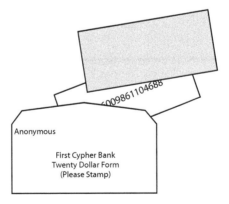

FIGURE 14.2 Preparing a cashier's check. Alice writes a random serial number on a form and encloses it together with carbon paper in an envelope. The bank stamps the envelope with its $20 stamp, which, thanks to the carbon paper, leaves a validating impression on the enclosed form. The bank does not see the serial number until Alice spends the $20.

This protocol can be carried out digitally using the digital signature protocol. The bank has a public RSA key $\langle n, f \rangle$ for $20 bills.

1. Alice selects a serial number S and a blinding factor k. Both of these are large numbers. The first several digits of this serial number identify the bank. She selects the remainder of the digits randomly. She also calculates g as the mod n inverse of k; that is, $kg \equiv 1 \bmod n$.

2. Alice computes

$$y = Sk^f \bmod n.$$

The result y is a blinded version of the serial number S. The blinding is analogous to Alice placing the form with her chosen serial number in an envelope with carbon paper. She sends the blinded value y to the bank.

3. The bank debts her account by $20 and signs the blinded serial number with its private $20 signature key $\langle n, e \rangle$, forming

$$t = y^e \bmod n.$$

The bank returns this value to Alice.

4. Alice unblinds t by computing

$$\begin{aligned} z &= tg \bmod n \\ &= y^e g \bmod n \\ &= [Sk^f]^e g \bmod n \\ &= S^e kg \bmod n \\ &= S^e \bmod n, \end{aligned}$$

which is then the bank's digital $20 bill.

5. Alice purchases merchandise by using the bank's digital $20 bill. The merchant verifies that it is valid by computing

$$S = z^f \bmod n$$

and noting that the leading digits are those that identify the bank. The merchant submits S to the bank.

6. The bank verifies that the serial number S has not been used before and then credits $20 to the merchant's account.

14.4 Identification

An important component of modern information services is the ability to identify oneself to a remote party or computer system, as in automated banking, message retrieval, investment transactions, and use of private databases. Modern cryptography provides several ways to do this, and each method has special characteristics and features.

Password Protection

The most common method of identification is by means of a password or Personal Identification Number (PIN). The password is usually established by a secure line, by mail, or by personal appearance. In use, the host computer requests the user's name and password and the user sends them. The computer compares the password–name combination with those stored in a master list and verifies identity if there is a match.

The system has several vulnerabilities, one of which is that someone might break into the computer that stores the passwords, and thereby obtain a list of passwords and associated names. This weakness can be mitigated by use of our familiar tool, the one-way function. In this method, the host computer does not store passwords with names, but instead, in each case, stores the result of a one-way function applied to the password. For example, if your password is the number x, the computer might store $f(x) = g^x \bmod n$, where g and n are fixed large integers. Whenever you log in and report your password x, the host easily computes $f(x)$ and compares the result with what is stored with your name. If the results match, you are admitted to the system. If someone manages to obtain a copy of the master list of transformed passwords, that person cannot deduce your password from the list of $f(x)$'s without solving the discrete logarithm problem, which, as we know, is extremely difficult. Hence, your password is protected against theft of the password list.

The basic password system is, however, vulnerable to eavesdropping. If someone is able to tap into your conversation while you send your password, that person can later pretend to be you by using your password.

Simple Challenge Method

Public key cryptography, such as RSA, provides the basis for a protocol that is more secure than the simple password system. Suppose Bob asks Alice to identify herself before sending further communication. Bob initiates the identification protocol by selecting a random number x, sending it to Alice, and asking her to encrypt it with her private key. Bob receives Alice's response and decrypts it with Alice's public key. If the result is the original random number x, Bob considers that Alice has indeed identified herself. No one else could have encrypted x so that it faithfully decrypts with Alice's public key. One way to look at this protocol is that Bob asks Alice to sign the random number x with her private key. Bob verifies Alice's digital signature.

The method can also be considered as a challenge-and-response protocol. Bob challenges Alice to properly encrypt the random number x. The challenge-and-response method is a central feature of many identification protocols. The public key challenge method is secure from theft within the host computer because the only thing stored is Alice's public key, which is public anyway, and hence, it does not matter if someone steals it.

The public key challenge is also secure against eavesdropping. If Eve intercepts the message between Alice and Bob, she will know the value of x and Alice's encrypted version of it. But this will not enable Eve to pose as Alice in the future, for when she tries, Bob will send a different value of x for encryption, and Eve will be unable to properly respond to the challenge.

14.5 Zero-Knowledge Proofs

Suppose you want to prove to someone that you know something specific without actually revealing what it is. For example, you might wish to prove that you know the combination of a safe, without revealing the combination. An easy solution for that case is to ask everyone else to leave the room for a few moments while you open the safe. When they return and see the open safe, they must conclude that you know the combination. You have proved what you wished to prove while transmitting zero knowledge about the specific thing you know.

There is a famous example of zero-knowledge proofs from mathematics.

Example 14.2 (Tartaglia's formula). Any student of algebra is familiar with the general quadratic equation $ax^2 + bx + c = 0$ and the explicit formula for its solution, $x = [-b \pm \sqrt{b^2 - 4ac}]/(2a)$. This formula has been known since at least the ninth century. A similar formula for the solution of a cubic equation $ax^3 + bx^2 + cx + d = 0$ was not known until the early 16th century. It was in 1535 that the Venetian mathematician Niccolò Tartaglia claimed to have discovered such a formula, but believing it to be of value, he did not wish to reveal the formula itself. However, he *proved* that he possessed such a formula by responding to challenges. People would submit a cubic equation to him, and shortly thereafter he would announce the solution (consisting of the three roots). His solutions were easily checked, by merely substituting them into the equation. Soon the mathematical community was convinced that Tartaglia possessed the formula, but no one else could deduce the formula from the challenges.

Eventually, Geronimo Cardano, sworn to secrecy, convinced Tartaglia to show him the formula. And, as might be expected, Cardano later published the formula in his own book, although he did give credit to Tartaglia. However, to this day the formula is generally referred to as Cardano's formula.

Zero-knowledge proofs (that is, proving you have knowledge without revealing the specifics) is accomplished by responding to a challenge. In the case of the safe, you open it while no one is looking. Likewise, for a formula that solves equations, you apply it to challenge equations. The challenge method for identity verification based on private key encryption of a random number is another example.

In some cases the response to a single challenge is sufficient, as in opening the safe or encrypting the value of x with one's private key, while in other cases such as Tartaglia's demonstration that he had the formula, successful responses to several challenges are required. Digital protocols of zero-knowledge proofs commonly require several challenges, each challenge having a probabilistic

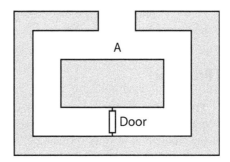

FIGURE 14.3 Tunnel with door. Alice wants to prove to Bob that she can pass through the door, but she does not want Bob to know in which direction she is able to pass.

component. The concept is illustrated by the standard textbook situation shown in figure 14.3.

Alice wants to prove to Bob that she can pass through the door in the tunnel, but she does not want to reveal how she does it or even in which direction she can pass. She could easily prove that she can pass through the door by asking Bob to wait at point A while she goes down one side and comes back the other. However, this reveals the direction by which she passes, which is more information than she wants to reveal. A more complex protocol using random challenges solves the problem.

Here is the method:

1. Alice tells Bob to wait outside while she goes down either the left or the right side. Bob is unable to observe which side she selects.

2. When Alice is inside, Bob comes to point A and shouts "left" or "right," selected randomly, as a challenge.

3. Alice reappears by coming from the direction that Bob calls out.

4. After several challenges and successful responses, Bob is convinced that Alice can indeed pass through the door.

Alice of course always goes down the side from which she is able to pass through the door. Suppose it is the left side. Then if Bob calls out "left," Alice simply returns up the left side, the way she came in. If Bob says "right," Alice passes through the door and returns up the right side. She can do this indefinitely until Bob is convinced, and Bob never learns the direction by which she passes. He obtains zero knowledge about the secret direction.

If Alice cannot pass through the door, she may be able to fool Bob into thinking she can—at least for awhile. Alice will randomly select a side to go down. If Bob calls out that side, she can return correctly; otherwise not. Alice has a 50 percent chance of meeting each challenge. Hence, Bob will want to issue several challenges, randomly calling out left or right, until Alice fails or until he is convinced that Alice can pass through the door.

If Alice meets the challenge N times in succession, there remains a probability of only $P = .5^N$ that she cannot pass through the door. For $N = 20$, for example,

$P \approx 1/(1 \text{ million})$. Hence, like Tartaglia and his formula, Alice can prove she knows her secret without revealing it.

A Numerical Protocol*

The concept of zero-knowledge proof has been implemented in digital form using concepts related to public key cryptography. One version is especially well suited to the problem of identity verification.

A central trusted agency selects an integer n as the product of two primes $n = pq$ and an integer s, $0 < s < n$. The agency also computes $y = s^2 \bmod n$. Alice is told n and s (and can compute y).

Suppose that Alice wishes to identify herself to Bob, her network provider. Bob has been given y and n by the central agency as the values to use when identifying Alice. Alice must prove to Bob that she knows the number s that is the square root (mod n) of y without revealing s.

Given only y and n, it is virtually impossible to deduce s. Indeed, doing so is equivalent to finding the message s that has been encrypted with an RSA public key of $\langle n, b = 2 \rangle$, producing the encrypted message $y = s^2 \bmod n$. This special case of going from $y = s^2 \bmod n$ to s is termed the **discrete square-root problem**, and when n is the product of two (unknown) primes, it is extremely difficult to solve, requiring the factoring of n, as in any RSA cipher.

Alice needs to prove that she has the discrete square root of y without stating what it is.

Here is a simplified version of the protocol designed by Fiat and Shamir:

1. Alice splits s into two pieces r_1 and r_2 such that $r_1 r_2 = s \bmod n$. (She can do this by first selecting r_1 with $\gcd(r_1, n) = 1$, and then setting $r_2 = s r_1^{-1}$, where r_1^{-1} is the mod n inverse of r_1, which can be found as explained in chapter 13.) Alice then computes $x_1 = r_1^2 \bmod n$ and $x_2 = r_2^2 \bmod n$ and sends x_1 and x_2 to Bob.

2. By simple calculation Bob verifies that $y = x_1 x_2 \bmod n$, which must be the case since

$$x_1 x_2 \bmod n = r_1^2 r_2^2 \bmod n$$
$$= (r_1 r_2)^2 \bmod n$$
$$= s^2 \bmod n = y.$$

Of course, Bob cannot see this detail, since he does not know r_1 or r_2.

Bob does know, however, that if Alice knew the discrete square root of both x_1 and x_2, she would know s since $s^2 = x_1 x_2 \bmod n$. Bob also would know s if he knew the discrete square roots of x_1 and x_2, so Alice will not tell him both of these.

Hence, Bob requests that Alice supply the square root of either x_1 or x_2, and he randomly selects which of them to request.

3. Alice sends Bob r_1 or r_2, corresponding to whether he requests the square root of x_1 or x_2, and Bob verifies that her response is a proper square root.

4. The procedure is repeated, with Alice selecting a new random pair r_1 and r_2, sending the new x_1 and x_2 to Bob, and responding to Bob's request for the discrete square root of either x_1 or x_2. This continues until Bob is convinced that Alice indeed knows s.

If Alice does *not* know s, then, like the door and the tunnel, she has a 50 percent chance of passing a challenge. She can select r_1 randomly and compute $x_1 = r_1^2 \bmod n$. She then determines $x_2 = yx_1^{-1} \bmod n$ where x_1^{-1} is the mod n inverse of x_1. Then $x_1x_2 = y \bmod n$, but it is extremely unlikely that x_2 is a square. That is, there is no r_2 such that $x_2 = r_2^2 \bmod n$.

When Bob receives x_1 and x_2 from Alice, he verifies that $y = x_1x_2 \bmod n$, but he requires proof that both x_1 and x_2 are squares. He cannot ask Alice to supply square roots of both x_1 and x_2, so he selects one. He has a 50 percent chance of selecting the one she did not prepare. Hence, each challenge has a 50 percent chance of exposing Alice's deceit if she does not know s. After sufficiently many challenges that are correctly met, Bob agrees that Alice knows s.

14.6 Smart Cards

A smart card is a wallet-sized plastic card that contains a microprocessor and memory so that it can carry out a limited amount of computation and store a modest-sized file. The card contains no internal power source, but gets power through contacts when inserted in a card-reading terminal. The computational capacity of a smart card is physically limited by the dimensions and physical characteristics of the card, but it is expected that the capacity of new cards will increase in coming years. Smart cards are used for personal banking, for the purchase of items from merchants or special vending machines, and for record keeping. Generally speaking, the computational aspects of the card are related to two main functions: authentication and financial payment. These operations illustrate how the protocols of this chapter are used in practice.

Authentication

We limit consideration to smart cards with authentication protocols that are based entirely on numerical data, as opposed to those employing photographs, hand-made signatures, fingerprints, or eye scans. The authentication process is divided into two steps: (1) the user is authenticated to the card, and (2) the card is authenticated to the computer of the bank or transaction agent. Thus for someone to fraudulently use an account, he or she must first obtain the card and be able to (falsely) identify himself or herself to the card. Then the card will identify itself to the computer.

The authentication of the user is generally based on a Personal Identification Number (PIN) known by the user. When the individual wishes to use the card, he or she inserts the card in a terminal and types the PIN into the terminal keyboard. The terminal relays this number to the card, which can compare it with the PIN stored internally. If there is a match, the card will place itself in a mode ready for additional service activity. Because this stage of verification is between the individual and the

Card   **Computer**

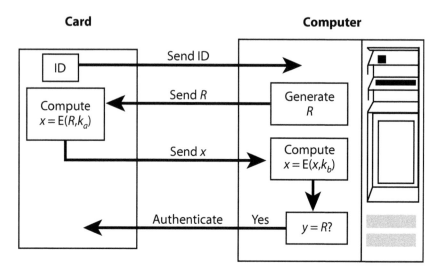

FIGURE 14.4 Smart card authentication protocol. The card sends its identification number to the computer. The computer responds by sending a random number as a challenge to the card. The card encrypts the random number with its private key and sends the result back to the computer. The computer decrypts the result with the user's public key. If this new result matches the random number, authentication is verified.

card, there is no reason for the terminal to store the PIN or compare it with a list. The terminal simply acts as a means for accessing the card.

The next phase is for the computer system to verify the card. This is carried out by a challenge-and-response protocol, which may be based on either a symmetric key cipher system or an asymmetric key system. In one of the simplest methods, the card and the computer share a common function f and a key k associated with that card. The computer generates a random number R and sends it to the card. The card computes $f(R, k)$ and sends the result back to the computer. The computer independently computes $f(R, k)$ and compares the result with the card's response. If they are equal, access is granted.

Authentication can be made more secure by use of public key cryptography as shown in figure 14.4. The card has a private key stored internally, and a public key stored in the computer. The computer generates a random number R and sends it to the card. The card encrypts R with its private key k_a to produce $x = E[R, k_a]$, which is sent to the computer. (Here again, $E[R, k_a]$ denotes the result of encrypting R with key k_a.) The computer decrypts x with the card's public key to verify that $R = E[x, k_b]$. If the verification is positive, the computer authenticates the card.

The authentication can also be based on a zero-knowledge proof, such as the Fiat–Shamir protocol. In this case the secret value s is hard-wired into the card and no one else need know it or store it. The card's identity is $y = s^2 \bmod n$, and the card proves its identity by proving that it knows s.

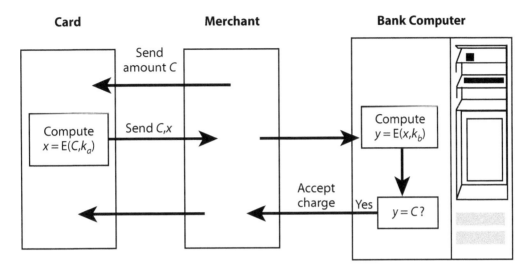

FIGURE 14.5 Smart card payment protocol. The merchant requests payment of an amount C from the card. The card encrypts this amount with its private key and sends the amount C and the encrypted version of C back to the merchant (who can check that C is still correct) and on to the bank. The bank verifies that C is consistent with its encrypted version (thus guarding against a change in C by the merchant). When verification is complete, the bank carries out the financial transaction.

Payment Protocols

In the context of a point-of-sale (POS) payment system, a smart card can authorize payment from a bank or can transfer digital cash. Again, either symmetric key or asymmetric key cipher systems can provide security for this process.

There are three parties involved in a payment system: the customer, the merchant, and the bank. All three must interact in the transaction, and each must be protected.

One way to carry out the transaction using an asymmetric key system is shown in figure 14.5. In this method neither the user nor the merchant can change the payment amount C as it is sent to the bank. The merchant can check what the card sends to the computer, and the bank can check that the merchant did not alter the amount.

14.7 EXERCISES

1. (Factor) Find Bob's key for example 14.1 and compute the number he would give to Alice if she changed her offer to $1,200.

2. (The birthday attack*) Suppose there are 23 people at a gathering. What is the chance that at least two people share the same birthday? This is known as the birthday paradox, for the answer is 50 percent, which at first seems surprisingly large.

   A variant of the puzzle is this: Suppose there are r people in each of two rooms. Each person holds a (random) number between 1 and n, where n is large compared to r. What is the probability that a pair of people, one from each room, possess the same number?

   (a) Argue that the probability that such a pair exists is $1 - (1 - \frac{r}{n})^r$.

   (b) Let $r = \sqrt{\lambda n}$ for some λ. Using the approximation $1 - x \approx e^{-x}$, find the probability that such a pair exists in terms of λ.

   (c) What is the probability of a matching pair if $n = 2^{50}$ and $r = 2^{30}$?

   (d) Alice is planning to digitally sign the hash of a contract. The hash will be a reduced document 50 binary digits in length. She feels safe because there are 2^{50} possible hash functions. Larry decides to launch a birthday attack by, first, preparing a fraudulent contract with terms unfavorable to Alice. Then he selects 30 places in each of the contracts where a slight change can be made (adding a comma for instance), changes that Alice is likely to accept. He prepares 2^{30} (approximately 1 million) versions of the contract by using all combinations either making a change or not at the 30 places. He then forms the hashed version of all of these, and looks for a match. What is the probability that there will be at least one pair of hashed versions, a good and a fraudulent, that match? If Larry finds a match, he can show Alice the corresponding fair contract and present the fraudulent hashed version for her digital signature.

3. (Bit commitment) Alice claims that she is able to predict what the stock market average will be one month from now. To prove she can do it without revealing the value ahead of time, she offers to give Bob now an encrypted version of the future average (say x) and in one month give Bob the inverse of the key she used so that he can verify the result. Bob is suspicious and wants better assurance that Alice does not have a number of different inversion keys that she could choose from to match the actual stock market value when it is known. Can he ask her to encrypt something else as well that will guarantee the security of her commitment?

4. (Secret sharing) Suppose you have a secret message m (a number) that you wish to divide among n people, so that the message can be determined by any t members of the group, but not by any subgroup of less than t members. This can be done as follows:

   (a) Select a prime number p larger than any m that might be used.

   (b) Select $t - 1$ coefficients $s_1, s_2, \ldots, s_{t-1}$ and form the polynomial

   $$s(x) = m + s_1 x + s_2 x^2 + \cdots s_{t-1} x^{t-1}.$$

   (c) Index people from 1 to n, and tell them all the number p. Tell person with index k the value $s(k) \bmod p$.

   Any group of t people can reconstruct the $(t-1)$th order polynomial from their information by using the Lagrange interpolation formula. Let $\mathcal{T}$ be the set of indices of t people. That group can determine m as

$$m = \sum_{k \in \mathcal{T}} s(k) \prod_{j \in \mathcal{T}/k} \frac{j}{j-k} \bmod p,$$

where $\mathcal{T}/k$ denotes every index in $\mathcal{T}$ except k. For example, for determination by three out of five people, suppose the message is 17, $p = 43$, and $s(x) = 17 + 24x + 36x^2$. Then the five individuals would be told p, and each would be given a result according to this list:

1. $17 + 24 + 36 \bmod 43 = 34$
2. $17 + 24 \cdot 2 + 36 \cdot 4 \bmod 43 = 37$
3. $17 + 24 \cdot 3 + 36 \cdot 9 \bmod 43 = 26$
4. $17 + 24 \cdot 4 + 36 \cdot 16 \bmod 43 = 1$
5. $17 + 24 \cdot 5 + 36 \cdot 25 \bmod 43 = 5.$

Suppose that persons 2, 3, and 5 share their information. They compute

$$m = 37 \left(\frac{3}{3-2}\right)\left(\frac{5}{5-2}\right) + 26 \left(\frac{2}{2-3}\right)\left(\frac{5}{5-3}\right) + 5 \left(\frac{2}{2-5}\right)\left(\frac{3}{3-5}\right) \bmod 43$$
$$= 37 \cdot 5 + 26 \cdot (-5) + 5 \cdot 1 \bmod 43 = 17.$$

Suppose on another day, the group is told $p = 57$ and persons 1, 3, and 4 are told the values 54, 48, 39 respectively. What is the message?

14.8 Bibliography

A readable discussion similar to parts of this chapter, including the story about Tartaglia, is contained in [1]. Numerous protocols are outlined in [2]. The concept of blinded signatures as well as several other signature and digital cash protocols were originated by David Chaum. See [3], [4]. For the discrete square root protocol, see [5]. The exercise on secret sharing is adapted from the general presentation in [6], which discusses several other protocols as well.

References

[1] Beutelspacher, Albrecht. *Cryptology*. Washington, D.C.: Mathematical Association of America, 1996.

[2] Schneier, Bruce. *Applied Cryptography*. 2nd ed. New York: John Wiley, 1996.

[3] Chaum, David. "Blind Signatures for Untraceable Payments." In *Advances in Cryptology: Proceedings of Crypto '82*, ed. David Chaum, R. L. Rivest, and A. T. Sherman. New York: Plenum Press, 1983.

[4] ———. "Achieving Electron Privacy." *Scientific American*, August 1992, 96–101.

[5] Fiat, A., and A. Shamir. "How to Prove Yourself: Practical Solutions to Identification and Signature Problems," *Advances in Cryptology—CRYPTO '86 Proceedings*, Springer-Verlag, 1987: 186–194.

[6] Trappe, Wade, and Lawrence C. Washington. *Introduction to Cryptography with Coding Theory*. Upper Saddle River, N.J.: Prentice-Hall, 2002.

Summary of Part III

Ciphers are designed to confound the efforts of outsiders wishing to read messages intended for others. Classical ciphers are based on a **key** known by both sender and intended receiver. The key associated with a **substitution cipher** is a transformation of the letters of the alphabet into other symbols, or it is a permutation of the alphabet itself. The key associated with a **transposition cipher** is a formula for changing the order of the letters sent. Complex ciphers may be combinations of both methods and may change with every letter sent. One measure of the complexity of a cipher is the number of its possible keys. A simple **Caesar cipher** has only 26 possible keys. The **Vigenère cipher** has 26^n possible keys, where n is the length of the key word. **Cryptograms** have $26! \approx 4 \times 10^{26}$ possible keys. The original **Enigma machine** had 7.6×10^{18} possible keys. **Jefferson's cipher wheel** had an astounding 3.72×10^{41} possible keys.

Shannon developed a comprehensive theory of classical encryption that includes a definition of **perfect security**, which can be related to entropy. It follows from this theory that the **one-time pad** method of encryption has perfect security if the symbols in the pad are truly random. The theory also quantifies the degree of security of a cipher system in terms of the **unicity point**, which is an estimate of the length of ciphertext that must be seen before an unknown key can be deduced.

Because a one-time pad is the most practical way to obtain perfect security, effort has been devoted to developing means for efficiently generating random numbers. One method employs **shift registers**, and another uses **linear congruences**. Both have the advantage that **pseudorandom sequences** with good properties can be generated from a few initial parameters. A disadvantage of these methods is that these few parameters generally can be inferred from observation of a long sequence generated by them.

Shannon's theory is based on the degree of uncertainty of the key. According to this theory a greater number of keys produces greater security. However, the Enigma machine has on the order of 10^{18} keys, while cryptograms have on the order of 10^{26} possible keys. Yet a cryptogram can be solved rather quickly by an average citizen, whereas the initial Enigma was breached only after a great deal of work and use of a computer. Computational complexity plays an important role in determining actual security.

The **public key** approach to encryption is directly based on computational complexity rather than absolute key size. The **discrete logarithm problem** of solving the equation $y = a^k \bmod n$ for k, given y and a, provides no security from a purely probabilistic standpoint because indeed it can be solved. There is no ambiguity. However, for large values of n it is extremely difficult from a computational standpoint.

The original motivation for public key systems was the difficulty of managing key distribution in a classical system. The original Diffie–Hellman paper gave a method, using the discrete logarithm problem, in which two parties jointly agree on a secret key even though their communication was open to anyone. Other public key systems follow that approach, finding functions that are hard to solve unless one knows a key. Some of the main tools used in the development of such systems are the **Euclidean algorithm** and **Fermat's theorem**.

A standard public key system is **RSA**. The security of this method rests on the computational difficulty of factoring a large number that is the product of two unknown primes. A person, say Alice, generates her **public key** $\langle n, b \rangle$, which can be openly published, and a corresponding **private key** $\langle n, a \rangle$ that is not published. Anyone wishing to send a message to Alice encrypts the message using Alice's public key. Alice can decrypt it with her private key. This method and its variants are now in common use—in Internet correspondence, financial transactions, and military operations.

Public key concepts form the basis of **protocols** designed to accomplish various secure operations. A digital document can be signed using a **digital signature** protocol that reverses the role of public and private keys. The document is encrypted with one's private key, and the signature is verified by application of the public key that decrypts the document. A more complex protocol provides a **blinded signature** whereby a document can be confusingly modified so that a notary can sign it without being able to read it. Public key protocols are embedded in **digital cash** systems and in some **smart cards** used for financial transactions.

PART IV

EXTRACTION

Information from Data

15

DATA STRUCTURES

Information often resides in data, but data is not always the same as information: certainly data is not the same as *useful* information. Nevertheless, often data—collected in experiments, gleaned from transactions, offered by individuals in questionnaires, or downloaded from other sources—does contain information of great value. The challenge is to extract the useful information from all that is available.

Data sources can become enormous and unwieldy. The first step toward transforming data into useful information is to organize the data so that it can be readily accessed, searched, manipulated, updated, simplified, and sometimes generalized. Data structures provide the basis of such work. Many of these structures were originated to facilitate the programming of complex data manipulations, but the principles underlying data structures are useful more generally, for constructing databases or data warehouses, for building efficient data retrieval systems, and ultimately for assisting with the processes of extracting information from data.

Basic data structures include lists, arrays, and trees. From these basics, more complex structures can be built.

We shall find that these basic structures often are used in concert and that one may be converted to another. For example, an effective way to sort a list is to transform it, either explicitly or implicitly, into a tree; then sort the tree and transform the results back to a list. In the process, the tree might be represented as an array. Data structures are fluid and adaptable, and come in numerous variations.

15.1 Lists

An obvious way to store data is sequentially, as a list, one item after another. Employee names might constitute a list, for example.

Abstractly, a **list** is an ordered set of objects (or items) of a given type. A list of length n can be represented as $(a_1, a_2, \ldots, a_n)$ where the a_i's are the objects. The

241

position of an object is its index i in the list. The objects themselves may be numbers, book catalog records, patient health profiles, gene descriptions, or names of state capitals. The objects in a list need not be numeric or alphabetic. A row of automobiles in a parking lot can be regarded as a list.

The objects in a list may be multidimensional. For example, items in a library catalog may include book title, author, Library of Congress catalog number, date of publication, publisher, date of acquisition, and availability status. Such an object is termed a **record** with individual portions of the record being **fields**.

For a list to be most useful, it must be possible to carry out certain basic operations on the list. For example, one may wish to **insert** additional items in the list, **delete** some items, **locate** an item or items that meet certain criteria, or move to the **next** or **previous** item. The ease with which such basic operations can be performed may depend on how the list is represented.

Two of the most important operations on a list are **sorting** and **searching**. Sorting is the process of arranging the items according to an ordering of the items. For example, it may be desired to sort items numerically or alphabetically. If an item is a record with several fields, the ordering is usually carried out with respect to a single field termed the **key** that uniquely identifies the entry in the list. Searching is the process of finding an item that meets a specific criterion, or concluding that no such item is in the list. Data structures facilitate sorting and searching, carried out by the basic operations mentioned in the previous paragraph.

Lists Represented by Arrays

Normally one thinks of a list as an array, the items being placed at successive locations. For example, the items might be written on successive rows of a sheet of lined paper or at successive locations in computer storage.

As a physical example, imagine the parking lot of a rental car agency located at an airport. The parking lot spaces are numbered consecutively. The rental cars also have identification numbers that serve as their keys. Only a small fraction of the total inventory of cars is in the lot at any one time.

Suppose the agency keeps track of available cars by storing them in numerical order in parking spaces, with the car of lowest identification number going in space number 1. When a car is returned to the lot, it must be put in its proper place between cars of lower and higher identification numbers. To make room for this car, all cars with higher identification numbers must be moved one space down the list to provide an opening (figure 15.1)

Likewise, when a car is rented and leaves the lot, all cars beyond it on the list must be moved up one space to close the gap. Clearly this is not an efficient way to store cars.

Inserting or deleting an element in a list that is implemented this way requires, on average, $n/2$ movements of items in the list, and even when the items are entries in a computer, this can be time consuming for large lists. On the other hand, if a particular item is found, finding the next or previous item is simple. One simply goes to the next or previous location.

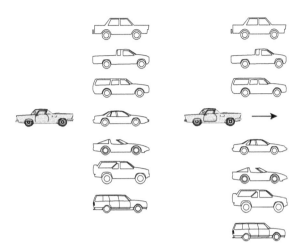

FIGURE 15.1 To insert a new vehicle in the parking lot, several others must be moved.

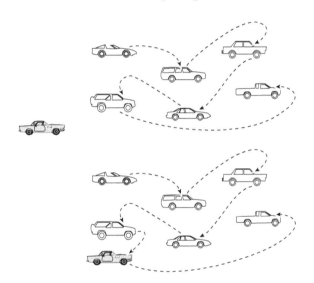

FIGURE 15.2 Each pointer gives the location of the next car. When a new car arrives, it can be inserted by updating the pointers, as shown in the bottom part of the figure when the entering vehicle is sixth on the list.

Linked Lists

The items of a list can be stored in arbitrary locations, provided that a record of their locations is kept. The simplest way to do this is with a **linked list**. In such a list the items are stored arbitrarily in the spaces available, but accompanying each item is a **pointer** to the location of the next item on the list.

In the parking lot example, cars can be stored in arbitrary spaces if on each car there is a pointer sign giving the location of the next car on the list (figure 15.2). To find, say the fifth car (the car fifth on the list of available cars sorted by identification number), you start at the first car, read its pointer telling where the second car is, go to the second car and learn where the third car is, and so forth, until you reach car 5.

When a new car arrives, it can be parked in any available space. The pointer of the car preceding it in the sorted list is then changed to indicate the location of this new car, and the new car is given the pointer that the preceding car had, pointing to the next car. Hence, by changing one pointer and adding another, the list is updated to include the new car, and the ordering is preserved.

This procedure works identically for lists stored in a computer. Items can occupy arbitrary memory locations, with each item appended with a pointer indicating the location of the next item. By following the pointers, one can traverse the entire list in order.

It is also simple to remove an item. To do so, it is only necessary to change the pointer of the preceding object to be the pointer of the object being removed. This deletes the object from the list even if the object is not physically removed. With no pointer pointing to it, the object is essentially nonexistent.

Although it is easy to move forward through a linked list, it is not easy to move backward. The pointers only point forward. This difficulty is solved by a **doubly linked list**, in which each object is accompanied by two pointers: one pointing to the successor item and the other to the predecessor.

Special Lists

Lists often have special uses that dictate a particular form of updating. One of these is the **stack** in which objects are entered one by one at the top of the list, each new addition causing the others to be pushed down one place. Objects are removed from the top as well, causing the other objects to move upward. The scheme is termed FILO, for "First In, Last Out." For example, if you make changes in a word processor, these changes are saved one at a time in a stack. Then if you decide to undo a change, the first change restored is the last that was made, since it comes off the top of the stack.

The sister to the stack is the **queue** in which objects are entered one by one at the top of the list, and removed from the bottom of the list. This is termed FIFO, for "First In, First Out." It simulates a queue of people waiting for service at the bank, or program steps waiting to be executed. The first in line is the first served.

15.2 Trees

Trees are valuable structures used in formal and informal representation, analysis, and manipulation of data. Trees represent structures such as organizational charts, genealogy (family trees), contest standings (as in a tennis ladder), and various other hierarchical structures.

In fact, a tree is basically a hierarchical arrangement of **nodes**. One node is designated as the **root**, and (although it is called a root) it is usually visualized as being at the top of the hierarchy. The simplest nonempty trees consist of a single root and no other nodes.

Generally, a node has a number of **children** nodes directly connected to it but one level further down the hierarchy. Every node i, except the root, has a single **parent**,

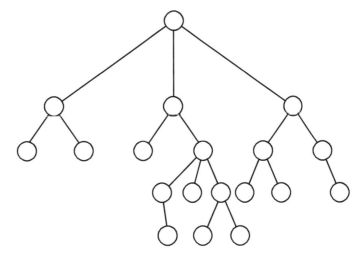

FIGURE 15.3 A tree. The root has three children, each of which has two children. There are a total of ten leaf nodes.

such that i is a child of this parent. A node without children is termed a **leaf**. The parent–child relation is described pictorially by lines connecting the corresponding nodes as shown in figure 15.3.

In a **binary tree** every node has at most two children. It is conventional to refer to a child in a binary tree as either a **left child** or a **right child**, where naturally the left child is the one located below and to the left of the parent and the right child is located below and to the right of the parent. Clearly, in a binary tree each node may have either no children, a left child, a right child, or both a left and right child.

Ordered Trees

It is often convenient to number the nodes systematically. In one simple method, the root is assigned number 1. Then at the next level numbers are assigned sequentially, starting from the left and working across to the right. This is continued through successive levels. The version of the tree of figure 15.3 numbered this way is shown in figure 15.4.

Other numbering strategies, useful in certain computational procedures, are discussed in the next section in the context of transversal.

Representation of Trees

One of the simplest ways to represent a tree is with a set of pointers that point down the tree. The position of the root is specified first. At every node, pointers to the locations of each of its children are listed. It is then possible to move from the root

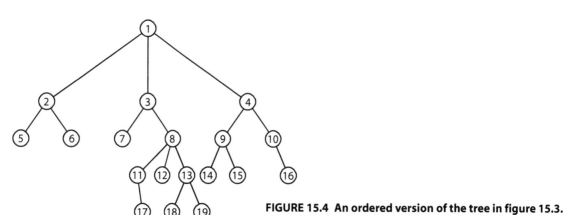

FIGURE 15.4 An ordered version of the tree in figure 15.3.

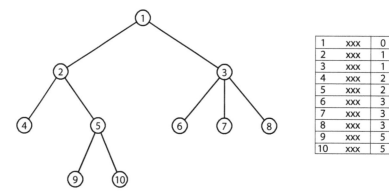

1	xxx	0
2	xxx	1
3	xxx	1
4	xxx	2
5	xxx	2
6	xxx	3
7	xxx	3
8	xxx	3
9	xxx	5
10	xxx	5

FIGURE 15.5 A tree and an array representation. The record of each node is placed in the array, followed by a pointer to the parent of the node.

1	xxx	2	3	
2	xxx	4	5	
3	xxx	6	7	8
4	xxx	N		
5	xxx	9	10	
6	xxx	N		
7	xxx	N		
8	xxx	N		
9	xxx	N		
10	xxx	N		

FIGURE 15.6 A list representation of the tree of fig. 15.5. Each node is represented by its contents and a list of its children. The symbol N denotes that the node has no children and is therefore a leaf node.

down a variety of paths, following one of the pointers at each node encountered. The entire structure of the tree is embodied in this pointer structure. Alternatively, each node may contain pointers to its parents. This too is sufficient to describe the tree.

A tree can be represented concretely as an array, in any one of several ways. Figure 15.5 shows a tree of record locations and the tree's representation as an array, which contains the records as well as the node numbers and pointer to the parent. The entire tree structure is embodied in this array.

A tree can also be represented as a series of lists. For example, a tree can be described by listing the children of each node. Figure 15.6 shows this representation for the tree of figure 15.5.

15.3 Traversal of Trees

Frequently it is desirable to traverse a tree, visiting every node to find one that satisfies a search criterion or to modify the contents of the nodes. If the tree is represented only by its parent–child relations, such a traverse must move systematically node by node: from parent to child, or from child to parent.

As an analogue, imagine that the connecting lines of the tree are pathways. To traverse the tree, one must walk on the paths in a route that visits every node. Some duplication is likely to be necessary—some nodes will be visited more than once—but we seek a systematic strategy. Such a strategy is illustrated in figure 15.7. The route indicated by the dotted line goes through every node, and it stays on the connecting paths. From the figure it is clear that if a complete cycle from the root back to the root is made, each leaf node will be visited only once, but others will be visited at least twice.

This traversal route can be used to order the nodes in one of several ways. The most direct numbering system is termed **preorder**. In this method, the nodes are numbered sequentially as they are passed the first time in the traversal that cycles the tree in the counterclockwise direction. The resulting node ordering for the tree of figure 15.7 is given by the numbers indicated in the nodes.

A special ordering for binary trees is termed **inorder**. In this method, the tree is again traversed according to the counterclockwise cycle, just as before. The leaf nodes are numbered the first (and only) time they are passed. However, other nodes are numbered the *second* time they are passed. The root, for example, will usually not be assigned number 1.

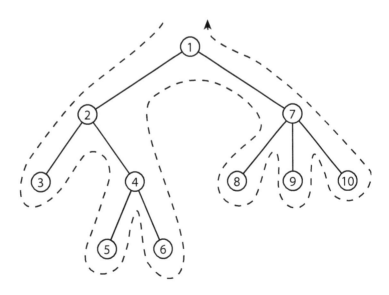

FIGURE 15.7 Traversal of a tree. A counterclockwise cycle defines a traversal that goes through every node at least once. The numbers in the nodes in this tree are those defined by the preorder method of ordering.

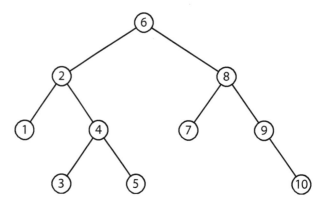

FIGURE 15.8 Inorder of a binary tree. Leaf nodes are numbered the first (and only) time they are passed. Other nodes are numbered the second time they are passed. A node without a left child is numbered before its right child.

There is a special case of this method that must be treated carefully. If a node has a single child and that child is a right child rather than a left child, then an artificial left child must be assigned to that parent. This artificial node does not get a number, but its existence insures that the parent will be visited twice before the traverse reaches the single right child. For example, in a tree consisting of a root and a single right child, the root would be numbered 1, because the root would be visited twice (the second time being after the visit to the artificial left child) before the right child is reached.

Another way to characterize the inorder order is that it numbers trees in LNR order; that is, in Left, Node, Right order. A left child is numbered first, followed by the node, followed by the right child. If there is no left child, then the node is numbered first.[1]

An example of a binary tree in inorder order is shown in figure 15.8. The numbers can be verified by making a counterclockwise cycle of the tree. Notice that node 9 has a single child, which is a right child. Hence, following the LNR rule, node 9 is numbered before node 10. This is the same ordering as would be obtained by appending an artificial left child to node 9, but not assigning it a number as the tree was traversed.

15.4 Binary Search Trees (BSTs)

The **binary search tree** is one of the most powerful of the basic data structures. Such trees lead to simple, yet highly efficient representations for searching and sorting data. It employs the inorder method of ordering.

A binary search tree is applicable when the objects to be processed possess key values that can be ranked (such as alphabetically or numerically). Construction of the

---

[1] The process can be described recursively by defining the general step at a node: visit left child and carry out the process, then number the current node, then visit right child and carry out the process. Start at the root.

tree and the search through it are governed by the order inherent in the key values. A binary search tree is built in such a way that an inorder transversal leads to a sorted ordering.

The process is simple. The first object becomes the root of the tree. The next object is compared with the root and becomes a left or right child of the root depending on whether its key value is less than the root or greater than (or equal to) the root. Subsequent objects are entered by comparing them first with the root, determining whether to go left or right, then continuing down the tree, making similar comparisons at every node encountered, until it becomes either the first or second child of a parent.

An example should clarify the procedure. Suppose we wish to alphabetize the following names: **Linda**, **Joanne**, **Carl**, **Robert**, **Jenna**, **Steve**, **Marion**, **Nancy**, **Ian**, **Jill**, **Susan**, **Fred**. The tree is built by taking the first name, **Linda**, as the root. The next name, **Joanne**, is then compared with the root. If it is lower in the alphabet, it becomes the left child, otherwise the right child. Thus **Joanne** becomes the left child of **Linda**. Then **Carl** goes left of **Linda** and left of **Joanne**. The complete binary search tree is shown in figure 15.9.

In the example, artificial nodes are adjoined as left children of **Carl**, **Marion**, and **Steve** to remind us how to define the inorder. Indeed, **Carl** is the first item in the inorder. The other items can be quickly ordered by traversing through the tree counterclockwise, leading to **Carl**, **Fred**, **Ian**, **Jenna**, **Jill**, **Joanne**, **Linda**, **Marion**, **Nancy**, **Robert**, **Steve**, **Susan**.

To search for a name, say **Nancy**, it is only necessary to follow the path downward. **Nancy** must be to the right of **Linda**, to the left of **Robert**, and to the right of **Marion**. Bingo! There she is. Alternatively, if a search is instituted for a name such as **Ralph** that is not on the list, one will progress down to a leaf node with no place farther to go, and hence conclude that **Ralph** is not on the list.

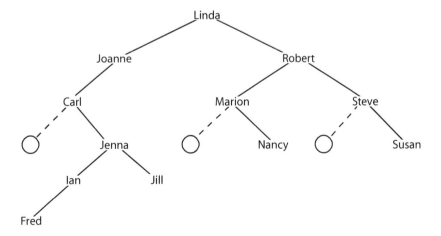

FIGURE 15.9 A binary search tree of names. The tree automatically puts the names in inorder as it is constructed.

Another example of a binary search tree is the tree of figure 15.8. It is the BST that would result from construction based on the unordered sequence 6, 2, 8, 4, 1, 3, 5, 9, 7, 10.

Binary search trees are used in many practical applications, such as airline reservation systems where individuals' names are entered sequentially as they book flights.

Average Path Length

Searching for an object in a BST entails traveling along the unique path from the root to the object. The total search time is proportional to the total number of comparisons required, and hence proportional to the length of the path.

The length of such a path in a BST with n nodes can vary widely, depending on the particular tree. The best case is when the tree is **balanced**, with each node having two children. In this case the total number of nodes n is of the form $n = 2^k - 1$ for some integer $k \geq 1$. The maximum length of a path, in terms of the number of nodes visited, is then $L_{max} = k$; or in terms of n, $L_{max} = \log_2(n + 1)$.

The worst case is when each node, except the last (which is a leaf node), has only a single child. The length of a path from top to bottom is n. In general, therefore, the maximum path length varies between $\log(n + 1)$ and n.

TABLE 15.1
Worst and Best Path Lengths to a Leaf. In the worst case the path length of a tree with n nodes is n. For a balanced tree the path length is $\log(n + 1)$.

n	$\log(n + 1)$
7	3
127	7
1,023	10
16,383	14
131,071	17
1,048,575	20
16,777,215	24
134,217,727	27
1,073,741,823	30

As shown in table 15.1, there is a tremendous difference between these two bounds for even modest values of n. For a tree of about 1 billion nodes, the length from the root to a leaf node is at most 30 if the tree is balanced. On the other hand, if the tree is completely unbalanced, the length is a billion.

It is of great importance to know what length might be expected in actual application. For $n = 1$ billion, is the number of required comparisons for a search closer to 1 billion or to 30?

This question can be addressed by considering a binary search tree with n objects, under the assumption that the ordering of the keys is initially random. Let $P(n)$ be the average path length to a random object where now the object is not necessarily at a leaf node. The following important result characterizes $P(n)$.

Theorem 15.1. The function $P(n)$ satisfies

$$P(n) \leq 1 + 2 \ln n \leq 1 + 1.386 \log n.$$

Proof: Define $Q(n)$ as the expected total number of node visits required to construct the entire binary search tree. The average number of visits to a particular node $P(n)$ is then $Q(n)$ divided by n.

The nodes are referred to by node numbers 1 though n, which are taken to be identical to the ranking of their keys. Hence the proper ordering is 1 through n. A step occurs when two elements are compared. Two elements i and j are compared at most once. We shall find the probability that i and j are compared, and for this purpose it can be assumed that $i < j$.

Consider the chain of values $i, i + 1, i + 2, \ldots, j$ that has $L = j + 1 - i$ members. The elements arrive randomly for placement. If any element k with $i < k < j$ arrives before i or j, then i will never be compared with j, for i will be sent left of k and j will be sent right. Hence, i is compared with j only if i or j occurs before all other elements in the chain. The probability of i or j occurring first among the L elements is $2/L$.

The expected total number of comparisons is therefore $n_L \times 2/L$, where n_L is the number of chains of length L. This number is $N_L = n + 1 - L$, for $L = 2, 3, \ldots, n$. In addition, each node is considered to visit itself. Hence the total number of visits is

$$Q(n) = n + \sum_{L=2}^{n} n_L \frac{2}{L} = n + \sum_{L=2}^{n} 2 \frac{(n+1-L)}{L}.$$

Using the standard approximation to the harmonic sum of $1/L$'s (see exercise 2),

$$\sum_{L=2}^{n} \frac{1}{L} \leq \int_{x=1}^{n} \frac{1}{x} dx = \ln n, \qquad (15.1)$$

gives $P(n) = Q(n)/n$ as

$$P(n) = 1 + \sum_{L=2}^{n} 2 \frac{n+1-L}{nL}$$

$$= 1 - 2\left(1 - \frac{1}{n}\right) + 2\left(1 + \frac{1}{n}\right) \sum_{L=2}^{n} \frac{1}{L}$$

$$\leq 1 + 2\ln n + 2\left[\frac{1}{n}(1 + \ln n) - 1\right]. \qquad (15.2)$$

The term in brackets is always less than or equal to zero. Hence

$$P(n) \leq 1 + 2\ln n \leq 1 + 1.386 \log n. \quad \blacksquare$$

The actual values of $P(n)$ are extremely close to the bound given by the theorem. If the bracketed term in equation (15.2) is included, a tighter upper bound $P^u(n)$ is obtained that is at most two steps less than $1 + 2\ln n$. A lower bound $P^l(n)$ can be constructed, by using a lower bound on the sum in equation (15.1) (see exercise 3) that is only about two steps less than $P^u(n)$. Hence the actual value of $P(n)$ is within two steps of either of these strong bounds. For example, the value of $P(1$ billion$)$ is between the bounds of 39.0602373 and 40.44653173.

Measures of efficiency as a function of the problem size n usually focus on the performance for large n. Typically, this asymptotic behavior is expressed in "big O" notation. A statement that the number of steps is $T(n) = O(n^k)$ means that there is a constant $c \geq 0$ such that $T(n) \leq cn^k$ for sufficiently large n. Thus if $T(n) = 47 + 2n + 19n^2$, then $T(n)$ is $O(n^2)$.

A stronger notion is defined by Θ notation. A statement that $T(n) = \Theta(n^k)$ means that there are positive constants c_1, c_2 such that $c_1 n^k \leq T(n) \leq c_2 n^k$ for sufficiently large n. Hence $\Theta(n^k)$ implies $O(n^k)$, but the reverse implication is not necessarily true. With this notation, the path length of BSTs is at worst $\Theta(n)$, but on average $\Theta(\log n)$.

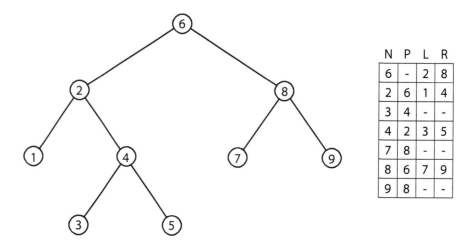

N	P	L	R
6	-	2	8
2	6	1	4
3	4	-	-
4	2	3	5
7	8	-	-
8	6	7	9
9	8	-	-

FIGURE 15.10 A binary tree and its representation as a table. The table shows node, parent, left child, right child.

Representation

A table representation of a binary tree listing the left and right children of various nodes facilitates rapid search through the tree. An example is shown in figure 15.10. To search for node 3, for instance, one begins at the root 6 and moves to the left child 2, then to the right child 4, then to the left child to arrive at 3. The parent pointers are not needed for this type of search, but they are useful when traversing a tree. For instance, after arriving at node 3 in the figure, which is seen to be a leaf because it has no children, the pointer to the parent makes it possible to move back up to node 4.

15.5 Partially Ordered Trees

A **partially ordered tree** is a binary tree that is balanced as much as possible and has all of its leaf nodes at the lowest level as far to the left as possible. Furthermore, the key value of any node is less than or equal to that of its children.

The first requirement implies that if there is a total of h levels, then the $(h - 1)$-th level is full with 2^{h-1} nodes, and the h-th level has all of its nodes to the left. The second requirement means that as one moves down the tree along any path, the key value never decreases. An example of a partially ordered tree is shown in figure 15.11.

Partially ordered trees are sometimes used to represent **priority queues**, ordering the service of various customers or jobs. The first customer in the queue is represented by the root. When served, that node is eliminated and the tree is reconfigured to a new priority queue.

To reorder the tree when the root is eliminated, the root is replaced by the node in the tree at the lowest level and at the rightmost position. The tree is then still balanced as much as possible but with one less node than before at the lowest level. To restore the partial order, the new root is pushed down the tree, exchanging it with its child of smallest key value until its key value is no smaller than that of either of its children

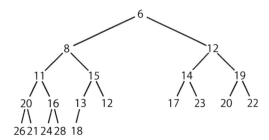

FIGURE 15.11 A partially ordered tree. Each node has a lower key value than its children, and the tree is balanced as much as possible, with all levels except the last full, and all nodes in the bottom level located to the left as far as possible.

or until it becomes a leaf node. Figure 15.12 shows the process of restoring the tree of figure 15.11 after the root node has been eliminated and replaced by the rightmost node at the bottom level.

The importance of partially ordered trees is derived from the efficiency of the restoration (push down) process. The maximum number of necessary exchanges is equal to the depth of the tree. This number is equal to at most $\log(n + 1)$. Hence, the restoration process is a $O(\log n)$ process. We will later see how this can be used to advantage when sorting large lists.

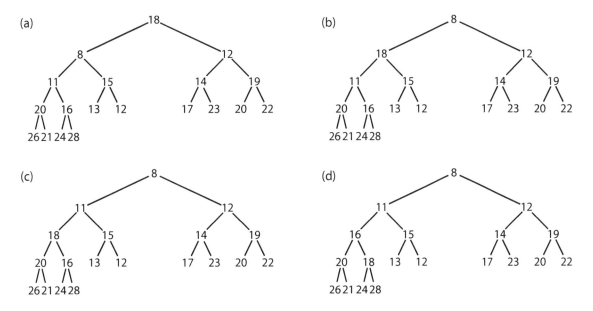

FIGURE 15.12 Pushing down a node. From figure 15.11 with the original root 6 dropped, the new root 18 is shown in (a). This root is exchanged with its smallest child 8 in (b). Then 18 is further exchanged with its new smallest child 11 in (c). Finally, 18 is exchanged with the new smallest child 16 as shown in (d). No further exchanges are necessary, and the tree is again partially ordered.

Heaps

Another advantage of partially ordered trees is that they can be stored efficiently in array form. This feature depends only on the balanced nature of the tree rather than its order, but the term **heap** usually refers to the partially ordered version.

Generally, the nodes of a partially ordered tree are numbered consecutively across each level. This numbering is independent of the key value. The root is number 1, its left child is 2, and this level is numbered up to 4. Because the tree is balanced as much as possible, the children of any node, say number i, are at node numbers $2i$ and $2i + 1$. Hence it is easy to move through the tree in array form. Suppose the tree of figure 15.11 is numbered that way. Then it can be represented by the following array.

node number	1	2	3	4	5	6	7	8	9	10	11	12	13	14	15	15	16	17	18	19
key value	6	8	12	11	15	14	19	20	16	13	12	17	23	20	22	26	21	24	28	18

No pointers need be appended to the list, since the children of node i are located systematically at $2i$ and $2i + 1$. Note, for example, that the children of node 5 (with key value 15) are nodes $2 \cdot 5 = 10$ and $2 \cdot 5 + 1 = 11$ (with key values 13 and 12). Manipulations such as the push-down process can be carried out directly on the array representation of the tree.

15.6 Tries[*]

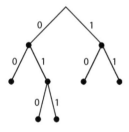

FIGURE 15.13 Codeword trie. Code words are read by beginning at the root and moving down a path to a leaf.

The term[2] **trie** is derived from re<u>trie</u>val and refers to a special type of tree useful for representing strings of data, such as words, codes, or numbers of several digits. The branches (or the nodes) of a trie are labeled with symbols in such a way that movement down the tree along a path from root to leaf defines an acceptable string. Tries were used in chapters 2 and 3 to study codes. Figure 15.13 is a trie representation of a Huffman code with codewords 00, 010, 011, 10, 11. The codewords are found by following the branches from the root to a leaf. Such a trie is a convenient way to verify that no codeword is a preface to another, for a preface would be found before reaching a leaf. Word tries are also used as dictionaries in some spell-checking programs.

In general, a valid string in the trie may in fact be a preface to other valid strings. THE is a preface to THESE, for instance. This situation is handled in a trie by introducing a special symbol, such as Δ, to indicate the end of a string. Figure 15.14 shows a partial trie for common English words. Note that there are three instances of the Δ symbol to signal that THE, TO, and TON are valid words even though they are prefaces to longer words.

When used as a dictionary words are checked by tracing a path down the trie, following the sequence of letters in a word being tested. If the word is in the trie, the path will end at a leaf node or a Δ. If a word is not in the trie, its path will reach a point where there is no appropriate branch or it will reach a leaf node before the word is complete.

---

[2]Tries are alternatively termed **digital search trees**.

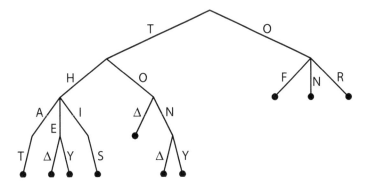

FIGURE 15.14 Word trie. Words are found by moving down a path. The △ symbol indicates the end of an acceptable word. For example, THE and THEY are valid words in this tree.

15.7 Basic Sorting Algorithms

One of the most important applications of the data structures studied in this chapter is to the sorting of lists. Sorting may seem to be a trivial or routine operation, but sorting is an integral component of many sophisticated data analysis procedures, and hence can be regarded as fundamental to the extraction of information from data. As much as 25 percent of computer time worldwide is devoted to sorting and searching. Improvements in sorting efficiency can accordingly pay large dividends. Good sorting methods are concrete illustrations of the importance and power of data structure theory.

Sorting is, of course, the ordering of a list of objects, with the order determined by a key. A sort can be made relative to numerical or alphabetical order, by date, by string length, or by any other key quantity that can be ordered. Usually ties are allowed, in which case the tied objects are placed together in the sorted list.

Two of the simplest sorting methods are presented in this section. Although they are useful only for relatively short lists, they are a preview of, and provide motivation for, the more efficient methods discussed in the next two sections.

Bubble Sort

The name **bubble sort** expresses the view that this sorting process bubbles up low-valued key items, floating them over the higher-valued key items.

Suppose there are n items in a list, and imagine that they are listed vertically. Suppose also that we want to sort this list according to key value, with the item with lowest key value at the top. To begin the process, the bottom two items, in positions $n - 1$ and n, are compared. If the bottom item has a key value less than the one above it, the two are exchanged; otherwise not. The two bottom items are then in order. The process then moves up one step, comparing the items at positions $n - 2$ and $n - 1$. They are exchanged if necessary to bring these two into proper order. This pair-wise comparison is continued up to the top of the list.

Math	**Art**	**Art**	**Art**	**Art**	**Art**
English	Math	**English**	**English**	**English**	**English**
History	English	Math	**Gymnastics**	**Gymnastics**	**Gymnastics**
Gymnastics	History	Gymnastics	Math	**History**	**History**
Language	Gymnastics	History	History	Math	**Language**
Art	Language	Language	Language	Language	Math

FIGURE 15.15 Bubble sort. Each successive column shows the result of an additional full pass. The boldface items have completed their upward bubbling.

After one complete pass, the lowest-valued item will be at the top, because once it is encountered in a comparison, it will be the lowest item in all subsequent comparisons and will thus be exchanged over and over again, bubbling up to the top. Another pass will bubble the second lowest item up to the second position.

Additional passes are made, although the k-th pass need not include the top $k - 1$ items since they are already in proper order. All items will be properly ordered after at most $n - 1$ passes. An example is shown in figure 15.15, where class titles are sorted alphabetically.

Measures of the efficiency of bubble sort focus on the number of comparisons or exchanges required. The best situation is when the list is initially in proper order, in which case $n - 1$ comparisons and no exchanges are needed. The worst situation is when the list is initially in reverse order. Then the comparisons and exchanges in the first pass are both $n - 1$ in number. Likewise, the k-th pass requires $n - k$ comparisons and exchanges. The total is $\sum_{k=1}^{n-1} (n - k) = n(n - 1)/2$ comparisons and exchanges. Therefore in the best case, bubble sort is a $\Theta(n)$ process, while in the worst case it is a $\Theta(n^2)$ process.

The average number of exchanges required in bubble sort can be deduced from the following clever observation. Consider a list L with n items ordered randomly, and consider the list $\overline{L}$, which is ordered in the exact reverse of L. Suppose bubble sort is applied to each list separately. Two items i and j will be out of order in exactly one of the lists, and so at some point they will be exchanged in that list. Since this applies to any two items, there must be exactly one exchange, in either L or $\overline{L}$, for every pair of items. Since there are exactly $n(n - 1)/2$ distinct pairs, sorting both L and $\overline{L}$ requires $n(n - 1)/2$ exchanges. This means that, on average, $n(n - 1)/4$ exchanges are required for a list of length n. Thus bubble sort is, on average, a $\Theta(n^2)$ process.[3] It can be shown that the average number of comparisons is also $\Theta(n^2)$.

Insertion Sort

In **insertion sort** items are inserted, one by one, into an incomplete list that is always properly sorted and that grows to full size. The result is the desired sorted list.

---

[3] It is assumed that all items have different key values.

Math	English	English	English	English	Art
English	**Math**	**History**	**Gymnastics**	**Gymnastics**	**English**
History		**Math**	**History**	**History**	**Gymnastics**
Gymnastics			**Math**	**Language**	**History**
Language				**Math**	**Language**
Art					**Math**

FIGURE 15.16 Insertion sort. Each successive column shows the result of an additional insertion of an item from the first column.

Again it is useful to imagine the list arranged vertically. The top item is considered, by itself, to be the single item in a short list of length 1; this short list is clearly in proper order. The second item in the main list is then inserted into the short list, and by an exchange if necessary, the new two-item list is properly ordered. Additional items are inserted one by one, keeping the partial list in order. When all items are inserted, the entire list is properly sorted. The details of an insertion sort applied to the same list used to illustrate a bubble sort are shown in figure 15.16.

The performance of insertion sort is similar to that of bubble sort. The number of exchanges is on average identical to the number required by bubble sort because the same symmetry argument applies. The average number of comparisons is, however, approximately one-half the number required by bubble sort, and for this reason insertion sort is considered superior to bubble sort. Both of these methods are $\Theta(n^2)$ processes on average.

The basic ideas of these algorithms, however, can be combined with tree structures to produce highly effective sorting algorithms, as discussed in the next section.

Information

From an information-theoretic viewpoint, the entropy associated with knowledge of the permutation embodied in the initial order of n items is $\log(n!)$. Since $\log(n!) \approx n \log(n/e)$, about $n \log(n/e)$ bits of information are needed to sort a list of length n.

Comparison of the order of two items constitutes a single bit. Hence, it might reasonably be inferred that there are sorting algorithms that on average require $\Theta(n \log n)$ comparisons. Furthermore, it is clear from the information-theoretic argument that this is the best that can be done. Two algorithms that achieve this average are presented in the following sections.

15.8 Quicksort

The sort algorithm considered most effective overall is **quicksort**. Its strategy is best understood as a practical implementation of the binary search tree discussed in section 15.4.

Tree Version of Insertion Sort

Imagine an insertion sort that inserts items one by one into a BST rather than into a linear list. When the tree is complete, the items can be read out in inorder to construct an ordered version of the original list. This is illustrated in figure 15.17.

This method can be extremely effective, with the one drawback that a tree must be constructed outside the original list. In other words, unlike bubble sort or insertion sort, this BST method does not take place within the list itself, but must build another structure as well.

The Quicksort Algorithm*

Quicksort provides a strategy that takes advantage of the BST structure but carries out the sort within the original list. The list is visualized as being laid out horizontally. To start, a root is selected and then, rather than processing each item in turn, all items with key values less than that of the root are placed to the left of the root, and all items with key values greater than or equal to that of the root are placed to the right. Those items now on the left can then be handled separately, using the same procedure, by selecting a lower-level root (called a pivot) for the left. Likewise, those items now on the right can be handled by a similar process. These processes are continued in each subgroup, leading to smaller subgroups, until the resulting subgroups contain only a single item or items that have equal key values.

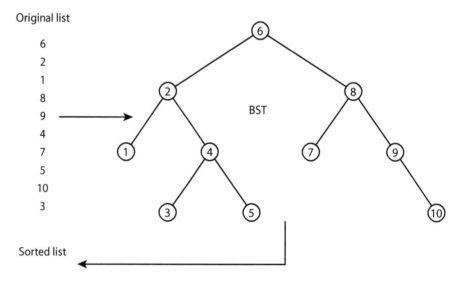

FIGURE 15.17 Insertion into a binary search tree (BST). Items from a list are inserted one by one into the BST; then the sorted version is read out to construct a sorted list.

One way to select the appropriate pivot for each group is to examine the two leftmost items and select the one with the largest key value. This guarantees that the pivot is not the item with the smallest key value.

Once the pivot is selected, some items must be moved left or right to their proper section of the list. For this purpose, left and right cursors are introduced. The left cursor begins at the far left and moves right until it encounters an element with key value equal to or greater than that of the pivot. The right cursor moves left until it encounters an item of key value less than that of the pivot. If the cursors have not met, the items they have encountered are swapped. Then the cursors continue their progress until they reach another stopping point, where another swap is made. This process continues until the cursors meet. The result is that the list is divided into two segments: a left-hand portion with all elements having key values less than that of the pivot and a right-hand portion with all items having key values greater than or equal to that of the pivot. These two segments are then processed individually in the same way, producing smaller segments, and so forth. If at any stage a segment consists of a single element or equal elements, that segment need not be processed further. Eventually, all segments will be of that type, and the sort is complete. An example is shown in figure 15.18.

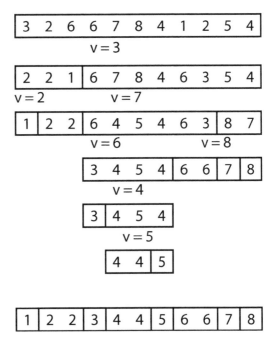

FIGURE 15.18 Quicksort. In the initial list, 3 is chosen as the pivot element (indicated by $v = 3$) since it is the larger of the first two elements. The left cursor is halted immediately at the 3. The right cursor advances leftward until it reaches 2, at which point the 3 and 2 are swapped. A further swap of 6 and 1 occurs. At that point the list is divided into two parts as shown by the separating bar in the figure. The individual portions are then processed in the same way. The final version of the list is shown in the last line.

Efficiency

Quicksort inherits its efficiency from the characteristics of the BST. The number of steps required in a path through a BST is in the worst case $\Theta(n)$ and in the best and average cases $\Theta(\log n)$. Sorting n numbers can be expected to require about n times as many steps, and accordingly, the worst performance of quicksort is $\Theta(n^2)$. The best and average cases are $O(n \log n)$ and $\Theta(n \log n)$, respectively. The $O(n \log n)$ performance is a huge improvement over bubble sort and insertion sort, and is consistent with the best performance implied by entropy considerations. A strategy to improve worst-case performance is to select the pivot points randomly. Then, against any particular set of input data, the expected number of steps is on average $\Theta(n \log n)$.

15.9 Heapsort

Heapsort is another tree-based sorting method, but it uses the partially ordered tree data structure rather than the BST. It has the (theoretical) advantage that it is at worst, best, and average a $\Theta(n \log n)$ process. Thus unlike quicksort, which may require $\Theta(n^2)$ operations in the worst case, heapsort is a $\Theta(n \log n)$ process in all cases.

First imagine that the list is to be entered into a partially ordered tree. There are two ways to do this. The first way can be viewed as a tree version of bubble sort. In this method items are initially entered into the tree in any order while simply assuring that the tree is balanced as much as possible. Then each item in the bottom level is compared with its parent, and if the parent has higher key value, the parent is swapped with its lowest-valued child. After the lowest level is processed in this way, the next higher level is processed in the same way, and so forth up to the top. This entire process is then repeated, starting again at the bottom level. After at most $O(n)$ such passes, the tree will be partially ordered.

The second method for achieving the partially ordered form can be viewed as a tree version of insertion sort. Items are entered one by one at the bottom level and moved up level by level until the key value of its parent is less than or equal to the key value of the new item.

Once the tree is partially ordered, the items can be sorted by using the push-down process discussed in section 15.5. As items are extracted from the tree, they are placed in a sorted version of the original list.

Heapsort can be carried out without constructing a tree separate from the original list by using the heap structure. The list itself is viewed as a tree by considering each item i as having items $2i$ and $2i + 1$ as its children.

Although heapsort's worst-case performance is far superior to that of quicksort and both are $O(n \log n)$ on average, experience has shown that heapsort is in practice rarely as efficient as quicksort.

Going from a linear list to a two-dimensional tree structure reduced the sorting time from $O(n^2)$ to $O(n \log n)$. One might suspect that going to a three-dimensional structure will give further improvement. However, consideration of entropy shows that $O(n \log n)$ is the best that can be achieved on average.

15.10 Merges

Frequently it is necessary to merge two lists L_1 and L_2, each of which has been sorted, into a new sorted list L containing all of the items in both L_1 and L_2. If these two lists can be simultaneously accommodated in the internal memory of the computer, there is a simple and effective method to attain the desired result.

The method begins by comparing the first item in each list, and inserting the item with the lowest key value of the two into the master list and removing it from its original list. This procedure is repeated until all items have been inserted, or until one of the original lists is exhausted, in which case the remainder of the surviving list is inserted.

This method can be extended to the merge of any number of sorted lists: simply compare the first of each and insert the one with lowest key value into the master list, deleting it from its original list.

It may be a law of nature that people seem to need more data than can be handled conveniently in their existing computing systems, and certainly more than can be stored in internal memory. Thus, historically, computer systems have employed tape drives, magnetic drums, and magnetic disks as external storage to augment internal storage.

Sorting huge lists that cannot be accommodated in internal memory is usually carried out by dividing the list into a number of smaller sublists, each of which can be sorted in internal memory. These separate sorted lists are then merged. The overall sorting and merging strategy moves segments of data back and forth between internal and external memory. There are many such strategies, the advantage of each depending somewhat on the physical characteristics and performance of internal and external storage devices.

Linear Time Sorting

There are a number of sorting algorithms that sort items in $O(n)$ time, which of course is faster than the $\Theta(n \log n)$ algorithms discussed in the past few sections. The difference is that these algorithms apply to special cases—cases where the key values of the items have some known structure. For example, **counting sort** applies to sorting integers known to lie in a fixed range 0 to k. These algorithms take advantage of the special structure to reduce direct comparisons between elements. As a simple example, in counting sort the proper placement of the element 0 is known to be at the top of the list; no comparison with other numbers is required.

15.11 EXERCISES

1. (Alphabetize) Insert these items into a BST (using alphabetical order): **Hockey, Baseball, Football, Tennis, Swimming, Ice skating, Badminton, Hopscotch, Basketball, Water polo, Crew**.

2. (Harmonic inequality) By graphically comparing $\int (1/x)\, dx$ to $\sum (1/k)$, prove that

$$\ln(n+1) - \ln 2 \le \sum_{k=2}^{n} \frac{1}{k} \le \ln n.$$

3. (Lower estimate) Use exercise 2 to find a lower bound $P^l(n)$ for the average path length of a BST, and show that this average length is in fact $\Theta(\log n)$.

4. (Balanced) Consider a binary tree balanced as much as possible. Suppose that all elements at the bottom level are first considered for exchange upward, then the next level, etc. However, a child is bubbled up only if its key value is less than that of its parent and no greater than that of its sibling. Show that once an element bubbles upward (after processing its entire level), it never moves down again. Hence, argue that putting an n-node tree in partial order with the bubble up process is an $O(n \log n)$ process.

5. (Perfectly balanced) A perfectly balanced binary tree has 2^{k-1} nodes at level k for each level k. If there are m levels, the total number of nodes is thus $n = \sum_{k=1}^{m} 2^{k-1} = 2^m - 1$.
   (a) Argue that the average length of a path from the root to a random node is

   $$L(m) = \frac{\sum_{k=1}^{m} k 2^{k-1}}{2^m - 1}.$$

   (b) Show that

   $$L(m) = m - 1 + \frac{m}{2^m - 1} = m - 1 + \frac{m}{n}.$$

   That is, for large m, the average length is essentially equal to the length to the second-to-last level.
   Hint:

   $$\sum_{k=1}^{m} a^k = \frac{a - a^{m+1}}{1 - a}$$

   $$\sum_{k=1}^{m} k a^k = \frac{a + a^{m+1}[am - m - 1]}{(1 - a)^2}.$$

6. (Bubble count) Consider the list $L = (5, 3, 1, 2, 4)$.
   (a) Sort the list L with bubble sort and count the number of exchanges required.
   (b) Sort the list $\overline{L}$, which has the reverse order of L, and count the number of exchanges required.
   (c) Is the sum of these exchanges equal to $n(n-1)/2$, where n is the length of the list?

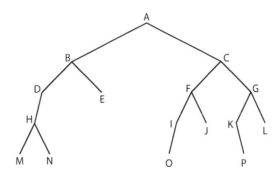

FIGURE 15.19 Tree for exercise.

7. (Order the tree) Given the tree in figure 15.19, order the elements in inorder and preorder.

8. (A quicksort) Do a quicksort of the following numbers: 3, 6, 7, 2, 9, 1, 4, using 6 as the initial root.

15.12 Bibliography

The material of this chapter is treated comprehensively in several good textbooks such as [2], [3], [4], and [5]. An especially concise and modern presentation that greatly influenced this chapter is [1]. The method for evaluating the average path length in a BST is adopted from the comprehensive text [5]. [6] is a valuable reference that provides depth and numerous extensions of the methods presented in this chapter.

References

[1] Aho, Alfred V., John E. Hopcroft, and Jeffrey Ullman. *Data Structures and Algorithms*. Reading, Mass.: Addison-Wesley, 1983.

[2] Lewis, T. G., and M. Z. Smith. *Applying Data Structures*. Boston: Houghton Mifflin, 1976.

[3] Reingold, Edward M., and Wilfred J. Hansen. *Data Structures in Pascal*. Boston: Little, Brown, 1986.

[4] Smith, Harry F. *Data Structures: Form and Function*. San Diego: Harcourt Brace Jovanovich, 1987.

[5] Cormen, Thomas H., Charles E. Leison, Ronald L. Rivest, and Clifford Stein. *Introduction to Algorithms*. 2nd ed. Cambridge: MIT Press, 2001.

[6] Knuth, Donald E. *The Art of Computer Programming*. Vol. 3, *Sorting and Searching*. Reading, Mass.: Addison-Wesley, 1973.

16 DATABASE SYSTEMS

Most people are familiar with the term **database**, for these repositories are practically everywhere, holding data both routine and critical, mundane and esoteric. Database systems are an essential element of modern information science, making available vast quantities of data for analysis, business operations, and ready access for a variety of purposes. A database is essentially a systematic organization of data, which through a database system can be queried to present data in useful form.

16.1 Relational Structure

Early database designs were based on standard data structures of lists, trees, and arrays studied in the previous chapter, but they proved to be rather unwieldy. A great revolution occurred with a publication by Edgar F. Codd in 1970. A mathematician working at the IBM San Jose Research Laboratory, Codd realized that the mathematical theory of relations could form a foundation for a rigorous theory of database design. His relational theory is indeed now the foundation of essentially all modern database systems.

A relational database system can be viewed as consisting of "tables and only tables," for a table is a natural representation of a relation. A table, of course, is a two-dimensional array. The top row of a relation table contains headings, termed **attributes**, and each other row of the table is an **instance** of the relation, and is defined by a **tuple** having components corresponding to the attributes. For example, you might own a relational database in the form of a small address book. It might have the attributes **name**, **house number**, **street name**, **city**, **state**, and **ZIP code**. Each entry would be regarded as a row or tuple and would consist of the corresponding information for one of your friends. Such an address book relation is shown in figure 16.1, with the name **AddressBook**.

AddressBook

Name	Number	Street	City	State	ZIP
Marge Davis	890	Water Street	Metropolis	WA	98202
Jane Doe	312	A Street	Smallsville	CA	94333
Ralph Flowers	314	A Street	Smallsville	CA	94333
Mary Green	8567	Baker Ave	Blue Field	AZ	85678
Harry Lincoln	786	Water Street	Metropolis	WA	98202
Peter Mallard	875	A Street	Smallsville	CA	94333
Cheri Stockton	7890	Viewpoint Ave	Smallsville	CA	94334
Nancy Strong	6589	Carpenter Ave	Blue Field	AZ	85678

FIGURE 16.1 The AddressBook relation. The relation is a table with columns corresponding to attributes and rows being instances.

A relation is just this kind of structure, with the one restriction that no two instances can be the same. If they are (temporarily) identical, they must be coalesced into one.

You might look down the **Name** column of **AddressBook** to find a particular person and then scan across that row to find the person's city. You could also scan down the **City** column looking for all friends that live in Smallsville, or the **Street** column to find all friends that live on Baker Street. It is a simple but useful database—and it is relational because it consists of tables and only tables (in this case just one table).

An abstract relation is referred to by its name, which might be simply R. If it has attributes $A_1, A_2, \ldots, A_m$, the fuller name $R(A_1, A_2, \ldots, A_m)$ is frequently used. Thus the full name of the address book is **AddressBook(Name, Number, Street, City, State, ZIP)**.

Another example relation **Clinic** is a (simplified) version of a database of a small medical clinic, keeping track of patient visits as shown in figure 16.2. Both the **AddressBook** and **Clinic** relations will be used to illustrate basic database design concepts.

Clinic

ID	Name	Birth year	Doctor	Specialty	Date
1106	Peter Mallard	75	Ralston	Allergy	Jan 14
4023	Alan Baker	37	Hartwood	Rheumatology	Jan 14
2469	Cheri Stockton	88	Elwood	Dermatology	Jan 14
3487	Robert Casent	80	Ralston	Allergy	Jan 14
2469	Cheri Stockton	88	Elwood	Dermatology	Jan 15
5602	Jane Doe	78	Vister	Dermatology	Jan 15
3671	Ralph Flowers	52	Hartwood	Rheumatology	Jan 15
4023	Alan Baker	37	Vister	Dermatology	Jan 15
2776	Gavin Jones	82	Ralston	Dermatology	Jan 16

FIGURE 16.2 The Clinic relation. This full table can be reduced.

Although it is impractical to do so in most cases, any relational database can be viewed as a single large table. It would be used exactly like the **AddressBook** or **Clinic** tables, although it might have an enormous number of columns and rows. Such a single-table representation of a large database is likely to be a highly inefficient representation, containing a great deal of redundancy and being awkward to update. Inefficiencies can be seen in **Clinic**, since each doctor's specialty is repeated numerous times, and this is fundamentally unnecessary. Redundancy can be reduced by representing a single table as a combination of smaller ones. In general, a good relational database can be regarded as a virtual master table based on several smaller efficient tables. The overall table need not ever be built explicitly since the equivalent information is fully contained in the collection of small tables.

A new representation of the relation **Clinic** as two relations **PatientVisits(ID, Name, Age, Doctor, Date)** and **Doctors(Doctor, Specialty)** is shown in figure 16.3. In this version, the information about doctor specialties is contained in a separate relation (or table) and hence is not repeated in the relation about patient visits.

There is still some redundancy in the two-table version of **Clinic** because each patient's birth year is repeated in each of that patient's visit tuple. This redundancy can be eliminated by representing **ClinicVisits** as two relations, as shown in figure 16.4. This produces the representation of **Clinic** as the three relations **Patients**, **Visits**, and **Doctors**.

The following sections study relations systematically and develop a general procedure for decomposing a relation into several small efficient component relations.

PatientVisits

ID	Name	Birth year	Doctor	Date
1106	Peter Mallard	75	Ralston	Jan 14
4023	Alan Baker	37	Hartwood	Jan 14
2469	Cheri Stockton	88	Elwood	Jan 14
3487	Robert Casent	80	Ralston	Jan 14
2469	Cheri Stockton	88	Elwood	Jan 15
5602	Jane Doe	78	Vister	Jan 15
3671	Ralph Flowers	52	Hartwood	Jan 15
4023	Alan Baker	37	Vister	Jan 15
2776	Gavin Jones	82	Ralston	Jan 16

Doctors

Doctor	Specialty
Elwood	Dermatology
Hartwood	Rheumatology
Ralston	Allergy
Vister	Dermatology

FIGURE 16.3 Two-table version of Clinic. Specialties are separated from visit information.

Patients

ID	Name	Birth year
4023	Alan Baker	37
3487	Robert Casent	80
5602	Jane Doe	78
3671	Ralph Flowers	52
2776	Gavin Jones	82
1106	Peter Mallard	75
2469	Cheri Stockton	88

Visits

ID	Doctor	Date
1106	Ralston	Jan 14
4023	Hartwood	Jan 14
2469	Elwood	Jan 14
3487	Ralston	Jan 14
2469	Elwood	Jan 15
5602	Vister	Jan 15
3671	Hartwood	Jan 15
4023	Vister	Jan 15
2776	Ralston	Jan 16

FIGURE 16.4 Decomposition of PatientVisits into Patients and Visits.

16.2 Keys

Loosely speaking, a key in a relation is an attribute or group of attributes that can be used to uniquely define any instance in the relation. For example, in **AddressBook** shown in figure 16.1, the attribute **Name** serves as a key because it uniquely defines an instance in the relation. Each name occurs only once.

More generally, a **superkey** of a relation is an attribute or group of attributes whose values always uniquely define an instance in the relation. That is, once the values of the superkey attributes are known, the values of all remaining attributes are defined. There can be at most one tuple for every combination of superkey attributes. Mathematically, a set of attributes $\{A_1, A_2, \ldots, A_m\}$ is a superkey for a relation if no two instances have the same values of these attributes.

By definition, every relation has at least one superkey since no two instances are identical; hence the entire set of attributes always constitutes a superkey.

A **key** is a minimal superkey. That is, a key is a set of attributes $\{A_1, A_2, \ldots, A_k\}$ that uniquely defines all other attributes, and there is no smaller subset of $\{A_1, A_2, \ldots, A_k\}$ that has this property. Again, every relation has at least one key because starting with a superkey, redundant attributes can be eliminated one by one until no further reduction is possible, and the result is a key.

Your email program might contain a small database of names and email addresses that are referenced by a unique nickname for each person. One person on the list might be "jim," another "james," and still another "jimmy." The nicknames are part of the database, and form the key used to access any instance. Of course, this database has other keys as well, including the email address itself.

Although a relation may have more than one key, it is standard practice to select one as the **primary key**, which is used to access the various instances. It is extremely important that at least this primary key be unambiguous, and remain so even if additions are made to the database, for repetition in a key is not only ruled out theoretically, it is disaster in practical terms. Often considerable effort is devoted to the design of a primary key (which may include more than a single attribute). The attribute **Number** (the house number on the street) might be considered as a key for **AddressBook** as currently constituted, since no two people have the same house number. But it is a poor candidate for a primary key. As other people are listed, someone at a different location may have the same house number. A key should be unambiguous for all conceivable tuples.

The issue arises frequently in database systems containing information about people. The use of people's names is often fine for small systems, but in large systems there is a strong possibility that two people will have the same name. For this reason, the U.S. Social Security number is frequently used as a key attribute. But, this too has weaknesses, since many foreign nationals do not have Social Security numbers. To get around this, most universities, for example, assign special university ID numbers to all students, faculty, and staff. Some stores use telephone number as an attribute that can serve as a key or part of a key. Banks assign account numbers. Some organizations use name and birth date. All of these practices are motivated by the need to ensure that a unique key is defined. The relation **Clinic** uses an ID number for each patient for this reason. However, doctors are not given an ID number, for it is assumed that no two doctors in the clinic have the same name—if a duplication should occur, the doctors would be distinguished in the database by, for example, using a first name as well as last name.

There is no single attribute that serves as a key for the relation **Clinic**. However, the pair **ID**, **Date** does serve as a key under the assumption that a patient has only one appointment and sees only one doctor on any given day. If this assumption is not warranted, a new attribute **Appointment** can be introduced that assigns a unique number to every appointment or doctor consultation. Each tuple in the database would have a unique appointment number, and hence **Appointment** could serve as a key. (See exercise 5.)

16.3 Operations

Part of the power of the relational model is that it is fairly easy to perform useful operations on relations. Queries can be formulated and answered, relations can be decomposed into smaller subrelations, and two or more relations can be combined. The output of queries can themselves be treated as relations, allowing nested queries. These operations are intuitive, yet flexible enough to carry out important manipulation. Some of these are used in the following sections. They may be divided into two main classes: set operations and algebraic operations.

Set Operations

Some operations are based on the fact that a relation is a set of tuples. Familiar set operations such as union and intersection may then be applied.

1. **Union.** Suppose there are two address books, each with the same attributes. Their union is a larger address book that includes the information in either of the individual books. However, if a particular instance (tuple) appears in both books, only one copy is included in the master book.

In general, the union of relations R and S, denoted $R \cup S$, is the set of tuples that are in either of the two relations. For this to make sense, it is necessary that the two individual relations have the same set of attributes, listed in the same order.

2. **Intersection.** The intersection of two address books is the (probably smaller) address book that contains only those listings that occur in both books.

   The intersection of relations R and S is denoted $R \cap S$. Again, it is necessary that R and S have the same set of attributes in the same order.

3. **Difference.** The difference $R - S$ is the set of tuples that are in R but not S.

Algebraic Operations

1. **Selection.** Selection is the process of reducing a relation so that it includes only some of its instances. The tuples to be included are determined by various rules applied to the attributes of the instances, depending on the data that is required. For example, a reduced clinic relation with entries corresponding to patients of Dr. Elwood could be formed. This new relation would consist precisely of the instances from the original **Clinic** relation that had Elwood as the **Doctor** attribute. Sometimes, the selection criterion is complex and involves many of the attributes. For example, the clinic may want to find all patients that are either older than 50 or have seen a doctor whose specialty is dermatology. The selection operation is the basic operation used in queries, for it isolates the subset of information needed for a particular reason.

2. **Projection.** Projection reduces a relation by including only a subset of the original attributes. The result is termed the projection onto these attributes. Projection may have the added effect of reducing the number of instances because duplicates of the reduced-dimension tuples must be eliminated. Consider the relation $R(A_1, A_2, \ldots, A_M)$. The projection onto the attributes $A_1, A_2, \ldots, A_k$ consists of tuples whose attributes are $A_1, A_2, \ldots, A_k$ but with the duplicate tuples removed.

   For example, if **Clinic** is projected onto the attributes {**Doctor, Specialty**}, the relation **Doctors** of figure 16.3 is obtained.

   Projection is the basic operation used to reduce a complex relation to a series of simpler relations. The decomposition of **Clinic** shown in figure 16.3 into **PatientVisits** and **Doctors** is indeed formed by two projections.

3. **Natural join.** The natural join operation combines two relations to form a single relation that contains all the relevant information in the two original ones. It is the basic operation for reconstructing a full relation after it has been decomposed. The natural join of relations R and S is denoted $R \bowtie S$.

The set of attributes of the natural join of two relations (often termed simply the **join**) is the union of the attributes of each of the relations, with duplicates eliminated. For example, if the two relations are **PatientVisits(ID, Name, Birth year, Doctor, Date)** and **Doctors(Doctor, Specialty)**, the join of these two has attributes {**ID, Name, Birth year, Doctor, Specialty, Date**}. The common attribute **Doctor** appears only once.

The tuples included in the join are those made up by concatenating all combinations of tuples in the corresponding individual relations, but only if their values agree in the common attributes. Thus if the relations are $R(A, B, C)$ and $S(C, D, E)$, the join consists of all tuples of the form (a, b, c, d, e), where (a, b, c) is in A and (c, d, e) is in B and with the c being the same in A and B. The joined relation can be much larger than either of its two components.

The join **PatientVisits(ID, Name, Birth year, Doctor, Date)** ⋈ **Doctors(Doctor, Specialty)** is the original relation **Clinic**.

The natural join is a powerful operation, and as the name suggests, there are other joins as well, some of which are discussed in the exercises.

4. **Product.** The product of two relations R and S, denoted $R \times S$, combines R and S in a way that preserves every possibility.

The set of attributes of the product $R \times S$ consists of the concatenation of the attributes of R followed by the attributes of S, without eliminating duplicates. Thus, if R has m attributes and S has n attributes, the product $R \times S$ will have $m + n$ attributes. If R and S do have identical attributes with the same name, it is necessary to rename the attributes of at least one of the duplicates so that they are distinguished. One way to do this, when the relations are $R(A_1, A_2, \ldots, A_m)$ and $S(B_1, B_2, \ldots, B_n)$ with some duplication of attributes, is to denote the product by

$$[R \times S](R.A_1, R.A_2, \ldots, R.A_m, S.B_1, S.B_2, \ldots, S.B_n).$$

Suppose that R has M tuples and S has N tuples. The product $R \times S$ is constructed by listing the first tuple in R a total of N times and adjoining to each of these the entire list of N tuples from S. Then the next tuple from R is listed N times and the entire list of N tuples from S adjoined again, and so forth. When the list is complete, it will contain a total of $M \times N$ tuples, each of the M tuples from R being paired with each of the N tuples of S.

Various smaller relations can be obtained from the product. The natural join $R \bowtie S$ can be obtained from the product $R \times S$ by applying the selection operator, selecting those tuples that agree on the attributes common to R and S. The result is then projected onto the set of attributes that contains all attributes only once, thereby eliminating the duplication. An example of this operation is shown in section 16.6.

There are other set and algebraic operations, such as various types of joins, but all of these can be obtained from the operations outlined in this section, together with a possible renaming of attributes.

16.4 Functional Dependencies

An important characterization of a particular relation is the set of functional dependencies that it contains. A **functional dependency** in a relation R is a dependency among attributes, written as

$$A_1 A_2, \ldots, A_m \longrightarrow B_1,$$

where the A_i's and B_1 are attributes from R. This is interpreted to mean that if the values of the attributes $A_1, A_2, \ldots, A_m$ in a tuple are known, then the value of B_1 is determined uniquely. Another way to state this is that if two tuples agree on the attributes $A_1, A_2, \cdots, A_m$, they agree on the attribute B_1. This is also expressed by saying that $A_1, A_2, \ldots, A_m$ determine B_1. If the attributes $A_1, A_2, \ldots, A_m$ determine additional attributes $B_2, B_3, \ldots B_n$ as well, one may write

$$A_1 A_2 \cdots A_m \longrightarrow B_1 B_2 B_3 \cdots B_n,$$

which is also considered to be a functional dependency. If the A_i's are all distinct from the B_j's, the dependency is said to be **nontrivial**.

The term "functional dependency" implies that the dependency is valid for all conceivable tuples that may occur in the relation. It is not enough that the dependency holds for a particular sample or for the current entries. For example the relation **Clinic** contains the apparent dependency **Name** $\rightarrow$ **Birth year** because everyone in the relation has a different birth year. However, this is not a functional dependency because, clearly, there is the possibility that a new patient may have a birth year that duplicates one that is already in the relation. When searching for functional dependencies, one must spell out the assumptions about what tuples might possibly be entered.

Note that all attributes are functionally dependent on any superkey that is in fact a valid superkey for all possible entries, and indeed an alternative way to define a superkey is that it has this property.

Many relations have functional dependencies that are not based on keys. For example, **Clinic** contains the functional dependency **Doctor** $\rightarrow$ **Specialty**.[1] It is these functional dependencies that guide the decomposition of a relation into useful components.

16.5 Normalization

A database is rarely static. Rather, it is frequently changed, with new data being added, old data being removed, and corrections being made. If a large database has

---

[1] But note the implicit assumption that each doctor has at most one specialty. If there is the possibility that a doctor has more than one specialty, the apparent dependency would not be a functional dependency.

its relation represented in full form, it may be inefficient, but equally important, the redundancy and inefficiency may render the database susceptible to update errors.

Suppose that the **Clinic** relation also included patient addresses as an attribute **Address**. Then if a patient alerted the office that his or her address was incorrect, it would be necessary to go through the entire database and change every entry that involved that patient. However, in the three-relation version of **Clinic**, it would only be necessary to change the address once, in the relation **Patients**. If the database were large, the first method would be lengthy and prone to error.

Deleting tuples can also lead to undesirable effects, including the loss of potentially valuable information. For example, suppose a specialist doctor is brought into the clinic to see a patient. If the listing for that one visit is deleted, the information about the doctor is also lost. This phenomenon can be important in critical or complex database systems.

Normal Forms

It is said that in the early part of 1970 when Codd was working out relational database theory, there were news reports of President Nixon normalizing U.S. relations with the People's Republic of China. Codd thought that if Nixon could normalize those relations, then perhaps Codd could normalize his database relations.

Codd's normalization required that all subrelations in a representation of relation satisfy certain desirable properties. There is now essentially a hierarchy of normal forms that impose increasing restrictions on the nature of the relational description. The early normal forms have largely been superseded by a later form, termed the Boyce–Codd normal form. However, the elementary dictates of the first normal form, denoted 1NF, are worth pointing out. This form requires only that the attributes be all single values. For example, in your address book relation you might consider adding the attribute **Children** to list the children of the people in your database. A corresponding tuple entry for that attribute might be Todd, Martha, Jane. This multiple value form is not allowed in 1NF, for it can make it difficult to compare tuples, to extract information, and to update the database. One way around this[2] is to provide for several children attributes in the relation, with headings **Child 1, Child 2, Child 3, Child 4, Child 5**.

Boyce–Codd Normal Form

Update and inefficiency difficulties can be ameliorated by decomposing a relation into components that are in Boyce–Codd normal form. The definition of that normal form is straightforward in terms of the notion of functional dependency discussed in section 16.4.

Boyce–Codd Normal Form. A relation R is in Boyce–Codd normal form (BCNF) if for all nontrivial dependencies $A_1 A_2, \ldots, A_k \rightarrow B$, the attribute set $\{A_1, A_2, \ldots, A_k\}$ is a superkey for R.

---

[2] See exercise 1 for another (better) method.

This definition can be regarded as a kind of test. If a database relation fails the test, it is not in BCNF and something can be done about it. Specifically, the dependency can be used to decompose the relation into components. Then the test can be applied to each of these components.

Let us apply this test to the relation **Clinic** (in its full original form). There are several nontrivial dependencies, including **Doctor** → **Specialty** and **ID** → **Name, Birth year** and **ID, Date** → **Specialty**. In the first two of these the left-hand side of the dependency is not a key, and hence these dependencies indicate that the relation is not in Boyce–Codd normal form and can be decomposed. Indeed, as will be shown later these two dependencies can be used to decompose **Clinic** into the three relations **Doctors**, **Patients**, and **Visits** and this decomposition is in Boyce–Codd normal form. Doesn't that seem right—that a clinic is described by its doctors, patients, and visits?

A School Example

Consider a simplified database of a small high school that records student grades during one semester. The relation **School** is shown in figure 16.5. It is assumed that each class type is taught by a single instructor and always in the same room.

Let us analyze the structure of this relation. First, what are the keys? In particular, is there a single attribute that can serve as a key? Clearly not. There is no single attribute that will suffice, since every column has repeated entries. However, consider the pair {**ID, Class**}. It does form a key, for any row in the table is specified uniquely by giving the student ID and the class.

Let us now search for nontrivial dependencies. An obvious one is **ID** → **Student**. Can we find other nontrivial dependencies with the same left-hand side? No, nothing else is uniquely determined by the student ID number. This tells us that a component relation can be constructed from this dependency. To do so, the attributes on both sides of the dependency are used. Let us call this relation **Students**. The dependent

School

ID	Student	Class	Subject	Instructor	Room	Grade
10112	J. Banes	MA 23	Calculus	LaGrange	F 203	B
10112	J. Banes	Eng 40	Fiction	Moss	K 112	A
10112	J. Banes	Hist 62	Latin Amer	Felding	G 331	A-
12343	S. Johnson	Eng 40	Fiction	Moss	K 112	B-
12343	S. Johnson	Hist 62	Latin Amer	Felding	G 331	B
12343	S. Johnson	Art 76	Drawing	Picas	A 2	A
12343	S. Johnson	PE 45	Dance	Murray	W. Gym	A
23678	M. Walters	MA 23	Calculus	LaGrange	F 203	C
23678	M. Walters	Art 76	Drawing	Picas	A 2	A-
23678	M. Walters	Eng 40	Fiction	Moss	K 112	B

FIGURE 16.5 The relation School. Every student is assigned an ID number. Classes have numbers and subject names, and are taught by a single instructor, always in the same room.

attribute can then be eliminated from the original relation to obtain the slightly smaller relation called **StudentClasses**. These two component relations are shown in figure 16.6.

Next let us search for a nontrivial dependency in the remaining component relation **StudentClasses**. We find **Class → Subject, Instructor, Room,** and **Class** is not a superkey for the **StudentClasses** relation. (Recall that it is assumed that each class is taught by a single instructor, always in the same room, during the semester; hence **Class → Instructor, Room** is a functional dependency—always true.) Using the dependency **Class → Subject, Instructor, Room,** a corresponding component relation called **Classes** can be constructed, and the three dependent attributes can be dropped from what remains. Let us call the remaining relation **Grades**. The overall decomposition, consisting of three relations each in BCNF, is shown in figure 16.7.

Notice how easy it is to update the final representation of the relation. If instructor Lagrange moves to a new room, for example, only one entry must be changed. If there is an error in the spelling of a student's name, it need be corrected only once.

The BCNF is a good goal, but there are both weaker and stronger normal forms that are occasionally of practical value. See exercise 4. Also, suppose that Mr. Lagrange were to teach a class in algebra as well as calculus but in the same room. Then the **classes** relation would contain a nontrivial dependency **Instructor → Room** that could be the basis for further decomposition. But this minor refinement may not be worthwhile.

Students

ID	Student
10112	J. Banes
12343	S. Johnson
23678	M. Walters

StudentClasses

ID	Class	Subject	Instructor	Room	Grade
10112	MA 23	Calculus	LaGrange	F 203	B
10112	Eng 40	Fiction	Moss	K 112	A
10112	Hist 62	Latin Amer	Felding	G 331	A-
12343	Eng 40	Fiction	Moss	K 112	B-
12343	Hist 62	Latin Amer	Felding	G 331	B
12343	Art 76	Drawing	Picas	A 2	A
12343	PE 45	Dance	Murray	W. Gym	A
23678	MA 23	Calculus	LaGrange	F 203	C
23678	Art 76	Drawing	Picas	A 2	A-
23678	Eng 40	Fiction	Moss	K 112	B

FIGURE 16.6 The first decomposition of School. The attribute **Student** is determined by the small relation, and hence is not needed in the remainder.

Students

ID	Student
10112	J. Banes
12343	S. Johnson
23678	M. Walters

Classes

Class	Subject	Instructor	Room
MA 23	Calculus	LaGrange	F 203
Eng 40	Fiction	Moss	K 112
Hist 62	Latin Amer	Felding	G 331
Art 76	Drawing	Picas	A 2
PE 45	Dance	Murray	W. Gym

Grades

ID	Class	Grade
10112	MA 23	B
10112	Eng 40	A
10112	Hist 62	A-
12343	Eng 40	B-
12343	Hist 62	B
12343	Art 76	A
12343	PE 45	A
23678	MA 23	C
23678	Art 76	A-
23678	Eng 40	B

FIGURE 16.7 Final decomposition. The remaining relation **Grades** has no nontrivial dependency that is not based on a superkey. Hence the complete relation is decomposed into components that are in BCNF.

The BCNF Algorithm

There is a general process used to obtain the BCNF that was illustrated by the **School** example.

Beginning with a relation R, one looks for a nontrivial dependency of the form $A_1 A_2 \cdots, A_k \rightarrow B_1, B_2, B_r$ in which the set $\{A_1, A_2, \ldots, A_k\}$ is *not* a superkey. To minimize the number of components in the final decomposition, it is best to include as many B_j's as possible and as few A_i's as possible for those B_j's. That is, given a nontrivial dependency of the form $A_1 A_2 \cdots A_k \rightarrow B_1$, one should seek additional attributes also dependent on $A_1, A_2, \ldots, A_k$ and include them as well; and then minimize the number of A_i's. These last procedures are optional. It is only necessary to begin with a nontrivial dependency in which the left-hand side is not a superkey.

The original relation R is then decomposed into two relations R_1 and R_2. R_1 is the projection of R onto the attributes $A_1, A_2, \ldots, A_k, B_1, B_2, \ldots, B_r$. And R_2 is the projection of R onto $A_1, A_2, \ldots, A_k, C_1, C_2, \ldots, C_s$, where the collection of C_i's includes all other attributes of R aside from the A_i's and B_i's. It then follows (as soon will be shown) that R is equal to the natural join of R_1 and R_2.

Next R_1 and R_2 are each examined to determine if they can be decomposed using the same process. If either of them cannot be so decomposed, then by definition it must be in Boyce–Codd normal form. This process continues until all the subrelations are in Boyce–Codd normal form.

Verification of the Algorithm*

To verify that the process works, it is necessary to establish two things. First it must be shown that the algorithm always comes to a successful conclusion. To see that it does, we first prove that any relation with just two attributes must be in BCNF. Suppose the two attributes are A and B. There are only two basic possibilities: (1) if there is no nontrivial dependency, R must be in BCNF, (2) if $A \rightarrow B$, then A must be a key, for if there were two tuples with the same first element, say (a, b_1) and (a, b_2), the dependency would imply that $b_1 = b_2$, which means that the two tuples are identical, contrary to the definition of a relation. The situation is parallel if $B \rightarrow A$. Thus a two-attribute relation is always in BCNF.

Now notice that at every stage of the decomposition the newest subrelation contains fewer attributes than the one from which it was derived. Hence, eventually, all components must either be in BCNF or consist of only two attributes, and in either case the relations are in BCNF. Therefore the algorithm always concludes successfully.

The second thing that must be established is that the decomposition is faithful to the original in the sense that the original relation can be reconstructed from the decomposed form. Suppose for simplicity that R has only three attributes, A, B, C. Suppose also that there is a nontrivial dependency $B \rightarrow C$ with B not a key. Decompose this relation into $R_1 = R_1(A, B)$ and $R_2 = R_2(B, C)$. We will show that the natural join of R_1 and R_2 is R.

Suppose there is a tuple (a, b, c) in R. Then R_1 will contain (a, b) and R_2 will contain (b, c). The natural join $R_1 \bowtie R_2$ will contain all combination tuples in R_1 and R_2 with a common B element. Thus clearly (a, b, c) will occur in the join. Hence the join contains everything in R. It still must be shown, however, that that is all there is in the natural join; no new false tuples are created.

Suppose as before that (a, b, c) is in R. R_1 contains (a, b). There could be a problem if R_2 contained a tuple of the form (b, c'), for then (a, b, c') would be in the natural join. But there can be no such (b, c') in R_2, for that would violate the dependency $B \rightarrow C$. Hence no new tuples are created, and the natural join faithfully reconstructs the original relation R. This argument applies directly (but with cumbersome notation) to an arbitrary number of attributes.

Relation to Entropy*

Consider a relation $R(A, B, C, D)$. If the entries are considered to be random, the entropy $H(A, B, C, D)$ can be defined. Suppose now that there is a functional dependency $A \rightarrow B$. Then the relation can be expressed as the natural join of two simpler relations of the form $R_1(A, B)$ and $R_2(A, C, D)$. The entropy can be written as

$$H(A, B, C, D) = H(B|A, C, D) + H(A, C, D)$$
$$= H(B|A) + H(A, C, D) \quad \text{(because } C \text{ and } D \text{ add nothing}$$
$$\text{to } A \text{ about } B)$$
$$= H(A, C, D) \quad \text{(because } H(B|A) = 0).$$

Hence it is necessary only to consider $H(A, C, D)$, and one should keep a record of $A \rightarrow B$, which is the relation $R_1(A, B)$. In general, functional dependencies imply a zero-valued conditional entropy that simplifies the structure of H in the same way that it simplifies the structure of the relation.

16.6 Joins and Products*

The product operation can be used to manipulate data contained in two or more relations in complex ways. This section displays the result of a product operation, and shows how the natural join of two relations can be easily found from the product.

Consider the two relations **Students(ID, Student)** and **Grades(ID, Class, Grade)** of the school example. These two component relations are shown again in figure 16.8. The product of these two relations includes all attributes of both, even duplicating **ID**. To make clear that they are distinct, let us rename them **ID1** and **ID2**. The complete product is displayed in figure 16.9, which has a total of $2 + 3 = 5$ attributes and $3 \times 10 = 30$ tuples.

Students

ID	Student
10112	J. Banes
12343	S. Johnson
23678	M. Walters

Grades

ID	Class	Grade
10112	MA 23	B
10112	Eng 40	A
10112	Hist 62	A-
12343	Eng 40	B-
12343	Hist 62	B
12343	Art 76	A
12343	PE 45	A
23678	MA 23	C
23678	Art 76	A-
23678	Eng 40	B

FIGURE 16.8 The two relations Students and Grades. Their product is shown in figure 16.9.

Product

ID1	Student	ID2	CLASS	Grade
10112	J. Banes	10112	MA 23	B
10112	J. Banes	10112	Eng 40	A
10112	J. Banes	10112	Hist 62	A-
10112	J. Banes	12343	Eng 40	B-
10112	J. Banes	12343	Hist 62	B
10112	J. Banes	12343	Art 76	A
10112	J. Banes	12343	PE 45	A
10112	J. Banes	23678	MA 23	C
10112	J. Banes	23678	Art 76	A-
10112	J. Banes	23678	Eng 40	B
12343	S. Johnson	10112	MA 23	B
12343	S. Johnson	10112	Eng 40	A
12343	S. Johnson	10112	Hist 62	A-
12343	S. Johnson	12343	Eng 40	B-
12343	S. Johnson	12343	Hist 62	B
12343	S. Johnson	12343	Art 76	A
12343	S. Johnson	12343	PE 45	A
12343	S. Johnson	23678	MA 23	C
12343	S. Johnson	23678	Art 76	A-
12343	S. Johnson	23678	Eng 40	B
23678	M. Walters	10112	MA 23	B
23678	M. Walters	10112	Eng 40	A
23678	M. Walters	10112	Hist 62	A-
23678	M. Walters	12343	Eng 40	B-
23678	M. Walters	12343	Hist 62	B
23678	M. Walters	12343	Art 76	A
23678	M. Walters	12343	PE 45	A
23678	M. Walters	23678	MA 23	C
23678	M. Walters	23678	Art 76	A-
23678	M. Walters	23678	Eng 40	B

FIGURE 16.9 Product of Students and Grades. The product contains every combination of tuples from the two underlying relations.

Once the product is formed, the natural join can be obtained by selection and projection. The section criterion is that **ID1** = **ID2**. The tuples that meet this requirement are shaded in the product representation shown in figure 16.9. After selecting on these tuples, the result is projected onto an attribute set that eliminates either **ID1** or **ID2**; it does not matter which is dropped, since their values agree on all selected tuples. The ID attribute that remains in the projection is renamed **ID** for simplicity. The outcome of this selection and projection procedure is the natural join of **Students** and **Grades** (figure 16.10), which is the relation **StudentGrades** (from which the two components were originally derived).

StudentGrades

ID	Student	Class	Grade
10112	J. Banes	MA 23	B
10112	J. Banes	Eng 40	A
10112	J. Banes	Hist 62	A-
12343	S. Johnson	Eng 40	B-
12343	S. Johnson	Hist 62	B
12343	S. Johnson	Art 76	A
12343	S. Johnson	PE 45	A
23678	M. Walters	MA 23	C
23678	M. Walters	Art 76	A-
23678	M. Walters	Eng 40	B

FIGURE 16.10 Natural join of Students and Grades. The join can be obtained from the product by selecting tuples that agree on the duplicated attribute, and then projecting to eliminate the duplication.

16.7 Database Languages

The primary purpose of a database is to readily access portions of data that are requested and add or delete entries. These operations are carried out through a query language. The standard query language SQL (for Structured Query Language, and pronounced "sequel") is used in most database systems. This language allows for flexible formulation and execution of queries and for modifying the entries in a database. This section illustrates the use of this language in a generalized manner.

Queries

Most users interface databases through **queries**, asking for the return of certain tuples, restrictions of tuples, or combinations of tuples from one or more relations.

The most important query is that of selecting the tuples that satisfy a simple specification. A query to the **School** relation might be

```
SELECT   Grade, Subject
FROM     School
WHERE    Name = 'J. Banes'  AND  Class = 'Hist 62'
```

This will return the grade and subject of J. Banes in the History 62 class. Specifically, it will return "A−, Latin America."

A slightly more complex query might be

```
SELECT   ID, Student
FROM     School
WHERE    Class = 'Ma 23'  AND  Grade = 'B'
```

This will return the ID and name of all students who received a B in Math 23.

Queries can involve more than one relation. For example, consider the query

```
SELECT   Grade, Instructor
FROM     Classes, Grades
WHERE    ID = '10112'  AND Class = 'Eng 40'
```

This will return the grade and instructor of J. Banes in the English 40 class using the relations **Classes(Class, Subject, Instructor, Room)** and **Grades(ID, Class, Grade)**.

Queries can be evaluated internally in several different ways. One way is to convert the query to the evaluation of a combination of relational operations and evaluate the result. For example, an intersection query can be evaluated by actually forming the intersection of relations. Any query can be expressed in terms of the six fundamental operations of select, project, union, difference, product, and rename. The conversion of a query to the appropriate combination of operations is the job of the database system, and this operation can be complex. As users, it is not necessary to know the details, but understanding the basic structure as described in this chapter is valuable for anyone beginning to probe database theory.

Modification

From time to time new tuples must be adjoined to the relations of a database, and out-of-date tuples must be modified or deleted.

New tuples are adjoined by use of the insert command. For example, consider the instruction

```
INSERT INTO   ID, Students(ID, Student)
VALUES        ('3001', 'R. Smith')
```

This will add the student R. Smith with ID 3001 to the relation **Student**. Later, to add class information about R. Smith, one might use

```
INSERT INTO   ID, Grades(ID, Class, Grade)
VALUES        ('3001', 'Hist 62', 'B+')
```

If this were an error, we could delete the tuple with the instruction

```
DELETE FROM   ID, Grades(ID, Class, Grade)
WHERE         ID = '3001'  AND
              Class = 'Hist 62'  AND
              Grade = 'B+'
```

16.8 EXERCISES

1. (Children) Explain how the problem of including any number of children's names in **AddressBook** can be solved by defining an additional relation that has only single-valued entries (using perhaps **Name** and **Child** as attributes).

2. (Find the keys) Let $R(A, B, C, D)$ be a relation and suppose it contains the following dependencies: $A \rightarrow B, BC \rightarrow D$, and $D \rightarrow A$.
   (a) Find all superkeys of R.
   (b) Find all keys of R.

3. (Addresses) Are there any nontrivial functional dependencies of the relation **AddressBook** in which the left-hand side is not a superkey?

4. (Multi-attributes) Consider the relation **ClubMembers** shown in figure 16.11. Show that it is in BCNF, but find a simpler representation as two relations. (The representation that eliminates this kind of redundancy is termed the **fourth normal form**.)

5. (Appointment) Suppose **Clinic** is augmented to **ClinicPlus** by the addition of the attribute **Appointment**, which is unique for every tuple. This can serve as a primary key. Find the BCNF of **ClinicPlus** and compare it with the BCNF of **Clinic**.

6. (Intersection) Express $R \cap S$ in terms of differences.

7. (Closure) Let R be a relation and let $X = \{A_1, A_2, \dots, A_k\}$ be a subset of the attributes of R. Define X^+, the **closure** of X, as the smallest set X^+ such that $X^+ \supset X$ and $X^+ \rightarrow B$ implies $B \in X^+$.
   (a) Find the closure of the set $\{$**Instructor, Grade**$\}$ in **School**.
   (b) Show that X is a superkey if and only if X^+ is the set of all attributes of R. Hint! For "if" construct X^+ step by step.

ClubMembers

Name	Child	Hobby
Alice	Mary	Tennis
Alice	Sam	Tennis
Alice	Mary	Golf
Alice	Sam	Golf
Barbara	John	Choir
Barbara	Kate	Choir
Barbara	Nancy	Choir
Barbara	John	Photography
Barbara	Kate	Photography
Barbara	Nancy	Photography

FIGURE 16.11 A relation for club members, listing members names, their children, and their hobbies.

8. (Equi-join) The **equi-join** of two relations R and S over the common attribute B is the set of all pairs of tuples, the first from R and the second from S, such that the values of attribute B are equal. The equi-join can also be defined relative to several common attributes. How does the equi-join of R and S over all common attributes differ from the natural join of R and S?

9. (Theta-join) The **theta-join** of relations R and S, denoted $R \bowtie_\theta S$, is the set of all pairs of tuples, the first from R and the second from S, that satisfy a specified condition θ. The condition θ can be equality of a given attribute, for example, in which case the theta-join is the equi-join. But θ can be more general. The theta-join can be obtained by first forming the product $R \times S$ and then selecting only those tuples that satisfy θ.

Consider the relations R and S defined as

R	A	B	C		S	D	E	F
	2	2	3			6	4	3
	6	4	2			2	4	1
	3	5	7			3	1	2
	2	4	2			7	0	4

Find $R \bowtie_{A<F} S$.

10. (A new relation) Assuming that all apparent dependencies (such as $A \rightarrow B$) are actual functional dependencies, put the relation below in BCNF.

A	B	C	D	E	F	G
6	2	1	8	2	4	3
7	1	1	8	2	6	1
4	2	2	6	6	2	3
6	2	1	8	2	4	1
6	2	2	7	0	5	0
4	2	2	6	6	2	5
7	1	1	8	2	6	8
9	0	2	9	8	9	4

11. (Two ways) Consider the two relations R and S shown below.

R	A	B		S	A	C
	2	3			2	8
	4	7			4	6
	2	9			2	1

(a) Form the natural join $T = R \bowtie S$. Note that it has five rows.

(b) Form the Boyce–Codd decomposition of T. Note that one component has five rows, and note that the original description of T in terms of R and S is more compact than the BCNF decomposition.

(c) Does the number of rows in at least one of the relations in BCNF decomposition of a natural join always equal the number of rows in the original (composite) relation?

16.9 Bibliography

There are many texts on database systems. The three listed here cover the range from elementary [1], to comprehensive [2], to a concise presentation of theory and application [3].

References

[1] Harrington, J. L. *Relational Database Design Clearly Explained*. London: Academic Press, 1998.

[2] Date, C. J. *An Introduction to Database Systems*, 4th ed. Vol. 1. Reading, Mass.: Addison-Wesley, 1986.

[3] Ullman, J. D., and J. Widom, *A First Course in Database Systems*. Upper Saddle River, N.J.: Prentice-Hall, 1997.

17

INFORMATION RETRIEVAL

Vast amounts of data are today available to all of us who have access to libraries, the Internet, or computer files of many sorts. This profusion brings with it the challenge of finding the particular elements of data that are relevant for us—the information that we seek. There are many tools that aid in our search. Libraries that once relied exclusively on card catalogs now have computer files that can execute searches by author, title, keywords, or subject. Many libraries make available database directories for collections residing in remote locations or on computer files. The Internet is connected to a vast heterogeneous assortment of data as pictures, music, movies, and text. And it is relatively easy to access this data from almost anywhere in the world, even with wireless connections.

In concert with the exponential expansion of accessible data, there have been advances in the techniques and tools for finding the data that we seek, or finding data relevant to our need but of whose existence we were previously unaware. There has indeed been a great deal of thought and creativity devoted to the development of tools that assist the retrieval of relevant information from the enormous amount of data available.

This chapter surveys some of the main retrieval concepts and tools. In order to bound the range of discussion, the chapter focuses almost exclusively on retrieval of text, leaving aside pictures, music, and movies. In practice, it is text that is most often sought. However, many of the ideas presented in this chapter have been extended and specialized to treat the other forms of data.

The general approach to information retrieval is not mysterious. It is simply an extension of the retrieval methods used for years in visits to a library or when consulting a text or reference work. Retrieval is partitioned into two parts: indexing and query response. Most textbooks and reference works include a comprehensive index, alphabetically listing terms and the key pages on which they appear. Some libraries still maintain a card catalog, which alphabetically lists works separately by author, title, and subject, together with a call number that indicates where in the library the work is located. Indexes are fundamental to information retrieval. But as data collections expand to encompass many millions of text sources, indexing becomes a science as well as an art.

Query response is also important. When consulting a book, such as this one, the reader can ask a question—such as "On what pages is entropy defined?"—and the index will provide the answer. When a data collection is enormous, such as that of the Internet, it is important that the retrieval system be able to respond usefully to complex queries.

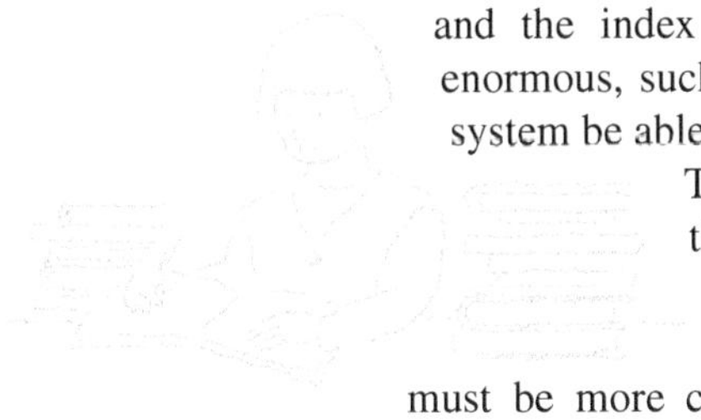

The vast majority of retrieval systems for huge collections do not search the collection in response to a query. Rather, they rely on a previously prepared index. Thus, to respond usefully to complex queries, the index itself must be more complex than that found in textbooks or card catalogs. Most retrieval systems are simply a combination of an index and a query response system—the two working together.

17.1 Inverted Files

Consider a body of text consisting of a collection of distinct documents. The terms **collection** and **document** are defined broadly. A collection might be a single book, with the documents being the individual pages. A collection may be a professional journal series (spanning several years) with documents being the individual articles in the series. A collection might be the entire set of web pages on the Internet with documents being the individual pages at various websites. Or a collection may be the set of books in the U.S. Library of Congress with documents being individual books. An important collection used often in retrieval experiments is the Bible, with the documents being the separate verses.

A simple index for a collection is a list of **terms** (usually in alphabetical order) with each term having a sublist giving the documents in which the term appears. This is the format of a textbook index, although book indexes generally list only important terms and report only important or defining instances of those terms. More complex indexes may indicate how often the term appears in each document and/or the positions of the term within a document.

And	2, 3
Is	1, 2, 4
Make	3
Now	1, 2, 3, 4
Place	2, 3
The	1, 2
Time	1, 3

Consider the small collection of four lines from the (fabricated) *Terrachrona*:

1 Now is the time
2 And now is the place
3 Time and place make now
4 Now is now

FIGURE 17.1 Index for Terrachrona. The index lists terms alphabetically and gives the document (line) numbers where each term appears.

A simple index for this collection is shown in figure 17.1.

A **lexicon** relative to a collection of documents is the set of words or terms contained in the collection. In the preceding example, the lexicon consists of seven words: and, is, make, now, place, the, time.

For purposes of indexing, a collection can be defined by an **incidence matrix**, also termed a **forward file**, which lists the words each document contains (although

Collection	Term 1	Term 2	Term 3	......	Term M
Doc1	1	0	1	......	0
Doc 2	0	1	1	......	1
Doc 3	1	1	1	......	1
⋮					⋮
Doc N	1	0	0	......	1

FIGURE 17.2 Forward file or incidence matrix. The terms in each document are indicated by the 1's in the corresponding row. In the case of a frequency matrix, the entries give the number of instances that the term appears in the document.

				Words			
Line No.	**and**	**is**	**make**	**now**	**place**	**the**	**time**
1	0	1	0	1	0	1	1
2	1	1	0	1	1	1	0
3	1	0	1	1	1	0	1
4	0	1	0	1	0	0	0

FIGURE 17.3 Forward file for *Terrachrona*. There are four documents (lines) and seven words in the lexicon.

it may only record important words, leaving out common words such as *the*). The general form of a forward file is displayed in figure 17.2. The forward file for the *Terrachrona* example is shown in figure 17.3.

A related but slightly more complex representation is a **frequency matrix**, which records the number of occurrences of the word in each document rather than simply indicating if there is at least one occurrence. The frequency matrix for the *Terrachrona* is identical to the incidence matrix except that the last row has a 2 in the "now" column.

The index of a collection is the transpose of the incidence or frequency matrix, for it lists the documents in which terms appear, rather than the terms that appear in documents. In this context the transpose is termed the **inverted file** of the forward file. The general form of an inverted file is shown in figure 17.4.

Finding the (forward) incidence or frequency matrix itself is straightforward. One merely scans each document in turn and lists the terms that it contains. The entries can be placed in an incidence matrix or written down in order. For example, a (not very practical) way to construct an index for a textbook is to scan each page and list alphabetically on an index card the important terms that that page contains. The index itself could then be constructed from these cards by selecting terms alphabetically and for each term, scanning through the cards to find entries.

Lexicon	Doc 1	Doc 2	Doc 3	......	Doc N
Term1	1	0	1	......	0
Term 2	0	1	1	......	1
Term 3	1	1	1	......	1
⋮					⋮
Term M	1	0	0	......	1

FIGURE 17.4 **General inverted file.** This is the transpose of the forward file.

A computer implementation for constructing an inverted file seems obvious. One first constructs the incidence matrix by scanning the documents sequentially. Then the inverted file is formed by scanning down the columns, listing the documents associated with a given term.

This simple method can in fact be used for small collections. A single 400-page book with 500 words per page may contain as many as 10,000 different words. In a simple implementation, four-byte integers may be used to represent the elements in the incidence or frequency matrix. The total storage requirement for the matrix would then be $400 \times 10,000 \times 4 = 16$ megabytes, which is quite manageable on present-day machines.

A concordance of the Bible lists the verses in which words appear. The Bible contains 31,101 verses and 8,965 different words. The same analysis as above gives $31,101 \times 8,965 \times 4 = 115$ gigabytes, which is too large to fit in the random access memory of most computers today. Of course, these estimates can be reduced by refined coding that may allocate only a few bits to an entry (one bit in an incidence matrix) or by compressing the matrix as discussed later in this chapter. On the other hand, it is clear that even modestly large collections can exceed available fast memory. It is therefore important to consider other approaches.

17.2 Strategies for Indexing

We shall describe four different strategies for indexing in this section. Each has its advantages depending on the size of the collection and the technology used.

Card File Indexing

Traditionally, authors indexed their books by hand using three-by-five index cards. In this method each card is headed by a term followed by a list of pages on which that term is found. The cards are placed alphabetically in a file box, or more often laid out alphabetically on the floor (as I can attest). The author progresses through the book a page at a time. When a term that should be placed in the index is found, the existing cards are searched for that term and the current page number is written on the card.

If no card exists for that term, a new card is inserted at the appropriate alphabetical position among the others. This is an effective time-tested method.

It is not easy to implement this method on a computer. A blank file card is essentially a block of free memory capable of storing several page references. Hence, whenever an author introduces a file card, a block of memory is being allocated to a particular term. If this strategy were used on a computer, it would be necessary to allocate a large block of memory to each term in anticipation that it may be filled, and this surely would be excessive.

Linked Lists

The computerized version of the file card method uses linked lists. A data structure for the terms is first established. A good choice is a binary sort tree, which is an efficient way to insert terms alphabetically and find them when needed. Each term occupies a memory slot large enough to contain the term, pointers for movement in the term data structure, and a pointer to the first element in a linked list for that term. The document numbers referenced by a term are located along this linked list.

The documents are scanned sequentially. When a term t to be indexed is found in document d, the data structure is searched for t. If t is in the structure, the links are followed to the end of the list for t. The document number d is then placed in the first available memory location (which may be some distance from the last member of the list) and the pointer of what was the last element is updated. If t is not already in the data structure, it is inserted together with a pointer to the first element of a linked list that is initiated with the document number d. The members of any particular linked list are likely to be scattered in memory because they arise erratically depending on when they are found in other documents. Hence, unlike the file card system, the only extra memory space is that of the pointers. The approach is illustrated in figure 17.5. (In the figure the terms have appeared in the order H E C M E K G M H H G E.)

Once scanning is complete, the data structure is traversed alphabetically. At each term, the linked list is followed (jumping from location to location) so that all references to that term can be written, in order, to construct the index entries for that term.

This method requires memory to store the linked lists as well as the data structure of terms. The total number of elements in all linked lists is equal to the number of document references in the final index. For large collections, this will exceed available random access memory, and hence most of the linked list storage will be in slower memory such as disk. Furthermore, since, as explained above, any particular linked list is scattered throughout memory, the list cannot be followed quickly.

The linked list method therefore requires either a huge random access memory or a great amount of time. Hence we again look for other methods.

Sort-Based Methods

Recall the author who indexes by hand. Suppose that, realizing that a blank index card is equivalent to a significant block of memory, the author instead uses small slips

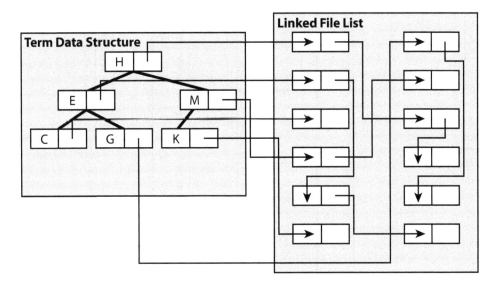

FIGURE 17.5 Linked list inverted file. In the figure, the possible terms are assumed to be the letters of the alphabet, and they appear in random order as documents are scanned. The term data structure (in this case a binary search tree) is used to store the terms for quick access. When a term is found, the corresponding document number is appended to the linked list for that term. The linked lists are stored outside of main memory. Since the terms appear in essentially random sequence, the linked list of a particular term is likely to be scattered throughout. The binary sort tree itself is stored as an array giving the term, node number, parent, and left and right child of each node.

of paper upon which only two entries can be made: a term and a page number. The author progresses through the documents (pages), filling out a slip whenever a term to be indexed is found. When all pages are scanned, there will be a large pile of these slips. It is only necessary to sort these slips alphabetically using term as the primary key and page number as the secondary key. The slips of a given term t will form an entry in the book index.

The sorting process is equivalent to constructing the inverted file. The slips are originally written in page (or document) order, and later sorted in term order.

This method easily translates to computerized form. Documents are scanned sequentially. When in document d the term t is encountered, the two-tuple (t, d) is stored. These two-tuples are then sorted by term and the common terms grouped to form the index, the inverted file.

This method requires roughly the same amount of memory as the linked list method, so for even modestly large collections the original list of two-tuples cannot be stored in fast memory, and hence a simple sort procedure cannot be used.

However, there is an efficient method of sorting, using an in-place merge sort. In outline: the entire list of two-tuples is partitioned into equal-sized segments that each fit in fast memory. Suppose there are R such segments. Each of these R segments is sorted in the fast memory using an efficient sort procedure such as quicksort, and then returned to the slower memory. Next the R segments are merged a few at a time. This is quite efficient compared with the linked list approach because data is accessed in blocks from the segments, rather than in the helter-skelter pattern associated with linked lists. The sort-based method is illustrated in figure 17.6.

Forward File

1	C	F	W	R
2	S	R	K	
3	T			
4	W	F		
5	R	K	V	
6	R	C		
7	S	T		
8	R	C		
9	K			

(a)

C	1
F	1
W	1
R	1
K	2
S	2
R	2
T	3
W	4
F	4
R	5
K	5
V	5
R	6
C	6
S	7
T	7
R	8
C	8
K	9

(b)

C	1
F	1
K	2
R	1
W	1
F	4
R	2
S	2
T	3
W	4
C	6
K	5
R	5
R	6
V	5
C	8
K	9
R	8
S	7
T	7

(c)

C	1
C	6
C	8
F	1
F	4
K	2
K	5
K	9
R	1
R	2
R	5
R	6
R	8
S	2
S	7
T	3
T	7
V	5
W	1
W	4

Inverted File

C	1	6	8	
F	1	4		
K	2	5	9	
R	1	2	6	8
S	2	7		
T	3	7		
V	5			
W	1	4		

FIGURE 17.6 Sort-based indexing. The forward file consists of terms (shown as letters) and document numbers. They are originally sorted by document number in (a). Next they are sorted by term. For large files, the sorting is carried out in segments (producing part (b) of the figure). The segments are merged to produce the file shown in (c). This is converted to the inverted file (in index form).

Text-Based Partitioning

Another strategy an author might use when compiling a book index is to index part of the book, say the first half, and then make a separate index for the second half. Finally the two indexes can be merged to form a completed full index.

In computer form this method is extremely practical. The collection of documents is partitioned into manageable groups, and an index constructed for each group using only fast memory, with for example, a linked list procedure or a simple sort-based procedure in fast memory. The separate indexes are then placed in the slower memory and merged in the same manner as separate partitions are merged with a full sort-based method.

Implementation

The methods outlined in the subsections above are general strategies rather than fully detailed methods. In actual implementations, the lists or tuples resulting from a scan of documents are usually compressed to save memory. These compression techniques are discussed in the next section. In addition, there are many details concerning pointers, storage allocation, and so forth that must be carefully designed and incorporated into a final system.

17.3 Inverted File Compression*

Return again to the view of indexing as the construction of an inverted file from an incidence matrix. If there are N total documents in the collection, any one row of the inverted file consists of N elements that are either 0 or 1, with the 1's indicating the various documents that contain the term. Although this view is theoretically useful, it is usually not practical to write the entire list of 0's and 1's. Instead, the row associated with a term is better represented as an ordered list of document numbers corresponding to documents that contain the term. A standard textbook index is of that form. It presents an ordered list of the page numbers where the term appears rather than an indication for each page as to whether the word appears there or not.

Generally, then, a row of the inverted file can be represented as a list $(d_{t1}, d_{t2}, \ldots, d_{tk_t})$ where the d_{ti}'s are the document numbers of documents that contain t. For example, a particular list may be $(3, 6, 15, 36, 44)$.

Note that the entries in such a list are increasing; that is, each document number in the list is greater than the previous one. Therefore, the list can be expressed using incremental rather than actual values, the elements in the list indicating how much to add to the previous document number to obtain the next in the list. For instance, the list $(3, 6, 15, 36, 44)$ can be expressed in incremental form as $(3, 3, 9, 21, 8)$. In traditional coding theory, this method is termed **run-length coding**. The advantage of this method is that it might lead to smaller entries, on average, and hence it may be possible to use fewer total bits than otherwise. To verify this, suppose the list of document numbers is $(d_1, d_2, \ldots, d_k)$. In run-length form the list is $(d_1, d_2 - d_1, d_3 - d_2, \ldots, d_k - d_{k-1})$. The average of the entries in the first list is $\text{Aver}_1 = (d_1 + d_2 + \cdots + d_k)/k$. The average of the entries in the second list is $\text{Aver}_2 = d_k/k$. Clearly $\text{Aver}_2 \leq \text{Aver}_1$. However, if there is only a single entry in the list, then $\text{Aver}_1 = \text{Aver}_2$. Hence the range of numbers that may be encountered by the two methods is the same. The number spread is equal in both methods, but the numbers associated with the run-length method tend to be bunched near small values. This feature can be used to advantage.

Unary Codes

TABLE 17.1
The Unary Code.

1	0
2	10
3	110
4	1110
5	11110
6	111110
7	1111110

A simple method of coding a series of numbers that tend to bunch up at the low end is the **unary code**. The codewords for the first seven integers are shown in table 17.1. The code for an integer i is a series of $i - 1$ ones followed by a single zero. The number zero is not coded, since it never appears in a run-length list.

This code can be recognized as the comma code of chapter 3. At first sight it may seem quite wasteful; after all, seven symbols can be coded with a code with words of length three, but actually it can be quite efficient.

Recall that the Shannon coding scheme sometimes achieves an average word length equal to the entropy of the source. The condition for this is that the lengths be chosen as

$$l_i = \log \frac{1}{p_i}, \tag{17.1}$$

where p_i is the probability of item i. Usually this formula is applied forward: starting with the p_i's and selecting the l_i's. It can be applied backward now to analyze the unary code. The lengths of this code are known; the p_i's for which this code is optimal can be found. Inverting (17.1) produces

$$p_i = 2^{-l_i}. \tag{17.2}$$

Using the fact that $l_i = i$ for the unary code, it follows that $p_i = 2^{-i}$. That is, $p_1 = \frac{1}{2}$, $p_2 = \frac{1}{4}, p_3 = \frac{1}{8}. \ldots$ These probabilities definitely favor small integers. If the entries of the incremental (or run length) list appear with these probabilities, then unary coding of the list achieves maximum compression.

This is a nice result, but unfortunately there is no reason to expect that entries appear with these probabilities. In fact, in most cases, the probabilities implied by unary coding emphasize small integers too strongly. We therefore seek alternative coding methods, but we have learned some useful ideas from study of this code.

Golomb Codes

Rather than invent a code arbitrarily, and then determine what probability distribution corresponds to it, it seems more logical to propose a probability distribution and then find the corresponding optimal code. This, of course, is the procedure used in the general study of coding theory presented in chapters 3 and 4.

A reasonable model for the probabilities of run lengths is the **Bernoulli** model. This model assumes that a given term to be indexed appears randomly in a document with a given probability, say p. That is, if one selects a document at random from the collection, there is a probability p that the document will contain the term t.

From this assumption the probability of runs of various lengths can be determined. The probability that t appears in document i and then not again until document $i + x$ is $(1 - p)^{x-1}p$, since there are $x - 1$ documents of no appearance and one where there is appearance. This resulting probability distribution is termed a **geometric distribution**.

Golomb devised a coding procedure that is very effective for this Bernoulli model and the resulting geometric distribution. The idea is that if a number $x - 1$ is divided by some integer $b > 0$, the result is a smaller quotient and a remainder. The quotient is encoded with a unary code and the remainder by a binary code, forming a two-part code. We consider the simplest case where b is selected to be a power of 2; that is, $b = 2^k$.

To construct the Golomb code for an integer $x > 0$, the quotient[1] $q = \lfloor (x - 1)/b \rfloor$ is computed and $q + 1$ is encoded with a unary code (which is possible because $q + 1$ is never zero).

The remainder $r = x - 1 - qb$ is coded in binary. There are $b = 2^k$ possible residue values: $0, 1, 2, \ldots, b - 1$, so since b is a power of 2, this number matches the number of values available in a pure binary code with k bits. The final code is formed by first using the unary code for $q + 1$ followed by the binary code for r.

As an example, let us encode $x = 118$ with $b = 2^4 = 16$. Here $q = \lfloor 117/16 \rfloor = 7$ and $r = 5$, since $117 = 7 \cdot 16 + 5$. Hence the code is 1111110 0101. The first part is the unary code for 7 and the last four bits are the binary code for 5. (In practice no space would be used.)

---

[1] For any number y, the expression $\lfloor y \rfloor$ denotes the largest integer less than or equal to y, and $\lceil y \rceil$ denotes the smallest integer greater than or equal to y.

These compression techniques can be used to compress the list of documents in the inverted file associated with a particular term. They can also be used in more complete inverted files that list the positions of all occurrences of a term in a specific document.

17.4 Queries

Data is retrieved in response to queries, and different data systems allow different types of queries and handle queries in various ways.

Tries

Systems for small collections may not use indexes. For example, the "Find" operation in a word processor simply scans a document for the string of characters that is requested. Some systems use tries, discussed in chapter 15, to facilitate string searches. These have the advantage that complex queries can be handled, but they are expensive in terms of memory.

Basic Index-Based Query Systems

Perhaps the simplest query is the request for all documents that contain a given word. Such a query fits perfectly with the structure of a standard inverted file index. For example, querying the word *God* in a Bible concordance will produce all verses in which *God* appears. When searching for a word on the Internet with a good search engine, the search engine can produce every web page in which that word appears (at least all web pages that are indexed by the search engine), although the search engine may report only the subset of these pages that are considered highly relevant. Such single-term queries are easily answered by direct consultation to the inverted file index.

A Boolean query is slightly more complex. Such a query incorporates logical operators of AND, OR, and NOT. For instance one might formulate the query

university AND **California** NOT **public**,

hoping to find all references to private universities in California. Such a query is processed by the addition of a logical computation in conjunction with an inverted file index. All documents with *university* are listed in a temporary file, then all with *California*, and all with *public*. The first two resulting files are merged, and then items are deleted if they appear in the list for *public*.

It must be remembered that such a query simply checks the word occurrence with no reference to placement or intent. Hence, the response to the above query would not report a document about the private university Stanford if the document contained the sentence, "Stanford University, located in California, serves the public interest."

Queries can be broadened by using more complex Boolean queries such as

(college OR **university)** AND **(California** OR **(Washington** NOT **D.C.))**.

Response to such a query requires an advanced Boolean processor, but such a processor represents a small part of a total retrieval system for a large collection of documents.

One difficulty with Boolean queries based on word indexing is that no account is made for word placement. Hence the query

University AND **of** AND **California**

could return many documents that contained those three words in any order. Some retrieval systems allow queries containing phrases, sometimes indicated by placing quotation marks around the phase, such as

University of California.

A proper response to such a query can be generated by retrieval systems that search strings. Proper responses also can be generated by inverted file systems that record the positions of a given word's occurrence in every document, for then it is only necessary to determine whether the words in the query phrase appear consecutively in a document. This, of course, requires a more complex index, but many indexes, including several Internet search engines, are structured this way.

17.5 Ranking Methods

Simple queries do not always adequately express the full intent of the inquirer. That realization motivates the idea of ranking query responses according to some measure. Most ranking methods applied to inverted file indexes are based on **vectors** associated with the forward file.

Consider the sample forward incidence matrix shown in figure 17.7. The rows, consisting of integers, can be regarded as vectors of dimension seven, the number of terms. The entire forward file can be regarded as a collection of these vectors, one for each document.

A query that specifies a word or set of words can itself be considered a vector of the same dimension (with ones corresponding to the terms in the query and zeros elsewhere).

	and	is	make	now	place	the	time
1	0	1	0	1	0	1	1
2	1	1	0	1	1	1	0
3	1	0	1	1	1	0	1
4	0	1	0	1	0	0	0

FIGURE 17.7 Forward file as collection of vectors. Each row is a seven-dimensional vector.

Suppose x and y are two vectors of length T. The **inner product** (or **dot product**) of x and y, denoted $x \cdot y$, is equal to

$$x \cdot y = \sum_{i=1}^{T} x_i y_i.$$

If the components of x and y consist only of zeros and ones, then the inner product is equal to the number of one-valued components that are in the same position in each vector. Hence, if x is a query vector and y is a document vector recording only the presence or absence of terms, the inner product gives the number of query terms present in the document. Thus the inner product can be used to rank documents by the number of queried terms they contain.

The naive way to compute the inner product is to scan each document and count the instances where a term matches one in the query. This procedure entails running through every document, which could take an enormous amount of time. Alternatively, the inner product can be computed by use of the inverted file. For each term in the query, the inverted file gives a list of documents that refer to that term. We credit a value of one unit to each of these documents. After processing all terms in the query, the total score assigned to each document is available, and the documents are then ranked accordingly.

Weighted Vectors

There are three obvious shortcomings of the simple counting procedure: (1) A document that contains many instances of a query term is likely to be more important than one that contains only a single instance. Hence, the frequency of the term's appearance should be incorporated. (2) Common words are likely to produce higher counts than uncommon words. Hence, the document ranking of a query such as **Eiffel** OR **Tower** will be dominated by documents that contain "tower" rather than "Eiffel." (3) On average, long documents are more likely than short documents to contain a given term. Hence, the counting procedure favors long documents, even though they may be less important to the query.

The first shortcoming can be ameliorated by recording the frequencies of terms, rather than mere presence. This is easily accomplished if the inverted file contains frequency information. The document vectors then consist of various integer components, rather than just ones and zeros.

The issue of common versus uncommon words has several facets. First, it is clear that common words will on average appear more frequently in documents and hence tend to skew rankings in favor of these terms, but this effect can be offset by applying weighting factors to terms. A common term such as *the* is given a small weight, while a term such as *tower* is given a larger weight. In practice, weights are usually based on the overall frequency of appearance in a collection. For example, if f_t is the number of documents that contain the term t and N is the total number of documents, then a reasonable weighting for t is

$$w_t = \ln\left(1 + \frac{f_t}{N}\right),$$

which increases with the frequency, but this increase flattens out for large f_t.

A second consideration relates to the intent of the query. Some words may be more important than others. For example, after an adjustment for frequency, the term *Eiffel* may be more important than *tower* in the query. This intent can be incorporated by allowing the query vector to have term weights other than zero or one.

Finally, the fact that long documents are likely to produce more total weight than short documents can be attenuated by normalizing the vectors. The most common normalization uses Euclidean length defined for a vector $x = (x_1, x_2, \ldots, x_T)$ as

$$||x|| = \sqrt{\sum_{t=1}^{T} x_t^2}.$$

The inner product of two vectors x and y is normalized by computing what is essentially the cosine of the angle between them, as

$$\text{cosine } \theta = \frac{x \cdot y}{||x|| \cdot ||y||},$$

which achieves a maximum value of 1 when x and y are proportional (that is, when $x = \alpha y$ for some $\alpha > 0$).

In a query system using weights, a fixed document d has an associated vector w_d with components w_{dt} for each term t. Likewise, a specific query q has an associated vector v_q with components v_{qt} for each term t. The cosine value of document d relative to query q is

$$\text{cosine } \theta = \frac{\sum_t v_{qt} w_{dt}}{||v_q|| \cdot ||w_d||}.$$

Of course, most of the components v_{qt} in the query vector v_q are zero, so the sum actually contains far fewer nonzero components than the general expression might indicate. The value of $||w_d||$ is independent of the query, and hence can be precomputed. The weights w_{dt} are found directly from the inverted file provided that the frequencies f_{dt} have been recorded.

The cosine ranking method forms the foundation for a fairly comprehensive ranking method. Other features can be added. For example, the document vectors can be expanded to record whether a given term appears in the title of the document.

17.6 Network Rankings

When documents contain links to other documents, as on the Internet, it is possible to use the structure of those links to formulate rankings based on popularity. If a document is referred to by many other documents, it is likely to be important and thus likely to be a good response to a query. This type of ranking mechanism was originally developed as PageRank[2] for the search engine Google.

One way to envision the ranking is to imagine a little program crawling through the web. It goes from document to document in a semirandom manner. Specifically,

---

[2] The "Page" in PageRank refers to Larry Page, one of the developers.

when at any given document, there is a fixed probability p ($p = 85$ percent has been suggested) that it will next follow one of the links originating at that document, choosing from among those available with equal probabilities. Alternatively, with probability $1 - p$ it jumps to an arbitrary document, again making the choice with equal probabilities.

By watching the behavior of this little program, one can record the relative frequencies with which it visits various documents. A document that has a lot of links pointing to it is likely to be visited relatively often, provided that the documents pointing to it are themselves visited often. The ranking of a document is defined as its relative visit frequency.

This procedure can be formalized in terms of the link structure of the network. Suppose that there are N documents, and suppose that document i contains a link to document j. Then define $a_{ij} = 1/n_i$, where n_i is the total number of links originating from document i. Otherwise, if there is no link from i to j, let $a_{ij} = 0$. Thus a_{ij} is the normalized link value. The sum of all link values a_{ij} from document i is 1, provided that there is at least one link (which for simplicity we assume). The matrix A is defined as the $N \times N$ matrix having entries a_{ij}.

Notice that a_{ij} can be interpreted as the probability that the crawling program will jump to document j if it arrives at document i and must follow a link. However, the crawler only follows some link with probability p; otherwise it jumps randomly. The probability that the crawler arrives at a node by jumping is $(1-p)/N$. Hence the total probability of the crawler moving to document j once it is on document i is $b_{ij} = (1-p)/N + pa_{ij}$. In matrix form the overall probability matrix is

$$B = (1 - p)\mathrm{E}/N + pA,$$

where E is the $N \times N$ matrix with all components equal to one.

Suppose a crawler starts at web page 1. Future positions will occur with various probabilities, expressed as a vector with N components. Initially this vector is $(r^0)^T = (1, 0, 0, \ldots, 0)^T$, indicating that it is with probability 1 at location 1 and with probability 0 at all other locations. For one move ahead, the probabilities change to $(r^1)^T = (b_{11}, b_{12}, b_{13}, \ldots, b_{1N})^T$. In general, if r^k is the probability vector (written as a column) for step k, then

$$r^{k+1} = B^T r^k, \tag{17.3}$$

where B^T is the transpose of the matrix B.

In equilibrium, after many, many steps, the probability vector will converge, so that $r^{k+1} = r^k$ in equation (17.3). The equilibrium probability vector will therefore satisfy

$$r = B^T r, \tag{17.4}$$

where the vector r is normalized with $\sum_{i=1}^{N} r_i = 1$.

The vector r is an **eigenvector** of the matrix B^T. It can be found computationally by iteration, or as in practice by actually carrying out the crawling procedure. For application to web searching the implied matrix B is huge, with the number of documents (web pages) N on the order of billions.

The popularity ranking of a particular site is its relative probability in the equilibrium vector r. An overall ranking is formed as a combination of the probability ranking and other factors such as key words and term frequency counts.

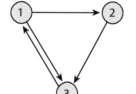

Example 17.1 (Three documents). Suppose that three documents have the link structure shown in figure 17.8. The corresponding link matrix is

$$A = \begin{bmatrix} 0 & \frac{1}{2} & \frac{1}{2} \\ 0 & 0 & 1 \\ 1 & 0 & 0 \end{bmatrix}.$$

FIGURE 17.8 A linked collection of documents.

Suppose p is set as $p = 0.7$. Then $(1 - p)/N = 0.1$ and the eigenvector equation is the matrix equation

$$\begin{bmatrix} .1 & .45 & .45 \\ .1 & .1 & .8 \\ .8 & .1 & .1 \end{bmatrix}^T \begin{bmatrix} r_1 \\ r_2 \\ r_3 \end{bmatrix} = \begin{bmatrix} r_1 \\ r_2 \\ r_3 \end{bmatrix}.$$

This equation can be solved together with the normalization condition $r_1 + r_2 + r_3 = 1$, yielding

$$r_1 = .375$$
$$r_2 = .231$$
$$r_3 = .393.$$

Document 3 has the highest rank, which is consistent with the fact that both other documents have links to it. Document 1 is next highest in rank since it has a link from the most highly ranked document.

17.7 EXERCISES

1. (An index) A scan through nine documents, searching for English letters, produced the table below:

1	F	K	L	A
2	B	L	C	F
3	A	F	L	
4	L	A		
5	K	C	F	H
6	C	A	L	
7	F			
8	K	F	A	B
9	A	H	K	

Make an alphabetical index for the terms by the following methods:
   (a) Construction of an incidence matrix followed by construction of the inverted file.
   (b) A sort and merge procedure.

2. (Golomb examples) Construct the Golomb codes for the following numbers using the indicated value of b: $(28, b = 2), (203, b = 32)$.

3. (Gamma codes*) Let x be a positive integer. The γ code for x consists of a unary code for $1 + \lfloor \log x \rfloor$ followed by the binary code for $x - 2^{\lfloor \log x \rfloor}$. Essentially, the first part of the γ code tells how many bits are required to code x in binary, and the second part is the binary code for x with all bits preceding the second 1 omitted. For example, the γ code for 13 is 1110 followed by 101, indicating the binary code 1101 of length 4.
   (a) Find the γ code for 20.
   (b) The length of the gamma code is approximately $1 + 2 \log x$ bits. Find the probability distribution of x that would make the γ code the one minimizing average length.

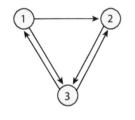

FIGURE   17.9 A   link structure for exercise 4.

4. Find the PageRank for the link structure shown in figure 17.9 using $p = .7$.

17.8 Bibliography

An excellent text on information retrieval is [1]. Another good reference is [2]. The Golomb code is presented in [3]. See [4] for a detailed analysis of the properties of the method. PageRank is described in [5]. There have been several other methods proposed for ranking based on links. One is described in [6].

References

[1] Witten, Ian H., Alistair Moffat, and Timothy C. Bell. *Managing Gigabytes*. 2nd ed. San Francisco: Morgan Kaufmann, 1999.

[2] Baeza-Yates, R., and B. Ribeiro-Neto. *Modern Information Retrieval*. Harlow, Eng.: Addison-Wesley, 1999.

[3] Golomb, S. W. "Run-Length Encodings." *IEEE Transactions on Information Theory* 12 (1966) 399–401.

[4] Gallager, R. G., and D. C. Van Voorhis. "Optimal Source Codes for Geometrically Distributed Integer Alphabets." *IEEE Transactions on Information Theory* 21, (1975): 228–30.

[5] Brin, S., and L. Page. "The Anatomy of a Large-Scale Hypertextual Web Search Engine." In *Proceedings of the 7th International World Wide Web Conference.* Amsterdam: Elsevier Science, 1998, pp. 107–17.

[6] Kleinberg, J. M. "Authoritative Sources in a Hyperlinked Environment." *Journal of the ACM* 46 (1999): 604–32.

DATA MINING

18

Organization or structuring of data can be considered an information operation, because a mass of completely random data has apparent high entropy, but when that data is organized or structured, its entropy is reduced. The associated decrease in entropy represents the informational effect of the structure. Effective data analysis decreases raw entropy, but the resulting information is more useful.

Data analysis is an enormous topic, consisting of dozens of basic techniques and procedures often individually tailored to one of a vast assortment of application areas. The popular catch phrase **data mining** loosely refers to methods of analysis that are more or less automatic, often computationally intensive, and designed for application to large banks of data. This focus on automatic methods has inspired researchers to carefully delineate and characterize proposed methods so as to unify several *ad hoc* techniques. However, it must be emphasized that the greatest success is achieved, even with automatic methods, when human insight contributes significantly to preliminary structuring, analysis, and choice of method. It is not enough to be expert in the technical aspects of data manipulation. One must also possess an understanding of the science and detail of the particular application area.

That having been said, this chapter does focus primarily on standard techniques of data mining, with the understanding that human intelligence and application understanding must be incorporated at every step.

18.1 Overview of Techniques

Data mining consists basically of three components: (1) the underlying data, (2) the questions that are posed with respect to the data, and (3) the types of functions or other mathematical structures used to answer the questions. We discuss these in turn:

1. The data. The data associated with a data-mining project can be considered to consist of **records**, each listing attribute values in various **fields**.

301

(See chapter 16.) A set of records may be symbolized by $R(A_1, A_2, \ldots, A_n)$ where each A_i represents an attribute field that can take on various values. A particular set of N samples has the form

$$R_1 = (a_{11}, a_{12}, a_{13}, \ldots, a_{1n})$$
$$R_2 = (a_{21}, a_{22}, a_{23}, \ldots, a_{2n}$$
$$\vdots \quad \vdots$$
$$R_N = (a_{N1}, a_{N2}, a_{N3}, \ldots, a_{Nn}),$$

where a_{ij} is a particular instance of the attribute A_i. In general, attributes may be numeric, categorical, alphanumeric, graphical, or any other data type that is defined consistently. The data therefore defines a **relation** as explained in chapter 16.[1]

Because of the diverse nature of data and methods of analysis, the notation for attributes varies in different applications and methods. When data are numerical, it is common to use X's, Y's, or Z's for attribute labels. A few specific attributes of categorical data may be labeled A, B, C, etc. And in specific examples is it clearest to refer to attributes by indicative names such as **Apples**, **Bananas**, and **Eggplant** or **Heart rate**, **Blood pressure**, and **Age**.

2. Questions. Data analysis attempts to discover relations, to classify, to simplify, to estimate, and to predict. Sometimes one expects unequivocal answers, such as how many instances of a particular combination of attribute values are contained in the data, but generally one settles for imprecise answers or answers in probabilistic form (such as whether it is likely that a certain value of A_1 implies a given value of A_2).

Here is a brief description of some of the most common types of questions that drive analysis:

(a) Estimation. Sometimes it is believed that the data is indirectly related to some underlying value x that one wishes to estimate. For example, suppose N measurements of the boiling point of water have been made with data attribute **Thermometer** from which an estimate of the true value of temperature is to be estimated. The estimation procedure is defined by a function f mapping the data into the estimate of x.

(b) Relation discovery. Often one seeks to discover relations that give insight or provide a means for simplifying or organizing a collection of data. A suitable relation is a function f of the attributes of the form $f(A_1, A_2, \ldots, A_n) = 0$. This may take a logical form, such as the following: If $A_1 = 3$, then $A_2 = 7$, which is expressed more simply as $(A_1 = 3) \Rightarrow (A_2 = 7)$. But such relations may hold only imperfectly. For example, in a collection of supermarket purchases it may be found that customers who purchase beer tend to purchase pretzels as well, but it is not a definite rule.

---

[1] Familiarity with methods explored in chapter 16 is not essential to the reading of this chapter.

(c) Classification. Sometimes one special attribute serves to classify data, in which case the possible values for that attribute are termed **classes** or **categories**. For example, in a database of patients' medical records, each record listing symptoms and medical history, one attribute may be **Heart disease**, having only the two possible values "yes" and "no," thereby classifying the patients as those with heart disease and those without. When a new patient who has not yet been diagnosed arrives, the other attributes (of symptoms and family history) may be used to roughly predict the heart disease classification.

Classifications derived from a finite sample of data are often imperfect and, similar to rules, are regarded as holding only with a certain level of confidence. Some classification techniques explicitly assign probabilities to categories.

(d) Clustering. Clustering is similar to classification except that the classes are not prespecified; that is, there is no designated classification attribute. Instead, clustering seeks to define classes by searching for disjoint (or nearly disjoint) clusters of data points in a multidimensional representation. Once these clusters are defined, the existing data can be interpreted and new data points can be categorized. For example, in a database of student grades, it may be found that one cluster of students does well in humanities and another cluster does well in science.

(e) Prediction. Most of the techniques discussed above lead to the possibility of predicting an event or condition. For example, once the parameter of a physical law is estimated, that value can be inserted into the law to predict outcomes and relations. Hence, if the velocity of an oncoming missile has been estimated, that velocity estimate can be used to predict the missile's future course. Or, the likely sales price of a particular house can be predicted from its age and square footage if a general relation between these variables has been determined from past sales data. Classification and clustering are often motivated by their potential for prediction.

3. Functions and structures. A wide assortment of functions and mathematical structures are used in data mining. Often they are interwoven within complex algorithms, but at root they draw on familiar constructs. They include linear and nonlinear functions, logical expressions and relations, tree structures, linear inequalities, probabilistic relations, and optimization algorithms. This chapter uses a sampling of these techniques with the objective of indicating the broad range of possibilities.

18.2 Market Basket Analysis

One of the simplest techniques of data mining, termed **market basket analysis**, is designed to analyze records where items either occur or do not. The standard example, responsible for the name of the method, is to the analysis of grocery purchases, where the records are customer purchases of various grocery items. The purpose of the analysis is to discover logical relations among attributes $A, B, C, \ldots$, such as $(A = \text{yes})$ AND $(B = \text{yes}) \Rightarrow (C = \text{yes})$. It may be found, for instance, that if a customer purchases apples (A), he or she is likely to purchase bananas (B) as well, leading to the rule $(A = \text{yes}) \Rightarrow (B = \text{yes})$. Such relations hold only approximately, in a percentage of cases, but in a higher percentage than would be expected by pure chance. Many such relations (such as the apples and bananas relation) may be obvious, but other less obvious and useful ones may be discovered. In some stores it was found that on Thursdays, beer and diapers were often purchased together. In hindsight it was deduced that shoppers were preparing for a weekend of undisturbed television. As a result of this discovery the store found it advantageous to place its own highly profitable brand of diapers near the aisle where beer was located.

In market basket analysis the attributes are typically items (apples, bananas, etc.) and the attribute values are binary "yes" or "no" or perhaps "true" or "false," and hence it is convenient to simply write A and $\overline{A}$ to stand for "A is yes" and "A is no," respectively. Then $p(A)$ denotes the probability that A occurs, and $p(A, B)$ denotes the probability that A and B occur together. The probability of joint events can be converted to conditional probabilities through Bayes' rule,

$$p(B|A) = \frac{p(A, B)}{p(A)}.$$

If this conditional probability is high, one might propose the rule $A \Rightarrow B$.

The notions of high probability and usefulness of a rule are formalized by the introduction of three quantities.

1. The **support** of a combination is the percentage of records (e.g. transactions) in which this combination occurs. For example, the support of the combination A, B is the number of times A and B occur together divided by the total number of records. In terms of probability, the support is an estimate of the probability $p(A, B)$.

2. The **confidence** of a rule $A \Rightarrow B$ is the number of times the pair (A, B) occurs divided by the number of times A occurs. In probability terms the confidence is an estimate of the conditional probability $p(B|A)$.

3. The **lift** of a rule $A \Rightarrow B$ measures how the proportion of B in a population changes when the population is restricted to samples in which A occurs. Formally,

$$\text{lift} = \frac{p(B|A)}{p(B)} = \frac{p(A, B)}{p(A)p(B)}.$$

This can be interpreted as the correlation between A and B. Lift greater than 1 is considered significant.

The *A Priori* Algorithm

Searching for pairs, triples, and general tuples of items that have significantly high support values can require astronomical levels of computation when the number of items is large. The *a priori* **algorithm** can significantly reduce that burden.

The algorithm progresses one level at a time, from single-item counts, to pairs, to triples, and so forth. The algorithm is based on the way support values propagate from one level to the next. In the algorithm, support is measured in absolute terms. That is, it is a count of the actual number of occurrences of multiple items, not a percentage. The support s of (A, B) is the actual number of occurrences of the pair (A, B).

The key observation is that if a tuple of length k is to have support of amount s, then each sub-tuple of length $k - 1$ must also have support of level s. For instance, if the triple (A, B, C) has support s, then the pairs (A, B), (B, C), and (A, C) must also have support s, for if one of these pairs occurs less than s times, then certainly (A, B, C) must occur less than s times.

Accordingly, the algorithm progresses through successive levels, keeping only those tuples having support of at least s for consideration as components of tuples at the next level.

Example 18.1 (Five items). A small sample of market basket data is shown in the left part of figure 18.1, with items $A =$ apple, $B =$ banana, $C =$ cantaloupe, $D =$ dates, $E =$ eggplant. The basic count of single items is shown to the right. (Only items that appear are shown; that is, the appearance of A means "$A =$ yes.")

Suppose that a threshold support count of 2 is established. Then all items pass through the first support test, since each count is at least 2.

The next step is to form all pairs of items that passed the first test, and count the occurrences of the pairs in the original data. The result is shown in the top left-hand corner of figure 18.2. Some of the pairs occur less than twice, and hence they can be pruned for the next phase. A list of the surviving pairs and their counts is shown on the right side of the figure.

Next, to evaluate triples, only those triples are considered that have the property that every contained pair is included in the previously pruned list of pairs. For example, the triple (A, B, C) is included because each of the pairs (A, B), (B, C), and (A, C) is in the pruned list. The triple (A, B, D) is not included because, although (A, B) and (A, D) are in the pruned list, (B, D) is not. The list of acceptable triples is shown at the bottom of figure 18.2. In general, this list would be pruned to include only those

Customer	Items
1	A, B
2	A, C, D
3	B, C
4	A, C, E
5	A, B, C
6	C, D
7	B, C, D, E
8	A, B, D
9	C, D, E
10	A, B, C

Items	Count
A	6
B	6
C	8
D	5
E	3

FIGURE 18.1 Market basket data. This is a record of the fruits and vegetables purchased by 10 customers. From these transactions, it is possible to infer some tentative rules. A preliminary step of the *a priori* algorithm is to count the number of times each item occurs.

Pairs	Count
A, B	4
A, C	3
A, D	2
A, E	1
B, C	4
B, D	1
B, E	1
C, D	4
C, E	3
D, E	2

Surviving Pairs	Count
A, B	4
A, C	3
A, D	2
B, C	4
C, D	4
C, E	3
D, E	2

Triples	Count
A, B, C	2
C, D, E	2

FIGURE 18.2 Market basket analysis. As a second step of the *a priori* algorithm, the number of occurrences of the pairs is determined. Those pairs that do not have adequate support (2 in this example) are pruned from the list, leaving the survivors shown the right. The next step examines triples, keeping those whose component pairs all appear in the previous pruned list.

triples that survive the support test, but in this case all these triples do. We conclude that the triples (A, B, C) and (C, D, E) have significant support.

It is possible to extract some rules from this data. From the pairs we can deduce, among others, the rules shown below:

Rule	Support	Confidence	Lift
$A \Rightarrow B$	40%	2/3	1.11
$C \Rightarrow D$	40%	1/2	1.0
$C \Rightarrow E$	30%	3/8	5/4

It might be concluded that $A \Rightarrow B$, which has good support and reasonable lift, is a good useful rule. The rule $C \Rightarrow E$ has better lift, so even though its support is modest, it is an interesting rule.

It is possible to extract rules such as $A, B \Rightarrow C$ from the triples. The lift of (C, D, E) is 1.67, and the lift of (A, B, C) is .694, but the support of each is only .2. Hence the associated rules such as $A \Rightarrow B, C$ are likely to be of little interest. However, in situations with more data, several interesting rules may be discovered.

18.3 Least-Squares Approximation

Probably the oldest and most widely used method of data analysis is least-squares approximation. In the simplest version of the method, the value of one attribute Y is approximated by a straight-line function of another attribute X, in the form

$$Y \approx a + bX.$$

The approximation is defined by the two constants a and b. As an example, the **height** Y of men might be approximated as a function of their **weight** X.

The approximation is based on data consisting of several actual pairs of x's and y's, which can be labeled as $(x_1, y_1), (x_2, y_2), \ldots, (x_N, y_N)$.

For any a and b the straight-line approximation will likely not match the data perfectly. The error associated with any data point (x_i, y_i) is $\varepsilon_i = y_i - (ax_i + b)$. In least-squares approximation, a and b are chosen so as to minimize the total sum of squared errors $E = \sum_{i=1}^{N} \varepsilon_i^2$. That is, the least-squares approximation solves

$$\min_{a,b} E = \min_{a,b} \sum_{i=1}^{N} (y_i - a - bx_i)^2. \tag{18.1}$$

An example is shown in figure 18.3, where the heights of 10 men are plotted versus their weights. We seek an estimate of the form **height** $= a + b\,$**weight**.

The best coefficients a and b can be found by setting the derivatives of equation (18.1) with respect to a and b equal to zero. It is algebraically neater to make a slight change in the parameters and write the approximations as

$$y_i = \alpha + \beta(x_i - \bar{x}), \quad i = 1, 2, \ldots N,$$

where $\bar{x} = \frac{1}{N} \sum_{i=1}^{N} x_i$ is the average of the x_i's. In this formulation the problem is

$$\min_{\alpha,\beta} \sum_{i=1}^{N} [y_i - \alpha - \beta(x_i - \bar{x})]^2.$$

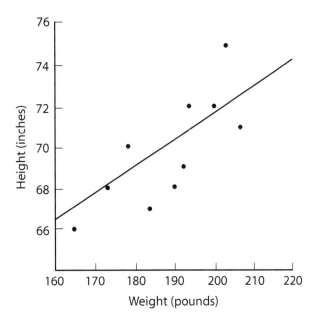

FIGURE 18.3 **Height versus weight.** A least-squares straight line is a standard method for approximating data. Here the straight line is **height** $= 50.7 + 0.1 \times$ **weight**.

The two equations that define the minimum are

$$0 = \frac{\partial E}{\partial \alpha} = -2 \sum_{i=1}^{N} [y_i - \alpha - \beta(x_i - \bar{x})]$$

$$0 = \frac{\partial E}{\partial \beta} = -2 \sum_{i=1}^{N} [y_i - \alpha - \beta(x_i - \bar{x})](x_i - \bar{x}).$$

If we recognize that $\sum_{i=1}^{N} (x_i - \bar{x}) = 0$, these equations can be solved separately, obtaining

$$\alpha = \frac{1}{N} \sum_{i=1}^{N} y_i$$

$$\beta = \frac{\sum_{i=1}^{N} y_i(x_i - \bar{x})}{\sum_{i=1}^{N} (x_i - \bar{x})^2}.$$

A useful measure of the closeness of approximation is the square root of the average squared error, which is, loosely, called the **standard deviation** since it is analogous to the standard deviation of a random variable. Formally, this measure is

$$s = \sqrt{\sum_{i=1}^{N} \varepsilon_i^2 / N}.$$

For the height–weight example, $s = 1.37$, which translates into an average approximation error of roughly 1.37 inches.

The method of least squares can be extended to include nonlinear functions of the independent variable. For example, approximations may take any of the following forms:

$$y \approx a + bx + cx^2 + dx^3$$
$$y \approx a + b \sin x + cx^2 + d \cos x$$
$$y \approx a \ln x + c \arctan x.$$

The method can also be extended to the case of more than one independent variable. For example, the attribute Z may be approximated by X and Y with the linear relation

$$Z = a + bX + cY.$$

Example 18.2 (Housing prices). In a certain neighborhood recent housing sales prices together with the ages and square footages of the houses have been recorded and are listed in the top table in figure 18.4. In order to visualize the data the price range is partitioned into the four intervals [0, 200], [201, 300], [301, 400], [401 and more], each with a different assigned icon. The houses with their icons are plotted on an age–square footage graph in the lower part of figure 18.4.

AGE	17	35	30	44	39	11	36	37	26	4	34	28
SQ.FT	36	32	17	23	19	12	17	11	40	25	22	22
PRICE.	474	397	222	278	285	210	195	140	504	377	334	312

AGE	10	33	26	27	31	26	6	19	38	34	29	39	17
SQ.FT	17	18	22	26	24	26	17	19	20	12	18	35	36
PRICE.	224	235	295	362	292	395	253	333	298	179	281	414	474

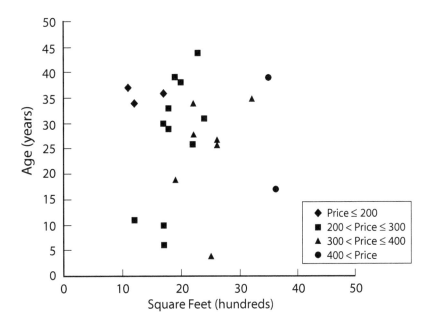

FIGURE 18.4 Housing price data. Prices are in thousands of dollars, square footage is in hundreds, and age is in years. The 25 data records are displayed visually using price intervals in the graph under the data.

The original data can be approximated by a relation giving price as a linear function of age and square feet. The coefficients that minimize the total squared error can be easily found with a simple spreadsheet program (by solving three linear equations or by directly minimizing the total squared error). The resulting approximation is

$$\text{Price} = 89 - 1.4 \times \text{Age} + 11.5 \times \text{Sq. feet}.$$

This result is in accord with intuition and even casual inspection of figure 18.4, which indicates that, roughly, prices decrease with age and increase with square footage.

The total sum of squares is 15,066. This translates to an average standard deviation of error in price of 24.5 thousand dollars, which is pretty good accuracy considering that house prices are in the hundreds of thousands of dollars.

18.4 Classification Trees

Suppose $A_1, A_2, \ldots, A_n$ are attributes of data records and there is one additional attribute, labeled C, that is a categorical class attribute. A classification scheme is a mapping from $A_1, A_2, \ldots, A_n$ to C. Classification rules for such data are designed by use of a **training set** of samples, in which all $n + 1$ attributes are available. Once the classification rule is established, it can be used to classify new data for which only the n independent attributes $A_1, A_2, \ldots, A_n$ are available.

Classification schemes are used in a wide variety of application areas. The attributes may be credit score, age, income, and debt figures for loan applicants, with the associated classification being high or low default risk. The attributes may be symptoms and medical test results, with the classification being diseased (or not). The attributes may be opinions about public policies, with the classification being political party disposition. Or the attributes may be pixel patterns on a screen derived from a handwriting sample, with the classification being the letter or number written.

One of the most popular methods for constructing the mapping from attributes to classification is with **classification trees**, also called **decision trees**. These trees are used in a manner similar to those used with binary search trees, discussed in chapter 15.

Classification trees are applied to situations in which there are a finite number of classes. The tree is constructed by beginning at a top **root node** that includes all possible combinations of attribute values. Then a special attribute is selected for that node and its possible values are partitioned into two or more subsets in preparation for **splitting** the node to improve class separation. Each of the subsets defines a lower-level node. All of the new nodes so created are **children** of the root.

Each child node is split using a special attribute for that node, leading to second-generation children and so forth. The process continues until a stopping criterion is satisfied. The final nodes, called **leaf nodes**, are labeled with the classification that best fits the attribute values that survive to that node.

An example of a simple hypothetical classification tree is shown in figure 18.5. There are two underlying attributes $T = $ **total cholesterol** (which is bad) and

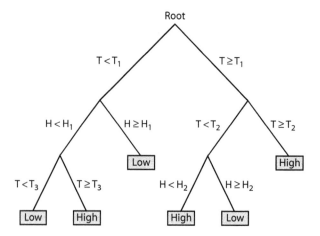

FIGURE 18.5 A (hypothetical) decision tree for heart disease. The splits are defined by critical values T_1, T_2, T_3, H_1, H_2.

H = **high density lipids** (which is good). There are fixed splitting values $T_1 >$ $T_2 > T_3$ and $H_1 > H_2$. The root node is split using the attribute T at the value T_1. At the next level there are two nodes, and one happens to be split using H at the value H_1, while the other again uses T. The leaf nodes are indicated by the shaded box and contain the classification value of **Low** or **High** risk of heart disease. When someone's blood is analyzed, the T and H values are used to progress down the tree to determine the associated risk. For example, $T < T_1$ and $H > H_1$ implies **Low** risk.

It is possible to construct a tree such that there are enough leaf nodes to cover practically every possible combination of attributes. For instance, if each attribute is categorical with k values and there are n independent attributes, a tree with k^n leaf nodes could represent every combination of independent attributes. However, there may be only a few test points in some of these nodes, implying that the tree may be an unreliable classifier of new data. Good classification trees have strong class separation with relatively few splits based on a large number of samples.

Node Splitting*

The quality of a classification tree is influenced by the procedure used to split nodes. This procedure must specify the attribute on which to base the split and how to partition that attribute's values.

The objective of a split is to reduce the diversity of the classes within the nodes. Equivalently, the objective is to make each node have a dominant class. For instance in cholesterol screening, splits of the attributes T or H are determined from a sample population for which the T and H values and the classification as low or high risk are known for each person in the sample. A node split is designed so that the two children nodes have higher proportions of low- or high-risk individuals, respectively, than the parent node. To formalize this, suppose there are K possible classes. The diversity of a node is characterized by the proportions $p_1, p_2, \ldots, p_K$ of the different classes in the sample population at that node. The objective is to minimize diversity as measured by a specific function $D(C)$ of these class proportions. Various functions are used for $D(C)$, all of which favor extreme probabilities (0 or 1). Some examples are

1. $D(C) = D(p_1, p_2, \ldots, p_K) = p_1 p_2 \cdots p_K$
2. $D(C) = D(p_1, p_2, \ldots, p_K) = -\max[p_1, p_2, \ldots, p_K]$
3. $D(C) = H(C) = -\sum_{i=1}^{K} p_i \log p_i$

Treating the observed proportions as probabilities, the last of these $D(C)$'s is entropy. With this choice the objective is to reduce entropy.

A diversity function is used by first computing the diversity of the node that is to be split. This is $D(C)$. Suppose the node is split along a certain attribute into J children. Define $p_{i|j}$ as the resulting proportion of class i in the j-th child node. The diversity of the j-th child is accordingly

$$D_j = D(C_j) = D(p_{1|j}, p_{2|j}, \ldots p_{K|j}).$$

The overall diversity is then the weighted average of these; namely,

$$D_{\text{new}} = \sum_{j=1}^{J} q_j D(C_j),$$

where q_j is the ratio of the number of samples in the j-th child to the number of samples in the original parent node.

The ideal splitting strategy selects the split, among those under consideration, that has the largest decrease in the diversity measure being employed. This can be computationally intensive since the effect of many different split possibilities must be worked out and compared. When the independent attributes are categorical, splits can be considered for every possible value of every attribute. When the attributes are continuous, as for example cholesterol levels, a finite number of split points must be proposed.

Proper termination of the process is important as well. A simple criterion is to terminate when no split reduces the diversity. But in practice it is advisable to terminate before the number of samples at a node becomes small. There are a number of procedures for termination, including the possibility of pruning nodes after a tentative tree has been constructed.

Regression Trees

A regression tree operates much like a classification tree except that the attribute to be predicted is continuous-valued rather than categorical. To treat the continuous variable, each leaf node is characterized by the mean value and variance of the sample values it contains, rather than simply a class designation.

The node-splitting procedure for regression trees seeks at each stage to minimize the total sum of squared errors (the total sample variance), in the prediction variable, among the child nodes. A perfect split produces children that have a variety of attribute combinations, but the same value for its predicted variable. The squared error (variance) for each such node would ideally be zero, and hence the total squared error would be zero.

Example 18.3 (The housing data). Let us construct a regression tree for the example of house sales, consisting of records of house price, age, and square feet studied in example 18.2. The data are shown again in the first four columns of table 18.1 but with the data sorted with respect to square feet.

To the right of the data are the running values of means and total squared error down from top to bottom and up from bottom to top. That is, record k in the list for "squares down" is the total of the sum of squares of price deviation from the mean, looking only at the first k records. Mathematically, it is $S_D(k) = \sum_{i=1}^{k} (p_i - \bar{p}_k^D)^2$, where $\bar{p}_k^D = \sum_{j=1}^{k} p_j / k$. The number for "squares up" is computed looking at only records from 25 up to k. The final column gives the sum of $S_D(k) + S_U(k+1)$, which is the sum of squares that would be obtained if a split were made at k, producing the two subsets $\{i \le k\}$ and $\{i > k\}$.

The prices themselves have a mean value of \$311.32 (thousands) with a total sum of square deviations (or equivalently, sum of variances) of 218,739.

By inspection, the best split is at $k = 17$, with square feet = 24, for a total sum of square deviations from the mean of 67,704 divided into 48,584 and 19,120. The mean values of the two subsets are $\bar{p} = 258$ and $\bar{p} = 425$, respectively. It turns out that this split is better than any split with the age variable, so this is the one to keep.

TABLE 18.1
Housing Data and Preparation for Split.

k	Price	Age	Square Feet	Mean Down	Squares Down	Mean Up	Squares Up	Sum of Vars.
1	140	37	11	140.00	0	311.32	218,739	188,166
2	210	11	12	175.00	2,450	318.46	188,166	178,341
3	179	34	12	176.33	2,461	323.17	175,891	156,621
4	253	6	17	195.50	6,869	329.73	154,160	154,862
5	244	10	17	205.20	8,751	333.38	147,993	148,355
6	222	30	17	208.00	8,986	337.85	139,605	134,463
7	195	36	17	206.14	9,131	343.95	125,477	111,190
8	281	29	18	215.50	14,034	352.22	102,059	110,722
9	235	33	18	217.67	14,372	356.41	96,688	95,398
10	333	19	19	229.20	26,344	364.00	81,026	106,345
11	285	39	19	234.27	29,174	366.07	80,001	102,134
12	298	38	20	239.58	32,897	371.86	72,960	99,982
13	295	26	22	243.85	35,732	377.54	67,085	95,437
14	312	28	22	248.71	40,045	384.42	59,705	94,029
15	334	34	22	254.40	46,834	391.00	53,984	97,244
16	278	44	23	255.88	47,356	396.70	50,410	82,111
17	292	31	24	258.00	48,584	409.89	34,755	67,704
18	377	4	25	264.81	61,958	424.63	19,120	78,488
19	395	26	26	271.47	78,065	431.43	16,528	93,044
20	362	27	26	276.00	85,850	437.50	14,980	93,989
21	397	35	32	281.76	99,794	452.60	8,139	104,069
22	414	39	35	287.77	116,486	466.50	4,275	117,086
23	474	17	36	295.87	149,659	484.00	600	150,109
24	474	17	36	303.29	180,067	489.00	450	180,067
25	504	26	40	311.32	218,739	504.00	0	218,739

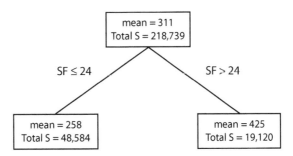

FIGURE 18.6 Start of house tree. A single split on square footage gives two nodes with different mean prices and different squared errors.

The single split into two parts (giving 67,704) has greatly reduced this sum of squares (from 218,739). The corresponding tree is shown in figure 18.6.

The same procedure is next applied to each of the two child nodes. The data for each are sorted according to the attribute considered as a basis for the split. It turns out that in this case both nodes should be split along square footage. This process is

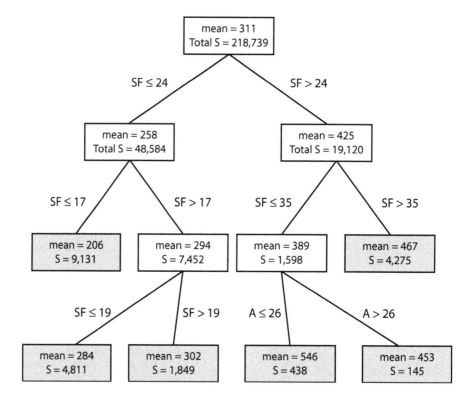

FIGURE 18.7 House tree. The leaf nodes are shaded, while intermediate nodes are clear. The mean prices and squared error of the data points included in a node are shown. The total squared error is the sum of the squared errors *S* of the leaf nodes. In this case the total sum of squared deviations is 20,649, which is considerably less than the 218,739 of the original.

continued, but a proposed split is rejected if one of the subsets contains only a single point.

The tree obtained after a number of splits is shown in figure 18.7. The leaf nodes are shaded, while the intermediate nodes are clear. The total sum of squared errors for the tree is the sum for the leaf nodes; namely: $9,131+4,811+1,849+438+145+4,275 = 20,649$, which is a huge reduction from the 227,854 of the original and from the 67,704 resulting from the first split. However, the result is not as good as the 15,066 obtained by the simple least-squares analysis of section 18.3. The data closely fits a linear model in this case. In other cases the regression tree may turn out to be superior.

18.5 Bayesian Methods

Classifiers are rarely perfect. Bayesian methods recognize this inherent imperfection and produce category probabilities rather than single category statements. There are two main types of Bayes' classifiers: **Naive Bayes classifiers** and **Bayesian belief networks**.

Naive Bayes

Suppose that there are attributes $(X_1, X_2, \ldots, X_n)$ and another attribute C termed the class with specific classes $C_1, C_2, \ldots, C_k$. The Naive Bayes method assumes that there is a conditional probability structure of the form

$$p(C_j | X_1, X_2, \ldots, X_n)$$

giving the probabilities of the various classes as a function of the other attributes. If it is desired to assign a specific class to a new observed attribute vector $(x_1, x_2, \ldots, x_n)$, the C_j with the greatest probability is selected. That is, C_j is chosen to maximize

$$p(C_j | x_1, x_2, \ldots, x_n)$$

with respect to the C_j's.

Unfortunately, it is not easy to evaluate the conditional probabilities

$$p(C_j | x_1, x_2, \ldots, x_n)$$

from data because although there may be only a few classes and several data samples, there are likely to be only a few samples with any particular attribute pattern $(x_1, x_2, \ldots, x_n)$ to average in order to estimate probabilities.

Instead, one uses Bayes' theorem to write

$$p(C_j | x_1, x_2, \ldots, x_n) = \frac{p(x_1, x_2, \ldots, x_n | C_j) p(C_j)}{p(x_1, x_2, \ldots, x_n)}.$$

Since $p(x_1, x_2, \ldots, x_n)$ is constant, independent of C_j, for purposes of maximization it is sufficient to maximize the numerator

$$p(x_1, x_2, \ldots, x_n | C_j) p(C_j)$$

with respect to C_j.

The probabilities $p(C_j)$ can be estimated from the data as

$$p(C_j) = N_j / N,$$

where N_j is the number of cases in which C_j appears in the N data points.

Greater difficulty arises when attempting to estimate the conditional probabilities $p(x_1, x_2, \ldots, x_n | C_j)$, and this is where the naive aspect of the naive Bayes method is employed. It is assumed (naively, for sake of computational simplicity) that the attributes are **class conditionally independent**. That is, given a fixed class, the probabilities of the other attributes are probabilistically independent. Mathematically,

$$p(x_1, x_2, \ldots, x_n | C_j) = p(x_1 | C_j) p(x_2 | C_j) \cdots p(x_n | C_j) = \Pi_{i=1}^{n} p(x_i | C_j).$$

The individual conditional probabilities $p(x_i|C_j)$ can be estimated quite readily; but there are two cases depending on whether the attributes are categorical (discrete-valued) or continuous-valued.

If attribute X_i is categorical, then the estimate is

$$p(x_i|C_j) = N_{ij}/N_j,$$

where N_{ij} is the number of instances of x_i when C_j occurred and N_j is the number of instances of class C_j.

If attribute X_i is continuous-valued, the conditional probabilities are represented by a parameterized family of continuous distributions, such as (Gaussian) normals. In the normal case the conditional probability density is of the form

$$p(x_i|C_j) = \frac{1}{\sqrt{2\pi}\sigma_{C_j}} e^{-(1/2)(x_i - m_{C_j})^2/\sigma_{C_j}^2}. \tag{18.2}$$

Example 18.4 (Football prediction). Your favorite football team has a good record in the current season, having won 8 out of 12 games. You would like to predict the probability that it will win the next game. A record of the games is shown below. The table shows whether the game was Home or Away, whether the opposing team is rated in the Top half or Bottom half in the region, and whether your team Won or Lost.

Field	H	A	H	A	H	H	H	A	H	A	H	A
Opponent	T	T	T	B	B	T	T	B	B	B	T	B
Win/Loss	W	L	W	W	L	W	W	L	W	W	L	W

The next game will be played at Home with a Top team. Inspection of the table yields these counts:

$$\text{Total games} = 12, W = 8, WH = 5, WT = 4, HT = 5.$$

These imply the probabilities

$$P(W) = 8/12, \ p(H|W) = 5/8, \ p(T|W) = 4/8, \ p(HT) = 5/12.$$

Using the naive Bayes method, you express the probability of winning as

$$p(W|HT) = \frac{p(H|W)p(T|W)p(W)}{p(HT)} = 1/2. \tag{18.3}$$

It looks like it will be a close contest.

However, since you went to all this work, it is not much harder in this small example to bypass the naive method and use the actual probability $p(HT|W)$ implied by the record. A simple count yields $HTW = 4$, implying $p(HT|W) = P(H, T, W)/p(W) = 4/8$. Using this instead of $p(H|W)p(T|W)$ in equation (18.3) produces $p(W|HT) = 4/5$. Now it looks like your team has a very good chance of winning!

Bayesian Belief Networks

Consider a given set of attributes $X_1, X_2, \ldots, X_n$ that may include categories. A complete probabilistic representation of them is defined by the joint probability density $p(X_1, X_2, \ldots, X_n)$. Given this density, it is possible to deduce various conditional probabilities. For example, if the values x_1 and x_2 of attributes X_1 and X_2 are known, the probability density of X_3 is, symbolically,

$$p(X_3|x_1, x_2) = \sum_{X_i = X_4}^{X_n} p(x_1, x_2, X_3, X_4, \ldots, X_n). \qquad (18.4)$$

Theoretically, this is a convenient and compact way to describe attribute relations and determine probabilities of unknown attribute values on the basis of observed attributes.

The difficulty, of course, is dimensionality.[2] If there are n attributes, each with m possible values, the joint density $p(X_1, X_2, \ldots, X_n)$ requires m^n values. This can be extremely large for even modest values of m and n. A number of data records several times m^n would be needed to even roughly estimate the m^n required values.

As said before, good models are based, in not insignificant part, on human intelligence. Selecting an all-purpose general model is seldom useful until it is simplified by the imposition of structure derived from science, common sense, intuition, or preliminary exploratory data analysis.

Bayesian belief networks (sometimes called simply **Bayesian networks**) provide a framework for describing a probability density structure in terms of probabilistic influences. Nodes in the network correspond to events or situations. A directed arc from one node to another implies that the event represented by the second node is probabilistically influenced by that at the first. An example (discussed below) is shown in figure 18.8.

A general Bayesian belief network consists of an **acyclic graph** (that is, a graph whose arcs are directed and for which it is impossible to find a path of arcs that forms a cycle) and a description of the conditional probabilities $p(X_i|X_{pa[i]})$, where $X_{pa[i]}$ denotes the set of parents of X_i. The fact that these conditional probabilities provide all the information that is needed is encapsulated in the basic formula for the entire joint density

$$p(X_1, X_2, \ldots, X_n) = \Pi_{i=1}^n p(X_i|X_{pa[i]}). \qquad (18.5)$$

Example 18.5 (Alarm system). In this standard example, it is assumed that you have installed an alarm system in your house to detect burglaries. If a burglary is attempted, there is a good chance that the alarm will be activated. This dependency is indicated in figure 18.8 by the line between **Burglary** and **Alarm**. However, if an earthquake occurs, there is a possibility that the alarm will be set off as well. This is indicated by the line between **Earthquake** and **Alarm**. There is no connection between **Earthquake** and **Burglary**, which implies that **Earthquake** and **Burglary** are independent events.

---

[2]Each X_i represents a number of specific values. To evaluate equation (18.4) at a specific x_3, it is necessary to substitute x_3 for X_3 and sum over all possible specific $x_4, x_5, \ldots, x_n$.

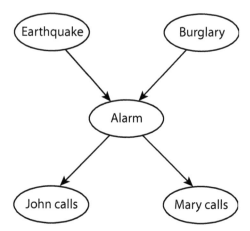

FIGURE 18.8 A Bayesian net for the alarm system. The alarm is influenced by an earthquake or a burglary. In turn, the alarm influences both John and Mary to call.

Two neighbors, John and Mary, agree to call you at work if they hear the alarm. Of course, either of them may call you at other times as well, and there is a chance that either of them may be away when the alarm sounds and hence will not call you. Nevertheless, the probability that either John or Mary calls is influenced by the state of the alarm, as indicated by a connection from **Alarm** to each of them.

The probability of any event is conditional on its parent nodes, and in fact knowledge of the state of the parents is all that is needed to determine the probability of a node.

Consequently, the probability that **Mary calls** is dependent on **Burglary**. But when conditioned on **Alarm**, **Mary calls** is independent of **Burglary** because Mary responds only to the alarm, not to burglars. Likewise, **Mary calls** and **John calls** are related, but when conditioned on **Alarm**, they are independent because Mary's calls do not influence John's for a given state of **Alarm**.

Associated with each node is a table of the probabilities of that node conditioned on its parents. The tables for the alarm system[3] are shown in figure 18.9. These tables are all that is required to define the complete set of probabilities for the network, as a special case of equation (18.5).

Let us use the notation E and $\overline{E}$ to mean that E is, respectively, True or False. The probability of the joint event $(E, \overline{B}, \overline{A})$ (there is an earthquake but no burglary and the alarm does not sound) is, according to equation (18.5),

$$p(E, \overline{B}, \overline{A}) = p(E)p(\overline{B})p(\overline{A}|E, \overline{B})$$
$$= .002 \times .999 \times .80 = .001598.$$

Bayesian networks are similar to the information channels discussed in chapter 5. Attributes that cannot be directly measured are akin to signals transmitted through the network structure, according to conditional probabilities, to the final observed

---

[3] These probabilities should be considered as referenced to a certain time period (say a week), since clearly the probability of an earthquake or burglary depends on the length of time considered.

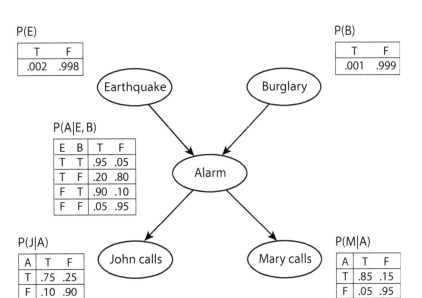

FIGURE 18.9 The conditional probability tables for the alarm system. For each node, there is a table giving the probabilities of the event at that node occurring (T for True or F for False) conditional on the status of its parent nodes. If there are no parents, the unconditional probability is given. From these tables, the entire probabilistic structure can be derived.

attributes. It is necessary to reverse the structure with Bayes' rule to deduce information about the original signals.

Although Bayesian belief networks are useful for expressing probabilistic relations, the computational demands can be enormous in large networks. It is important, therefore, that the model be as simple as possible consistent with the objective of getting good results.

18.6 Support Vector Machines

Support vector machines represent yet another powerful and versatile method for classification. The basic idea is rather simple and has been used for decades, but modern developments have greatly expanded the range of applications and power of the method. Today support vector machines are used to analyze textual documents, decipher handwriting, classify graphical images, filter email messages, and serve in many other classification applications.

The central idea is illustrated in figure 18.10. In this example, the data samples have two attributes X and Y, and there are two possible classes: square and round. The two classes are separated by a straight line in the two-dimensional attribute space, so that the squares are on one side and the circles on the other. The straight line is defined by a linear equation of the form $y = ax + b$. A new item is classified by plotting its attribute values and determining on which side of the separating line it lies.

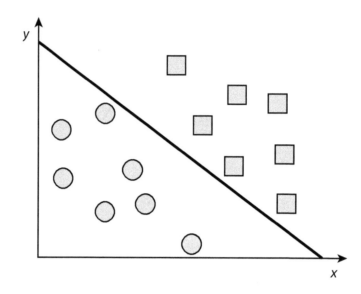

FIGURE 18.10 A separating line. In higher dimensions two classes are separated by a separating hyperplane.

The idea is easily extended to an arbitrary number of dimensions n. In the general case the linear separation is by a **hyperplane**, a flat surface of dimension $n-1$. In three dimensions, for instance, a hyperplane is a two-dimensional plane. A hyperplane in n-dimensional space is defined by solutions $\mathbf{x}$ to a linear equation of the form

$$\mathbf{x}^T \mathbf{w} + w_0 = 0,$$

where $\mathbf{x}$ is the vector of attribute components, $\mathbf{w}$ defines the coefficients, w_0 is a constant, and $\mathbf{x}^T\mathbf{w}$ denotes the inner (or dot) product of $\mathbf{x}$ and $\mathbf{w}$.

The separating hyperplane method was developed in the 1950s as a learning procedure using artificial neurons intended to mimic the functioning of the human brain. The highly influential book *Automata Studies*, of which Shannon was an editor, popularized the approach. The first separating hyperplane updating device was proposed by Rosenblatt and called a **perceptron**. An alternative with a different updating procedure was developed by Professor Widrow and Marcian (Ted) Hoff[4] at Stanford University and called **ADALINE**.

Depending on their distribution, it is not always possible to separate sample points with a hyperplane; but if it is possible, there is a simple step-by-step procedure for doing so. The method begins with a collection of just one sample point and adjoins additional sample points to the collection one by one and adjusts the hyperplane as each new point is adjoined. It is always possible to construct a separating hyperplane if the number of data points N is less than or equal to the number of dimensions plus one, that is, $N \leq n + 1$ (although the separation may not be strict). (See exercise 8.) This condition is often met when the method is applied to textual material, where the number of dimensions corresponds to the number of dictionary words, while the number of training samples is far less than that.

[4]Shortly after that, Hoff joined the small company Intel, where he invented the first microprocessor.

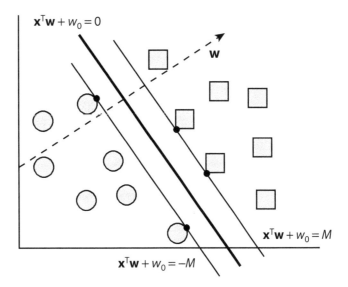

FIGURE 18.11 Maximum separating hyperplane and support vectors. By properly tilting a hyperplane, the separation between two classes can be maximized, as illustrated by the heavy dark line. The two translations of this that touch the classes define the separation distance, and the sample points where these lines touch are termed *support vectors*.

A difficulty with the simplest separating hyperplane method is that the separation it defines may be unnecessarily weak. That is, points of different classes may be close to the same hyperplane even though the points themselves are widely separated in terms of distance. A **support vector machine** (SVM) ameliorates this problem by selecting the hyperplane that produces maximum possible separation. The idea is illustrated in figure 18.11.

Comparison of the hyperplane indicated by the heavy line with the hyperplane of figure 18.10 shows that the distance from the heavy hyperplane to either of the two classified sets is substantially greater than the corresponding distances for the earlier hyperplane. This greater distance is likely to enhance the classification performance when applied to new data.

Consider the two translated hyperplanes of figure 18.11 that touch one of the classification sets. The distance between these two hyperplanes, measured in the direction $\mathbf{w}$, is the distance that should be made as large as possible. Suppose the points $\alpha\mathbf{w}$ and $\beta\mathbf{w}$ are points at the lower and upper hyperplanes, respectively. The distance between them is $d = (\beta - \alpha)||\mathbf{w}||$, where $||\mathbf{w}||$ is the length of $\mathbf{w}$. These points satisfy

$$\alpha\mathbf{w}^T\mathbf{w} + w_0 = -M$$
$$\beta\mathbf{w}^T\mathbf{w} + w_0 = M.$$

Subtracting the first equation from the second gives

$$(\beta - \alpha)||\mathbf{w}||^2 = 2M.$$

Or, equivalently,

$$d = (\beta - \alpha)||\mathbf{w}|| = \frac{2M}{||\mathbf{w}||}.$$

The separating hyperplanes are unchanged by multiplying their equations by any constant: that is, by multiplying $\mathbf{w}$, w_0, and M by a constant. Hence M can be specified as $M = 1$. The maximum distance d is found by minimizing $||\mathbf{w}||$ or equivalently minimizing $||\mathbf{w}||^2$. Hence the problem of finding the best hyperplane is mathematically

$$\min_{\mathbf{w}, w_0} ||\mathbf{w}||^2$$

$$\text{subject to } y_i[\mathbf{x}_i^T \mathbf{w} + w_0] \geq 1, \text{for all } i,$$

where for each data sample i, one sets $y_i = +1$ if i is of class 1, and $y_i = -1$ if i is of class 2. Optimization problems of this form are called **quadratic programs**, and efficient software packages are available to solve them.

As in figure 18.11, the solution will have at least one, and generally more than one, data point touching the lower boundary hyperplane and others touching the upper boundary hyperplane. The points that touch are termed **support vectors** because they are points where a boundary hyperplane visually supports one of the classification sets.

If the two sets cannot be separated with a hyperplane, it is still possible to define a best hyperplane as the one that minimizes the amount of overlap.

The most important extension of support vector machines is to nonlinear separation, illustrated in figure 18.12.

Suppose the dimension of the original space is N. A nonlinear mapping Φ is defined from N-dimensional space to a space $R^{\overline{N}}$ that may be of higher or lower dimension, and data points $\mathbf{x}_i$ are mapped as $\mathbf{x}_i \rightarrow \Phi(\mathbf{x}_i)$. For example, the transformation

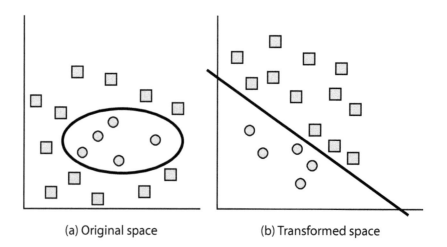

(a) Original space   (b) Transformed space

FIGURE 18.12 Nonlinear separation. A nonlinear transformation of the data may make it possible to separate two classes. Increasing the dimension may also be advantageous.

$\mathbf{x} = (x_1, x_2) \rightarrow \Phi(\mathbf{x}) = (x_1, x_2, x_1x_2, x_1^2)$ maps from two dimensions to four. A judicious choice of Φ often can produce excellent separation when none was initially possible.

18.7 Other Methods

The field of data mining is evolving quickly, fueled by the increasing availability of massive amounts of data and great computer power. There are now several other important methods beyond those described in earlier sections, and we briefly describe four of them here.

1. Artificial neural networks. The method of artificial neural networks has a long history in the field of data mining, and it is still widely used. As the name implies, these networks are modeled after the neural connections in the brain and consist of interconnected nodes (like neurons) and arcs (like neuron connections). The values of observed attributes enter the network as values to input nodes. These values then progress through interconnections (arcs) that apply various weights to them before they reach a node at a deeper layer of the network. A receiving node then applies a nonlinear function to the value it receives and sends the result on to other nodes at the next layer. The final nodes give the class or predicted value.

The main parameters of the network are the weights associated with the arcs. These are determined by adjusting the network with training data.

An advantage of neural networks is that they are widely applicable and software implementations are available. A disadvantage is that it is often difficult to train optimally and the results often are difficult to interpret as simple rules.

2. Fuzzy logic. Many classification systems assign a specific class to an occurrence (a combination of attribute values). For example, a system might classify people with a medical test result of 51–100 as high risk and those with result of 0–50 as low risk. However, it can be argued that someone with a result of 49 may in reality be of only marginally low risk. The method of fuzzy sets allows for the possibility of having less than full degree of membership in a set. In the case of two possible categories A and B, an instance would have membership values v_A and v_B that are nonnegative and sum to 1 (like probabilities) representing degrees of membership in the sets.

The same instance may be processed by another classification system to determine fuzzy membership in two other sets C and D. Then using the rules of fuzzy logic, final membership values for A, B, C, and D can be determined. Fuzzy logic methods have similarities to probabilistic methods.

3. Memory-based reasoning (MBR). In this method, the training instances from past situations are stored directly as tuples of attributes and their corresponding classification categories. Then when a new instance arises, its nearest neighbor (according to some measure) is found, and the new instance is assigned the category of this neighbor.

A disadvantage of this system is that it is sensitive to the parameters used for measuring distance. In addition, the method does not produce simplified rules for classification, and this means that categorization of a new instance may require a great deal of computational time.

4. Case-based reasoning (CBR). Case-based reasoning goes further than MBR by, in general, proposing solutions to difficult problems within a domain by looking at similar problems and their solutions. For example, a doctor may enter into the system the symptoms and test results of a current patient. The CBR system will search for similar situations and looking at their solutions, propose a solution for the current case.

In most advanced form, a CBR system will consider several similar cases and invent a unique solution for the current case based on the experience with previous cases. There have been successful applications of CBR, but this method is, as can be imagined, in an early stage of development.

18.8 EXERCISES

1. (Market basket entropy) For example 18.1 calculate the relative entropy $H(B|A)$ and the mutual information $I(A; B)$ based on probabilities estimated from the data, and hence give an entropy measure to the rule $A \implies B$.

2. (New rules) Find other rules for the market basket example discussed in the text.

3. (Zero mean) Show that the average value of the error in least-squares estimation of the form $y = a + bx$ is zero.

4. (Add a square*) For the housing data find a best fit of the form **Price** $\approx a + b$ **Age** $+ c$ **Sqft** $+ d$ **Sqft**^{2}.

5. (Linear estimation entropy*) Suppose that an unknown parameter x has a normal (Gaussian) distribution with variance σ^2. Independent measurements of the form $y_i = x + \varepsilon_i$ are made, where each ε_i is Gaussian with mean zero and variance ω^2 and is independent of x. The entropy of a multidimensional Gaussian variable is $H = \frac{1}{2} \log (2\pi e)^n |Q|$ bits, where $|Q|$ is the determinant of the covariance matrix corresponding to the variable. (See chapter 22 for the $n = 1$ case.)

   (a) Find $I(y_1, y_2, \ldots, y_n; x)$.

   (b) Let $\bar{y} = \frac{1}{n} \sum_{i=1}^{n} y_i$ and find $I(\bar{y}; x)$, showing that for purposes of estimation it is only necessary to consider $\bar{y}$ rather than keeping track of the individual y_i's.

6. (The alarm) Referring to example 18.5, find the probability that there is an attempted burglary if John calls.

7. (K-means clustering) Suppose that there are n records (or items) each with m attributes that can be expressed in numerical form, so that each item corresponds to a vector in m-dimensional space. These can be grouped into k clusters with the k-**means algorithm**: First, k records are chosen as **seeds**. Next, each other item is assigned to the seed closest to it in terms of Euclidean distance in the m-dimensional space. These assignments define k clusters. Next, the centroid of each cluster is formed (being the point that minimizes the total squared distance to all points in the cluster), and these centroids replace the original seeds. The algorithm then proceeds by reassigning each point to the centroid closest to it. This defines a revised clustering. The steps are repeated, calculating new centroids and obtaining new clusters, until the resulting change in clusters is small.
   Using this method, cluster the following items into two groups: [2, 4, 10, 12, 3, 20, 30, 11, 25]. Begin by assigning the first two items as seeds, and arbitrarily assign 3 to the seed 2 (rather than to 4 since there is a tie).

8. (Separation condition) Let N be the number of items and n the dimension of the space of items. Show that if $N \leq n + 1$, items of two different classes can be separated by a hyperplane. (For simplicity assume that $N = n + 1$ and that any n of the N items are linearly independent.)

18.9 Bibliography

There are a number of good survey texts on data mining. One of the most accessible and useful is [1]. Other good texts are [2], [3], [4], and [5]. A good discussion of data mining and web mining is [6]. A nice set of notes on Bayesian networks is [7].

A comprehensive book of statistical learning with emphasis on support vector machines is [8]. The book edited by Shannon and McCarthy [9] is interesting and was highly influential. The separating hyperplane devices were first presented in [10] and [11]. Exercise 7 is from [4].

References

[1] Berry, Michael J. A., and Gordon Linoff. *Data Mining Techniques*. New York: John Wiley, 1997.

[2] Hand, David, Heikki Mannila, and Padhraic Smyth. *Principles of Data Mining*. Cambridge: MIT Press, 2001.

[3] Witten, Ian H., and Eibe Frank. *Data Mining*. San Francisco: Morgan Kaufmann, 2000.

[4] Dunham, Margaret H. *Data Mining: Introductory and Advanced Topics*. Upper Saddle River, N.J.: Prentice-Hall, 2002.

[5] Han, Jiawei, and Micheline Kamber. *Data Mining: Concepts and Techniques*. San Francisco: Morgan Kaufmann, 2001.

[6] Baldi, Pierre, Paolo Fransconi, and Padhraic Smyth. *Modeling the Internet and the Web*. New York: John Wiley and Sons, 2003.

[7] Hauskrecht, Milos. "Bayesian Belief Networks." Lecture 1. http://www.ecs.pitt.edu/~milus/conrses/cs2001/.

[8] Vapnik, V. N. *Statistical Learning Theory*. New York: John Wiley and Sons, 1998.

[9] Shannon, Claude E., and J. McCarthy, eds. *Automata Studies*. Annals of Mathematical Studies 34. Princeton: Princeton University Press, 1956.

[10] Rosenblatt, F. "The Perceptron: A Probabilistic Model for Information Storage and Organization in the Brain." *Psychological Review* 65 (1958): 386–408.

[11] Widrow, B., and M. E. Hoff. "Adaptive Switching Circuits." *IRE Westcon Convention Record* 4 (1960): 96–104.

SUMMARY OF PART IV

Data is most useful when it is organized, and data structures provide the tools and concepts for that purpose. The simplest data structure is a **list**, but generally it is desirable for the list to be ordered according to a **key**, such as alphabetically or by date. If the list is stored in a computer and is dynamic, with additions and deletions occurring from time to time, some provision must be made for updating the list and arranging it so that it can be queried efficiently. This can be facilitated by the use of **pointers** associated with each entry, indicating the memory location of the next or previous entry. However, searching a list of length n requires an average of $n/2$ operations, which is excessive for long lists. More efficient data structures are based on **trees**. **Binary search trees** and **partially ordered trees** are especially effective. Both of these can be searched in an average time proportional to $\log n$, which is consistent with the entropy of n equally likely items. Information theory, in fact, implies that times proportional to $\log n$ are the best that can be done.

A related problem is that of sorting a list of items according to a key. Two simple methods are **bubble sort** and **insertion sort**, but these both require a number of operations proportional to n^2. The faster methods of **quicksort** and **heapsort** are based on binary search trees and partially ordered trees. These sort algorithms have ordering times proportional to $n \log n$.

A major breakthrough in data organization was the introduction of the **relational database** system. This system of organization, which is surprising simple and yet wonderfully elegant, consists simply of arrays, like spreadsheets. The algebraic theory of relations provides systematic methods for manipulating these arrays, decomposing them into simpler components and combining them again when needed. A fundamental concept used for decomposition is that of a **functional dependency** among attributes. Today most large database systems are relational.

How can a huge collection of textual and other material be searched quickly? This is the issue addressed by modern information retrieval systems. These systems typically regard a collection as a set of **documents**. Depending on the application, the term *document* may refer to a page in a book, a web page, a verse in the Bible, a book in the Library of Congress, a technical article in a collection of professional journals, and so forth. Most large-scale retrieval systems begin by constructing a **forward file**, which lists the terms that appear in each document. In simplest form such a file lists documents one by one down the first column. The other columns correspond to possible **terms**. A non-zero entry at document i and term j indicates that term j appears in document i. The file is most useful when it is transformed into an **inverted file**, listing for each term the documents that contain it, much like a book index that lists the pages on which various terms appear. Forward files can be enormous, so special strategies have been developed to invert them. The principles of linked lists, binomial search trees, and sorting algorithms are useful for this purpose.

Size is also managed by compression. Inverted files are typically sparse and thus can be significantly compressed using run-length codes.

Efficacious response to queries of document collections is facilitated by ranking methods that quantify the closeness of a particular document to the terms indicated in the query. Many of these methods are based on the dot product of the query and the document when each is considered to be a vector of 1's and 0's, with a 1 in a particular component indicating the presence of a corresponding term.

When documents contain links to other documents, as in book references and especially in Internet documents, it is possible to rank query responses by a measure of their popularity as measured by the link structure.

It is both important and challenging to extract useful information, beyond standard queries, from a collection of data. The term **data mining** refers to this activity. In a sense, data mining looks for structure in data when none is initially obvious. The process can be regarded as decreasing entropy while increasing usefulness; for a large mass of seemingly unstructured data is initially characterized as having high apparent entropy. Structure decreases apparent entropy. For example, if the data comprises two variables A and B that appear to be independent, the entropy of the collection is estimated as $H(A) + H(B)$. However, if a relation between A and B is known, then the perceived entropy is $H(A, B) \leq H(A) + H(B)$, the difference being the mutual information of A and B.

There are several data-mining techniques. They can be classified by the nature of the structure they seek to discover. These include logical relations such as A implies B, functional relations such as $y = f(x)$, and partitioning relations that divide the attribute space into regions. All of these can be modified to allow for probabilistic statements; for example, $A = a_1$ implies that $B = b_1$ with probability p. The probabilistic structure of **Bayesian networks** is explicitly based on conditional probabilities and hence is closely related to many concepts of information science. **Support vector machines** are designed to group samples into classes by establishing separate regions in the space of attributes.

PART V

EMISSION

The Mastery of Frequency

19

FREQUENCY CONCEPTS

requency has long been recognized as a fundamental part of nature—basic to music, all matters of sound transmission, mechanical vibrations, and the mathematics of differential equations. Centuries ago it was known that the frequency of vibrating strings or of resonant flutes can be controlled by adjusting physical parameters such as tension, length, or volume. But deeper understanding of frequency—its true mastery—developed rapidly with the advent of modern communication technology, beginning with the telegraph.

Indeed, frequency concepts play a major role in almost all forms of electrical and optical communication, but these concepts, although now clear and in many respects simple, were originally elusive and mysterious. They were discovered and honed only through patience, genius, and happenstance, tinged by human frailty and fierce competition. Today the theory of frequency is regarded as both beautiful and powerful, and its myriad applications are enjoyed by many.

The next few chapters study several theoretical concepts that comprise the mastery of frequency. This study emphasizes historical aspects more than do earlier chapters, for this history—of the telegraph, telephone, radio, and modern communication—is a significant part of our culture, and one can learn from the missteps as well as brilliant insights of key characters in this development. Through this history we will witness how frequency was mastered.

Here are five basic principles of frequency:

1. Invariance of sinusoids. Consider a continuous signal that passes through a linear medium such as an electric circuit. As a general rule, the result differs in form from the original. For example, perhaps you have listened to your voice spoken under water, and found it highly distorted, or noticed that if you tap on a musical string, the response is much more sustained than a tap on a piece of wood. Graphs of the entering and exiting signals in a linear system will, in general, differ in shape.

Signal in

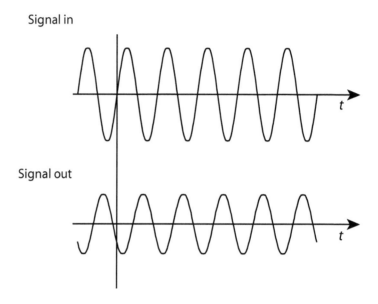

Signal out

FIGURE 19.1 Invariance of a sine wave. A sinusoidal signal passing through a linear medium or system preserves its sinusoidal character (of the same frequency). Only amplitude and phase are affected.

However, if the original signal is a sinusoid (a time function of the form[1] $A \sin(2\pi ft + \theta)$) of frequency f, the resulting output will also be sinusoidal of the same frequency f, although the amplitude and phase may differ from the original (that is, the peaks may be of different height and shifted in time from those of the original). See figure 19.1. The reason that the basic shape is preserved is that a linear system consists of elements that perform some or all of the following operations on a signal: multiplication by a constant, differentiation, integration, delay, and sums of these. From calculus, we know that sine and cosine waves remain sinusoidal of the same frequency under each of these operations.

This invariance of form allows a simple analysis of the behavior of a sinusoidal wave as it passes through a cable, a circuit, or a physical medium such as water or metal. It is not necessary to deduce the output shape, only its amplitude and phase. Much of modern systems analysis is based on this concept.

2. Fourier representation. In the early 1800s the great French mathematician Jean Baptiste Joseph Fourier deduced that any periodic function (within broad limits) can be expressed as a combination of sinusoidal waves. Although he developed this theory as early as 1807, publication was held up by controversy regarding the validity of his methods, and

---

[1] Here A is a constant, f is the frequency, t is a variable representing time, and θ is a phase shift with $0 \le \theta \le 2\pi$. For example, the function $\cos 2\pi t$ is a sinusoid of frequency $f = 1$ with $\theta = \pi/2$.

his work was not publicly available until 1822, when he published his *Théorie analytique de la chaleur*, which not only presented the Fourier series concept but also applied it to solve complicated partial differential equations related to heat transfer. His method was later extended to the Fourier transform that can be used to represent functions that are not periodic.

The fact that an arbitrary function can be represented by a series of sinusoids has profound implications when combined with the principle of invariance discussed above. It means that the linear transformation of an arbitrary signal can be analyzed in three steps: (1) represent the signal as a combination of sinusoids, (2) compute the effect on each of those separate sinusoidal waves using the invariance principle, and (3) combine the separate results to determine the final result. This idea is used, directly or indirectly, in almost all analyses of continuous signals.

3. Resonance. While the previous two concepts are mathematical in nature, the phenomenon of resonance is easily experienced physically. You witness it when you blow across the mouth of a soda bottle to produce a deep tone characteristic of the shape and volume inside, or when you hear a harpist pluck a string. Electrical circuits display resonance as well. Gradually it was learned that it is possible to manipulate resonance to create oscillations, tune radios, and filter out noise.

4. Modulation and the heterodyne principle. Eventually it was discovered that nonlinear transformations of signals, such as squaring or multiplying two signals together, produced important frequency transformations. From elementary trigonometry it is known that when a sine wave of frequency f_1 is multiplied by a sine wave of frequency f_2, the result can be expressed as the sum of two sinusoidal waves—one of frequency $f_1 + f_2$ and the other of frequency $|f_1 - f_2|$. This mathematical identity is manifested physically by the production of a wave that consists of two component frequencies. Purposeful control of this effect is inherent in amplitude modulation (AM) and it also underlies the more general heterodyne principle by which signal frequencies are changed.

5. Nyquist–Shannon sampling theorem. If a continuous signal is sampled at regular intervals, the resulting sequence of sample values can be regarded as an approximation to the actual signal, but obviously some detail may be lost. The Nyquist–Shannon sampling theorem states that if the frequencies contained in the original signal are all below some value W, and the signal is sampled at least every $1/2W$ seconds, then the original signal can be exactly reconstructed from the samples. This result plays a fundamental role in modern communications technology that relies on digital processing. The result also forms the basis of Shannon's most famous capacity calculation, which is presented in chapter 21.

These key principles along with other concepts were discovered gradually along the path to frequency mastery, and we shall follow that development during the course of the next few chapters.

19.1 The Telegraph

The nature of communication technology was fundamentally changed by the invention of the electric telegraph, for it made possible almost instant communication around the world—over continents and across the seas. Messages of national emergency, of family crises, of economic developments, of business relations, and of every aspect of human experience could be sent almost routinely. The world was profoundly different.

The electric telegraph also initiated the modern theory of information and communication. It was the first application of electricity to practical life, and as such it was a visible and friendly breeding ground for fundamental research, the results of which played a significant role in later inventions. Many aspects of electricity were tested and advanced in the course of telegraph development, including the understanding of electrical capacity, induction, batteries, long cables, measurement devices, relays, and so forth.

We study the telegraph and later inventions partly for their interest but mainly because the telegraph motivated a major step in the understanding and use of frequency.

A method for long-distance communication over land was initiated in 1791 when Claude Chappe and his brother transmitted messages visually by sequentially displaying black and white panels that could be observed at a distance of a few miles using a telescope. Their invention was coined *télégraphe*, meaning "far writer." The primitive system was improved by the development of mechanical semaphores between line-of-sight stations. Usually, a large building on a hill was equipped with a tower with long mechanical armlike appendages that could be rotated to various positions to represent alphabetical characters or codewords (figure 19.2). Messages could be sent for several miles, and even further by relaying messages from station to station. These telegraph systems were deployed widely throughout Western Europe and in parts of the United States over the next several decades. Even today their past is evidenced by the existence of a "Telegraph Hill" in many cities.

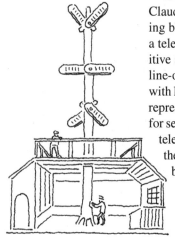

FIGURE 19.2 An early telegraph station. Stations used codes that could be observed from several miles away.

During that period, several individuals imagined an "electric telegraph" using the then newly discovered phenomenon of electricity, generated in those days as static electricity. Real progress was made by Alessandro Volta's invention of the battery, which generated electricity chemically using metals and acid configured into a "pile," and by the revolutionary discovery in 1820 of electromagnetism by the Danish professor Hans Christian Oersted. Oersted observed that a current in a wire influenced the direction of a nearby compass needle, thus establishing a connection between electricity and magnetism. Soon electromagnets and galvanometers were invented that responded to the presence of electric current. Some people attempted to create a telegraph based on these principles, but without significant success.

In 1832 on a return voyage to the United States from Europe, Samuel Morse, a fairly accomplished artist, heard passengers discuss the phenomenon of electromagnetism and the possibility of detecting a current at any point along a wire. He immediately saw that this could lead to the construction of an electric telegraph, and (incorrectly assuming that this idea was original) he determined to develop such a system.

With the help of the physicist Joseph Henry and others, he eventually produced a working telegraph. He also developed the **Morse code** using dots and dashes to correspond directly to letters of the alphabet and numbers. He assigned short code words to frequently used letters (using "dot" for "e," for example), determining the relative frequencies on the basis of a visit to a printer, where he observed the relative numbers of different letters available in a type box.

After four years of difficult negotiations, Morse eventually convinced a skeptical U.S. Congress to support the construction of a 40-mile telegraph line between Baltimore, Maryland, and Washington, D.C. On May 24, 1844, the line operated successfully, reportedly sending "What hath God wrought!" as the first transmission.[2] Over the next decade the telegraph system grew enormously, and its influence was felt in practically every area of life. By 1850 there were over 12,000 miles of telegraph line in the United States.

The main component of the telegraph was an electromagnet, made by wrapping a length of wire many times around an iron core. When current was passed through the coil, a magnetic field was produced. This actuated an armature that indicated the presence of the current.

Electromagnets also form the basis for relays. In this case the armature acts like a switch in a separate circuit: when the first circuit is active and pulls the armature down, the second circuit is completed. By this procedure a relatively weak current in the coil circuit controls what could be a much larger current in the second circuit. Morse understood the importance of the principle that something small could control something large. When his associate Gale worried that the telegraph signal would be too weak at even modest distances, Morse responded, "If a lever can be moved at any distance, it can operate a control point and send a strong signal to the next point, and so on around the globe if desired."

19.2 When Dots Became Dashes

As the telegraph system expanded in both the United States and Europe, it was natural to envision a telegraph crossing bodies of water, including the Atlantic Ocean. Once batteries, relays, and basic line configurations were established, telegraph over land did not seriously challenge the concepts of electric transmission. Over land, voltages could be periodically bumped up along a line by relay stations. And since dots and dashes were coded by hand with telegraph keys and interpreted by ear, the rates of transmission were limited to about 25 words per minute. This meant that dots and dashes were relatively long in duration, as compared to the capabilities of the line. However, transmission through undersea cables was found to be an entirely different matter.

The most obvious difficulties associated with underwater transmission were purely mechanical, but after a few failed attempts, it was found that sufficiently heavy cables insulated with the gummy material extracted from the gutta-percha tree[3] were reasonably reliable. Such cables were laid across the English Channel and other spans of water up to about 100 miles.

However, once the early cables were in place, a fundamental difficulty was discovered. The lines were electrically sluggish, and hence messages were garbled. Dots blurred and became indistinguishable from dashes. Consequently, it was necessary to transmit more slowly than on land-based systems. A typical resulting response to

---

[2] There is controversy over when this famous phrase was actually first transmitted by telegraph.
[3] This material was used in early golf balls and is currently used in dentistry to fill root canals.

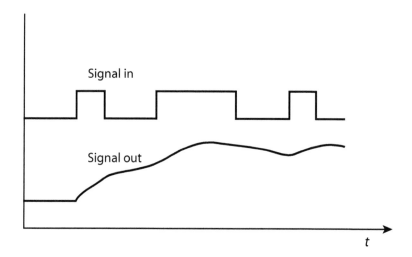

FIGURE 19.3 Response to dots and dashes. When a line has high capacitance, the signals tend to blur together, which necessitates that the rate of signaling be significantly slowed for clear reception.

a series of dots and dashes is shown in figure 19.3. The received signal is a smoothed version of the original transmission, making it difficult to distinguish dots from dashes.

Scientists understood that, because the conductor in the cable would be in close proximity to the ocean water, which is also a conductor, a large level of electrical capacitance would be associated with the line. However, it was not understood what the consequences would be.

William Thompson (later Lord Kelvin) was deeply involved in the undersea telegraph project, and developed much of the associated theory. He based his work on the Fourier series, for he was well aware of Fourier's analysis of the heat equation in his famous 1822 publication; after all, Thompson was an expert on heat.[4]

Thompson's analysis explained the effect of undersea line capacity. Capacity is the ability to store charge. Thompson explained the effect of capacity in cables by stating that the cable acted like a thin flexible tube through which water was to be sent. It would be sluggish because the tube would first have to be stretched. Or, as a scientifically more accurate analogy, he explained that sending a pulse along the cable was like sending a pulse of heat along a thin metal wire. The entire wire must be heated before the far end would register a temperature change.

The laying of the Atlantic cable was one of most spectacular and difficult engineering feats of modern times, and it was fraught with a series of disasters. Finally, after several failed attempts at laying a cable, the *Great Eastern*, a huge ship, three times as large as any other ship in service at that time and capable of storing the entire length of nearly 3,000 miles of cable in its hold, was commissioned. On the first attempt with this vessel, in 1865, cable was laid extending about 70 percent across the Atlantic from Ireland to Newfoundland before the cable broke and the venture was

[4]Fourier also was an expert on heat. It is reported that he was in fact obsessed with heat, keeping his rooms unusually warm, perhaps as a result of the three harsh years he spent on a mission to Egypt with Napoleon.

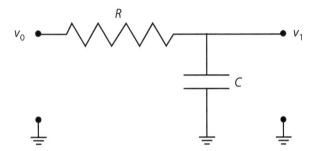

FIGURE 19.4 Highly simplified transmission cable. A transmission cable has resistance R due to the resistance of the wire conductor. There is capacitance because the cable is in close proximity to the sea, which serves as the other side of the circuit. This version is simplified in that it does not reflect the fact that resistance and capacitance are present continuously along the line.

terminated. The second attempt, a year later in 1866 again with the *Great Eastern*, was successful, and another mission was immediately launched that recovered the cable of 1865 and completed it as well. Thompson traveled on these ventures, and provided invaluable technical support based on his theory.

When the two cables were in place, an engineer in Ireland requested that the two ends be connected in Newfoundland. Then from a borrowed silver thimble, some acid, and a piece of zinc he improvised a tiny battery and sent a signal across the Atlantic and back. The returning pulse was detected by a mirror galvanometer after traveling over 4,000 miles.

Thompson's fundamental analysis can be simplified by approximating a telegraph line by the simple circuit shown in figure 19.4, consisting of line resistance R (in **ohms**) due to the resistance of the wire, and capacitance C (in **farads**) resulting from the proximity of the wire to the seawater through the insulation.[5]

The circuit is analyzed by finding its response to a general sinusoidal signal voltage.[6] For example, suppose $v_0(t) = \sin 2\pi ft$. It can be shown that if this voltage wave is applied to the circuit of figure 19.4, the output will be

$$v_1(t) = \frac{1}{\sqrt{1 + (2\pi f)^2 (RC)^2}} \sin(2\pi ft - \theta),$$

where $\tan \theta = 2\pi fRC$. This shows that the magnitude of the signal is sharply attenuated at high frequencies. Thompson understood that the sharp edges of a pulse represented high-frequency components of the signal. Their attenuation explains the

---

[5]Thompson's analysis actually took account of the propagation of voltage v along the continuous line. He developed the partial differential equation $\partial^2 v/\partial x^2 = RC\partial v/\partial t$ for the voltage at distance x and time t, which is still regarded as the fundamental equation for a line with capacity and resistance. This equation is, as he understood, identical in form to the equation for propagation of heat along an iron rod.

[6]The equations governing the circuit of figure 19.4 are $v_0 - v_1 = IR$ and $CI = dv_1/dt$, where I is the current. Elimination of I gives

$$\frac{dv_1}{dt} = -\frac{1}{RC}(v_1 - v_0).$$

If a constant voltage v_0 is applied at time $t = 0$, then the solution is $v_1(t) = (1 - e^{-t/RC})v_0$.

blurring effect of the transmission, illustrated in figure 19.3. From this blurring effect he estimated that the first transatlantic cable could sustain a transmission rate of 4,000 words per day (that is, about 3 words per minute), which was just enough to render the project economically feasible.

Thompson at least partially discovered and used two of the most important properties of frequency: the invariance property of sinusoids and the Fourier representation. His work was proof that even the early theory had enormous practical power, which could transform the world of communication.

19.3 Fourier Series

A function $x(t)$ is **periodic** with period T if $x(t + T) = x(t)$ for all t, $-\infty < t < \infty$, and if T is the smallest positive value with this property. Two periodic functions are the sine and cosine functions $\sin 2\pi t/T$ and $\cos 2\pi t/T$. These are said to be sinusoidal functions with period T and frequency $f_0 = 1/T$. If, as we almost always suppose, t represents time, frequency has units of cycles per second (or equivalently, units of hertz). One complete cycle of the sinusoidal function is completed every $T = 1/f_0$ seconds.

Frequency is alternatively expressed in terms of radians per second. In these units the frequency associated with a period of duration T is $\omega_0 = 2\pi/T$ radians per second, and the basic sine function with period T is $\sin \omega_0 t$. Radians are often used because the notation is somewhat cleaner without the factors of 2π.

The **harmonics** of the basic sinusoidal functions of period T are the sine and cosine functions with frequencies that are integer multiples of the basic frequency f_0. For example, $\sin 2\pi(2f_0)t = \sin 2\pi(2t/T)$ completes two cycles every T seconds.

The amazing theory of Fourier is that any[7] function of period T can be decomposed into a series of sinusoidal functions of period T and harmonics of these sinusoids. Specifically, the **Fourier series** of a function $x(t)$ with period T uses sines and cosines, and is written as

$$x(t) = \frac{a_0}{2} + \sum_{n=1}^{\infty} a_n \cos(2\pi n f_0 t) + \sum_{n=1}^{\infty} b_n \sin(2\pi n f_0 t). \qquad (19.1)$$

The coefficients a_0, and a_n, b_n for $n \geq 1$ can be found by simple formulas as well, as shown in exercise 2.

Imaginary Exponentials

A shorthand way of expressing combinations of sines and cosines is to use the special identity
$$e^{i\theta} = \cos \theta + i \sin \theta,$$
where i is the imaginary number $i = \sqrt{-1}$. It follows, likewise, that $e^{-i\theta} = \cos \theta - i \sin \theta$. Complex exponentials obey the rule $e^{a+i\theta} = e^a[\cos \theta + i \sin \theta]$.

[7] The function must satisfy certain technical assumptions concerning boundedness and continuity.

Fourier series can be expressed in terms of complex exponentials as

$$x(t) = \sum_{n=-\infty}^{\infty} c_n e^{i2\pi n f_0 t}, \tag{19.2}$$

where again $f_0 = 1/T$. In this expression the coefficients c_n may be complex numbers.

This more compact representation of the Fourier series is, of course, equivalent to the earlier one. In fact the coefficients are related by

$$a_0 = 2c_0$$
$$a_n = c_n + c_{-n}$$
$$b_n = \frac{c_n - c_{-n}}{i}.$$

19.4 The Fourier Transform

The Fourier series can be extended to nonperiodic functions by letting the period T tend to infinity. The result is termed the **Fourier transform**, and because of its generality, it is today used more often than Fourier series. Indeed, it is the workhorse of almost all frequency analyses.

The Fourier transform representation of a function $x(t)$ is

$$x(t) = \int_{-\infty}^{\infty} X(f) e^{i2\pi f t} \mathrm{d}f. \tag{19.3}$$

Since $e^{i2\pi f t} = \cos 2\pi f t + i \sin 2\pi f t$, the Fourier transform representation of the function $x(t)$ is in terms of sinusoids. The function $X(f)$ is itself termed the **Fourier transform** or alternatively the **frequency spectrum** of $x(t)$ and is analogous to the coefficients of a Fourier series, for $X(f)$ gives the weight to be applied to the sinusoidal term $e^{i2\pi f t}$. A Fourier transform may have $|X(f)| > 0$ for almost all frequencies, since all frequencies can be regarded as harmonics when the period approaches infinity and the corresponding basic frequency approaches zero.

The Fourier transform can be found explicitly from the equation

$$X(f) = \int_{-\infty}^{\infty} x(t) e^{-i2\pi f t} \mathrm{d}t. \tag{19.4}$$

The relationship between the original function and its transform is therefore defined by the symmetric pair

$$X(f) = \int_{-\infty}^{\infty} x(t) e^{-i2\pi f t} \mathrm{d}t$$
$$x(t) = \int_{-\infty}^{\infty} X(f) e^{i2\pi f t} \mathrm{d}f.$$

Tables of such pairs for various functions are available in reference texts.

Example 19.1 (A pulse). Consider the unit pulse $x(t)$ of width T centered at $t = 0$ shown in the top portion of figure 19.5. The Fourier transform of this pulse is

$$X(f) = \int_{-T/2}^{T/2} e^{-i2\pi ft} dt$$
$$= \int_{T/2}^{T/2} [\cos 2\pi ft - i \sin 2\pi ft] dt. \qquad (19.5)$$

Since the sine is antisymmetric about $t = 0$, the imaginary part of the integral is zero. The remaining integral is easily found to be

$$X(f) = \frac{\sin(\pi fT)}{\pi f}.$$

This function of f is shown in the bottom portion of figure 19.5. This example is useful in many applications. Notice that as the width of the pulse is decreased (T made smaller), the Fourier transform widens. This is a general characteristic of functions and their Fourier transforms: narrower functions produce wider transforms.

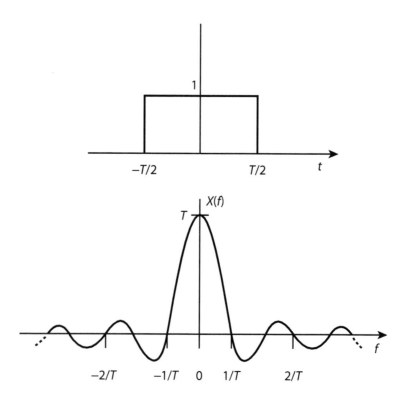

FIGURE 19.5 A square pulse and its transform. A single pulse of finite duration has a Fourier transform that extends over the entire range of frequencies.

Energy Distribution

An important quantity in physical communication systems is the energy represented by a signal, for that energy must be supplied by some physical source. Generally, energy over a short period is proportional to the square of the signal, and hence the total energy required by a signal function $x(t)$ is proportional to the integral

$$E = \int_{-\infty}^{\infty} x(t)^2 dt.$$

This energy is distributed among frequencies, and that distribution is directly represented by the Fourier transform of the signal. This connection is established by a series of simple steps, manipulating the formula for the total energy.

$$\begin{aligned}
E &= \int_{-\infty}^{\infty} x(t)^2 dt \\
&= \int_{-\infty}^{\infty} x(t) \left[\int_{-\infty}^{\infty} X(f) e^{i2\pi ft} df \right] dt \\
&= \int_{-\infty}^{\infty} X(f) \left[\int_{-\infty}^{\infty} x(t) e^{i2\pi ft} dt \right] df \\
&= \int_{-\infty}^{\infty} X(f) X^*(f) df \\
&= \int_{-\infty}^{\infty} |X(f)|^2 df.
\end{aligned} \tag{19.6}$$

where X^* denotes the complex conjugate of X. This expression for energy is known as **Parseval's theorem** or **Rayleigh's energy theorem**. It shows that the energy distribution among frequencies is given by the function $|X(f)|^2$, which is called the **energy spectral density** of $x(t)$.

For the pulse of example 19.1, the energy spectral density is

$$|X(f)|^2 = \left[\frac{\sin(\pi Tf)}{\pi f} \right]^2,$$

which indicates that the energy is concentrated around $f = 0$ but falls off with higher frequencies.

Since $X(-f) = X^*(f)$, it follows that $|X(-f)|^2 = |X(f)|^2$. Hence, the energy can alternatively be expressed as

$$E = 2 \int_{0}^{\infty} |X(f)|^2 df,$$

where the factor 2 is due to writing the integral from 0 to infinity rather than from minus infinity to infinity.

The Fourier transform is a basic tool of frequency analysis. Its power was exhibited by its fundamental role in the development of the telegraph, and it is central in the remaining chapters.

19.5 Thomas Edison and the Telegraph

The discovery of electricity, and especially electromagnetism, spawned a flurry of invention activity beginning in the mid-1800s. The telegraph, being highly visible and commercially successful, attracted young inventive minds like a magnet attracts iron filings.

Among those attracted by the promise and technical excitement of the telegraph was the energetic young man Thomas Alva Edison, who began work as a telegraph operator in 1863 when he was sixteen. He was an unusual operator, however. Frequently he played practical jokes on coworkers (such as wiring a water bucket, from which workers drank, to a high-voltage source), and his energies often drifted away from routine keying and interpretation of messages to the study of telegraph technology. Edison did not stay long in any one location, but rather became a transient operator, traveling from city to city, typically playing jokes on his new colleagues and irritating his supervisors with his inventions. Yet he was an excellent telegrapher, certainly one of the fastest Morse code operators in the country.

One project that fascinated him was the possibility of designing a duplex, which would allow simultaneous transmission of messages in both directions along a single telegraph line. Most of his supervisors dismissed his excursions into the duplex idea as a fruitless waste of time.

Eventually, he gave up telegraph operation to become a full-time inventor. His initial laboratory consisted of space in the shop of Charles Williams, Jr. on Court Street in Boston. Mr. Williams sold electric equipment and often rented laboratory space to young inventors. It was here that Edison first constructed a working duplex using polarized electromagnets (although he was not the first to do so).

Edison's first commercially successful inventions were the quadruplex, improved gold price tickers, and improved stock tickers—all related to the telegraph. As we shall see, his contributions, built on the experience gained while working on the telegraph, played a fundamental role in the advancement of communication.

19.6 Bell and the Telephone

Alexander Graham Bell was born March 3, 1847, in Edinburgh, Scotland. He was exposed to voice training and the synthetic reproduction of voice during his entire youth. Both his father and his grandfather made their livings by the study and teaching of speech and elocution. In 1864 Bell's father, Alexander Melville Bell, developed a universal phonetic alphabet called "Visible Speech," the symbols of which were representations of the configurations of the mouth and tongue that produced the corresponding sounds. This was a major achievement, for several leading phoneticians had failed in their attempts to create such a universal alphabet. It also led to a unique and effective method for teaching deaf people to speak, and Alexander Graham Bell became an excellent instructor in the method.

It was a year later, in 1865, that Alexander Graham Bell, at nineteen, began to experiment with tuning forks. He found that by holding a tuning fork in front of his mouth as he spoke a vowel, the tuning fork of proper pitch would resonate. He soon discovered that each vowel corresponded to a few specific tones. His friend Alexander Ellis told him that these experiments were duplications of earlier experiments by the great German physicist Hermann von Helmholtz, and so Bell studied Helmholtz's book *On the Sensations of Tone*.

Based on his knowledge of Helmholtz's experiments, Bell conceived the idea of sending several messages simultaneously over the telegraph by means of a "harmonic telegraph" in which each of several messages would be sent in Morse code but each using a different pulse tone. To pursue this idea, he began to duplicate Helmholtz's experiments with oscillations derived from tuning forks. He soon replaced the tuning forks with metal reeds, which had the advantage of being smaller and being tunable by adjusting the length of the free part of the reed. A working harmonic telegraph was not easily produced, although Bell tried numerous variations.

In 1874 Bell met Thomas Watson at Williams's electric shop on Court Street, where Thomas Edison had set up his laboratory three years earlier. Watson assisted Bell in his attempts to perfect a workable harmonic telegraph using space in the attic of Williams's shop.

A major breakthrough occurred to Bell on June 2, 1875, while he was engaged in his work on the harmonic telegraph. It was important that the transmitter reeds and receiver reeds be tuned identically. Bell typically held a receiver reed to his ear while with a small instrument he adjusted it to match the tone that was being sent by the transmitter in another room. On that particular day, Watson interrupted the vibration of the transmitter reed to make an adjustment, and Bell, with his ear to the receiver reed, heard sounds associated with Watson's manipulations of the transmitter reed. Bell immediately knew that that was of fundamental significance. He understood then that the wires could transmit, not only tones, but complex sounds. He knew then that a telephone could be built.

What had occurred was the demonstration of the converse of the familiar principle of electromagnets. If a base current flows though the coil, and the armature is forcibly moved (as by impressed sound waves on a reed armature), the current in the coil will respond in accord with the armature movements. This is a manifestation of the dual aspect of electromagnetism: changing current causes armature movement, and armature movement causes changing current. In this context it was a brilliant discovery.

Bell and Watson set about to construct the telephone. The receiver and the transmitter were identical, each consisting of a stretched parchment drumhead with the free end of a harmonic transmitter reed attached at the center of the drumhead. When Watson spoke toward the transmitting drumhead, the sound would be transmitted electrically to another room, where Bell could hear Watson's voice reproduced by the drumhead of the receiver. At least that was the concept. Unfortunately, it did not work well because the feeble current generated by the transmitter was not sufficient to vigorously drive the receiver, so that only a weak and vague response to Watson's voice could be heard. Nevertheless, this primitive telephone launched a transition in understanding. Arbitrary sounds could be sent as electrical signals, simply by electromagnetism.

Bell filed for a patent on his device on February 14, 1876. A diagram in the patent, shown in figure 19.6, makes clear how his first telephone was constructed. The transmitter reed's vibration induces a corresponding vibration in the transmitter coil, which then proceeds to the receiver coil, where it causes the receiver reed to vibrate in sympathy to the original reed. Another inventor, Elisha Gray, filed a caveat (a preliminary patent filing) for a telephone two hours after Bell filed his patent application.

The famous telephone conversation in which Bell entreated, "Mr. Watson. Come here. I want to see you," did not occur until about a month later, on March 10 of the

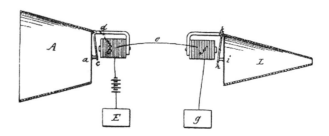

FIGURE 19.6 A diagram from Bell's patent application. Sound waves from a speaker move the transmitting diaphragm and induce a feeble varying current in the coil of the transmitting electromagnet. This current is carried to the receiving electromagnet, which moves the receiver diaphragm.

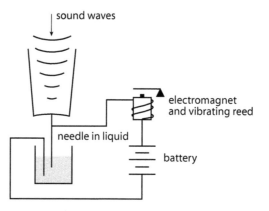

FIGURE 19.7 Bell's variable resistance. The diaphragm moves a needle up and down in the acidic liquid, causing the resistance between the needle and a fixed brass rod to vary according to the sound patterns impressed on the diaphragm.

same year, and it was made over a telephone whose transmitter was constructed on an entirely different principle—that of a microphone.

It was evident that a major problem with the original design was that the current generated by an electromagnet directly connected to a diaphragm was not sufficient to produce intelligible speech at the receiver. The breakthrough modification connected the transmitting diaphragm to a device whose resistance varied as the diaphragm moved. This resistive device was then part of a high-voltage circuit that drove the receiving electromagnet. The setup is shown in figure 19.7. Bell's variable resistance device (or microphone) consisted of a small dish of dilute sulfuric acid into which a brass tube was inserted to serve as one side of the circuit. A needle was connected to the transmitting diaphragm, and the tip of this needle was also inserted into the liquid. As the needle mimicked the vertical vibrations of the diaphragm and slid up and down in the liquid, the needle surface exposed to the liquid varied, and consequently the resistance between the needle and the brass tube was continuously modified. Thus the tiny vibrations of the diagram controlled a large current through an electromagnet that vibrated a reed or other diaphragm. It was with this arrangement that the famous first phone conversation between Bell and Watson took place. Strangely, however, the incident was not reported by either Bell or Watson for more than 10 years.

Controversy and intrigue surrounded the original patent application. The main body of the application says nothing about the use of variable resistance; however, a relatively short paragraph, almost an afterthought, mentions the possibility of using a liquid variable resistance. On the other hand, Elisha Gray's filing fully describes this possibility. Yet there is some evidence that Bell had conceived the concept much earlier. The controversy and numerous other patent disputes, by inventors who apparently had working telephones prior to Bell's, were the subject of much litigation, with one case being finally resolved in the U.S. Supreme Court, which found in favor of Bell in a split decision. A logical inference from these favorable judgments is that Bell was likely very charming and very lucky, as well as hardworking. As an interesting side note, Bell and Gray had earlier each filed patent applications for the harmonic telegraph, with Gray being two days ahead of Bell, and Gray eventually won that patent.

The liquid variable resistance was not commercially practical, and other inventors soon devised superior devices. Indeed it was Thomas Edison who developed the carbon microphone that is still in common use. In the basic design as a diaphragm moves in response to sound waves, it compresses a little space that is filled with small granules of carbon. As the granules are pressed more tightly together, the resistance through them decreases.

19.7 Lessons in Frequency

The saga of Bell's researches, and those of his contemporaries working on similar paths, brought forth an improved mastery of frequency. Two major ideas seem to have emerged. First was the implicit understanding that the shape of a sinusoidal signal is preserved as it passes through electric circuits. Indeed this was the principle underlying the harmonic telegraph.

A second principle—that of a tuned circuit—also was only partly known to the early telephone researchers. These researchers generated sinusoidal signals mechanically, and detected them mechanically as well, with tuning forks or reeds tuned to match those at the transmitter. These practical inventors had not yet discovered that tuning itself could be done with electrical components.

Perhaps the greatest leap in understanding was the recognition that complex sounds, such as those of the human voice, can be thought of as patterns of a single quantity whose amplitude varies; and this pattern can be transcribed as mechanical movements or fluctuations of electrical current, and hence transported over great distances and then converted back to duplicate the original sound. It is perhaps difficult for us to imagine a state of knowledge in which this simple fact is not known. Certainly musicians and physicists knew that sound comes from mechanical movements, but although it was known for a single tone, it was apparently unclear that the complexity of sound could be regarded as a single pattern. This lack of understanding perhaps explains Bell's first conception of a telephone—the harp phone—consisting of a series of reed transmitters with different closely spaced frequencies. A voice projected onto the reeds would stimulate each according to the frequency makeup of the sound, and then be transported to the corresponding array of tuned receiver reeds, where the voice would be reasonably reconstituted as the combination of their separate vibrations. Bell knew that all of these separate tones could be transported over a single wire to the array of tuned receivers, but he did not understand that they did not have to be impressed in their separate identities. However, because of its complexity. Bell never built the harp phone. The original telephone of figure 19.6 is so simple, it is hard to imagine that most of Bell's investigations were attempts to separate and then later gather the individual frequencies.

Shortly after the introduction of the telephone, Thomas Edison created what came to be his favorite invention: the phonograph. In original form, it was not even electric. Edison had been experimenting in his Menlo Park, New Jersey, laboratory with modifications to the telephone, including his carbon microphone, and hence had telephone components readily available. One day in 1877, Edison attached a pin to a diaphragm, and while shouting the word "halloo" onto the diaphragm, he pulled a strip of waxed paper past the pin to record the vibrations as a groove on the paper. He then placed the pin in the groove and pulled the paper through again. He heard a faint "halloo," and the concept of the phonograph was born. Edison was amazed that his voice was captured with that thin wavy scratch. This was one of the first observations of a human speech waveform.

The phonograph was immediately popular, due in part to that little wavy groove that amazed, mystified, and delighted average citizens as much as it did Edison. The breakthrough in understanding could be appreciated and enjoyed by everyone. Indeed, it was upon the introduction of the phonograph that Edison was fondly referred to as the "Wizard of Menlo Park."

In terms of its pure technology, Bell's telephone represents an interesting perplexity. The fundamental observation that electromagnetism works both to generate current and to respond to it enabled his first telephone. That was the essential idea described in his patent, and it clearly was an outstanding achievement. Yet the first practical telephone, constructed a month or so later and capable of producing truly audible sounds at the receiver, did not use that principle; and neither do modern telephones. Instead, a different principle was used—one we find in almost every major technological idea that has advanced the mastery of frequency—the principle of a small signal controlling a larger one (as with Morse's electromagnetic relays). The liquid microphone embodies that idea in Bell's telephone. The tiny vibrations of the transmitting diaphragm are used, by way of the variable resistance, to control the large current in the receiver circuit. It is not certain when Bell conceived of that idea or when he understood its true significance (for he continued to experiment with the earlier idea), but it was that principle that was most profound.

Later, secure with monies generated by the telephone, Bell continued to invent. He invented the hydrofoil ship, the box-shaped airplane wing, and several other important innovations. His favorite, however, which he considered "his most important invention," was what he called the "photophone," by which voice signals were transmitted on a light beam.[8] The device relied on the element selenium, which changes resistance according to the amount of light to which it is exposed. The variable resistance is used to convert the small light signal to a large current in the receiver!

---

[8] The intensity of the beam was changed by reflecting the beam from a diaphragm controlled by voice. As the diaphragm became more convex, the reflected beam was spread and hence was less intense over any given receiving angle.

19.8 EXERCISES

1. (Alternate form) An alternate version of the Fourier transform uses radians per unit time (ω) rather than cycles per unit time (f). In general, $\omega = 2\pi f$. The Fourier transform of $x(t)$ based on ω is

$$X(\omega) = \int_{-\infty}^{\infty} x(t)e^{-i\omega t}\,dt.$$

(a) Show that

$$x(t) = \frac{1}{2\pi}\int_{-\infty}^{\infty} X(\omega)e^{i\omega t}\,d\omega.$$

(b) Show that

$$E = \frac{1}{2\pi}\int_{-\infty}^{\infty} |X(\omega)|^2 d\omega.$$

2. (Fourier coefficients) It is easily seen from figure 19.8 that the integral over a period of the product of a sinusoid and its first harmonic is zero. More generally,

$$\int_0^T \sin 2\pi m f_0 t \, \sin 2\pi n f_0 t \, dt = \begin{cases} 0, & m \neq n \\ T/2 & m = n. \end{cases}$$

$$\int_0^T \cos 2\pi m f_0 t \, \cos 2\pi n f_0 t \, dt = \begin{cases} 0, & m \neq n \\ T/2 & m = n. \end{cases}$$

$$\int_0^T \sin 2\pi m f_0 t \, \cos 2\pi n f_0 t \, dt = 0, \quad \text{all } m \text{ and } n.$$

By using these orthogonality relations, show that the coefficients of the Fourier series

$$x(t) = \frac{a_0}{2} + \sum_{n=1}^{\infty} a_n \cos (2\pi n f_0 t) + \sum_{n=1}^{\infty} b_n \sin (2\pi n f_0 t)$$

are

$$a_0 = \frac{2}{T}\int_0^T x(t)\,dt$$

$$a_n = \frac{2}{T}\int_0^T x(t) \cos (2\pi n f_0 t)\,dt$$

$$b_n = \frac{2}{T}\int_0^T x(t) \sin (2\pi n f_0 t)\,dt.$$

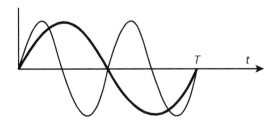

FIGURE 19.8 Orthogonality of a sinusoid and its harmonic.

3. (Square wave series) Find the Fourier series of the square wave with period T defined on $0 \le t \le T$ by

$$x(t) = \begin{cases} 1, & 0 \le t < T/2 \\ -1, & T/2 \le t < T, \end{cases}$$

and extended periodically for all t.

4. (Exponential decay) Find the Fourier transform of

$$x(t) = \begin{cases} 0, & t < 0 \\ e^{-at}, & t \ge 0. \end{cases}$$

5. (Fourier identities) Suppose that the Fourier transforms of $x(t)$ and $y(t)$ are $X(f)$ and $Y(f)$, respectively. Find the Fourier transform of the following signals.
   (a) $ax(t)$
   (b) $x(t) + y(t)$
   (c) $x(at)$
   (d) $x(t + a)$
   (e) $z(t) = \int_{-\infty}^{\infty} x(t - \tau)y(\tau)d\tau$.

6. (Cosine pulse) Consider the cosine pulse that is zero except in the interval $-T/2 < t < T/2$, where it is $x(t) = \cos 2\pi f_0 t$ with $f_0 = 1/2T$.
   (a) Sketch the pulse shape.
   (b) Write the integral expression for the Fourier transform of $x(t)$.
   (c) Can you infer that you only need to integrate the product of two cosines?
   (d) Find the Fourier transform. Hint: You may find it useful to use the identities

$$2\cos A \cos B = \cos(A + B) + \cos(A - B)$$

$$\sin(A + B) = \sin A \cos B + \cos A \sin B$$

$$\sin(A - B) = \sin A \cos B - \cos A \sin B.$$

7. (Pulse energy) Using Rayleigh's energy theorem applied to the pulse of example 19.1 show that

$$\int_0^{\infty} \frac{\sin^2 x}{x^2} dx = \frac{\pi}{2}.$$

8. (More Fourier identities) If $X(f)$ is the Fourier transform of $x(t)$, find the transform of
   (a) $\dfrac{dx(t)}{dt}$
   (b) $tx(t)$.

9. (An integral*) Evaluate the integral

$$\int_0^{\infty} \frac{\cos \omega}{1 + \omega^2} d\omega$$

by considering the Fourier transform of

$$x(t) = \begin{cases} 0, & t < 0 \\ e^{-t}, & 0 \le t \le 1. \end{cases}$$

Hint: Use exercise 1 and first consider $y(t) = e^{-t}$ for $t \ge 0$, $y(t) = 0$ for $t < 0$.

19.9 Bibliography

The colorful history of the early telegraph is presented in [1], [2], [3], [4]. An excellent popular textbook treatment of the Fourier transform, including a brief biography of Joseph Fourier, is [5]. A general history of early electrical discovery and engineering is [6]. Histories of Bell's telephone work are [7] and [8]. See also [9] for an introduction to telephone technology and its history. A detailed study of the controversy surrounding the Bell telephone patent is contained in [10]. There are several biographies of Edison; two that were especially useful in preparation of this chapter are [11] and [12].

References

[1] Standage, Tom. *The Victorian Internet*. New York: Berkley Books, 1998.

[2] Wilson, Geoffrey. *The Old Telegraphs*. London: Phillimore, 1976.

[3] Clarke, Arthur. *Voice across the Sea*. London: William Luscombe, 1958.

[4] Coe, Lewis. *The Telegraph: A History of Morse's Invention and Its Predecessors in the United States*. Jefferson, N.C.: McFarland, 1993.

[5] Bracewell, Ronald N. *The Fourier Transform and Its Applications*. 3rd ed. New York: McGraw-Hill, 1986.

[6] Skilling, H. H. *Exploring Electricity: Man's Unfinished Quest*. New York: Ronald Press, 1948.

[7] Bruce, Robert V. *Bell: Alexander Graham Bell and the Conquest of Solitude*. Boston: Little, Brown, 1973.

[8] Frailey, Jarrell D., and James M. Velayas. *In the Spirit of Service: Telecommunications from the Founders to the Future*. St. Louis: Columbia Creek Publishing, 1993.

[9] Noll, A. Michael. *Introduction to Telephones & Telephone Systems*. 2nd ed. Norwood, Mass.: Artech House, 1991.

[10] Aitken, William. *Who Invented the Telephone?* London: Blackie and Sons, 1939.

[11] Adair, Gene. *Thomas Alva Edison: Inventing the Electric Age*. New York: Oxford University Press, 1996.

[12] Israel, Paul. *Edison: A Life of Invention*. New York: Wiley, 1998.

20

RADIO WAVES

One of the greatest scientific achievements of the nineteenth century—indeed, perhaps of all time—is the unified theory of electromagnetism developed by James Clerk Maxwell. His general equations, presented in his masterful *A Treatise on Electricity and Magnetism*, published in 1873, capture the phenomenon and partial theories of Oersted, Faraday, and Ampère. Maxwell was largely inspired by the theory of light as a wave phenomenon, and his theory implied that if electromagnetic waves existed, they would move at the speed of light. Using his theory, he in fact measured the speed of light indirectly by making electrical measurements.

Unfortunately, Maxwell died of cancer at the age of 48 while he was still working on the second edition of his great *Treatise*. He was therefore unable to spend much time working out the implications of his theory or fully testing it empirically. That work, then, fell to a small group of scientists who called themselves the "Maxwellians."

One of that group was Oliver Heaviside. Another was Heinrich Hertz, a student of von Helmholtz. Heaviside contributed greatly to the theory, recasting it in a modern vector approach that is easily understood. Hertz's experiments were as important to development and acceptance of Maxwell's theory as was Heaviside's theoretical work. It was understood by physicists familiar with the theory that Maxwell's equations implied the possibility of generating electromagnetic waves that would radiate indefinitely away from their source, carrying energy outward. It was Hertz who first generated and measured such waves, providing major experimental verification of Maxwell's theory, as described below in section 20.3.

20.1 Why Frequencies?

To discuss radio it is useful to understand a bit more about frequency—especially how it is treated by circuits. The introduction to chapter 19 mentioned the invariance property of sinusoids in linear systems. This connection is made more explicit here.

Homogeneous Solutions

Consider the differential equation

$$\frac{d^n x(t)}{dt^n} + a_{n-1}\frac{d^n x(t)}{dt^n} + \cdots + a_0 x(t) = 0. \tag{20.1}$$

This is termed a **linear differential equation** or **order** n. It is linear because all functions of the unknown $x(t)$ appear linearly. The equation is defined by the coefficients $a_0, a_1, \ldots, a_{n-1}$. In (20.1) these are **constant coefficients** since they do not depend on time. It is these two properties, linearity and constant coefficients, that together lead to sinusoids. As an example, the low-amplitude motion of a pendulum is described by such an equation of order two.

The particular equation (20.1) is called a **homogeneous equation** because the right-hand side is zero.

If x or one of its n derivatives is nonzero at $t = 0$, there will be a nonzero solution to the homogeneous equation. For example, if a pendulum bob is held up and then released, the pendulum will swing.

Equation (20.1) can be solved by assuming a solution of the form $x(t) = e^{\lambda t}$ for some value of λ. Substituting this solution into the equation produces

$$\lambda^n e^{\lambda t} + a_{n-1}\lambda^{n-1}e^{\lambda t} + \cdots + a_0 e^{\lambda t} = 0.$$

Canceling the common factor of $e^{\lambda t}$ yields

$$\lambda^n + a_{n-1}\lambda^{n-1} + \cdots + a_0 = 0, \tag{20.2}$$

which is the **characteristic equation** of equation (20.1). This polynomial equation can be solved for n roots. Some roots may be repeated, and some may be complex. If they are complex, they occur in complex conjugate pairs. In general, the real part of a root describes a growing exponential (if the real part is positive) or a decaying exponential (if the real part is negative). The imaginary part of a root indicates the presence of sinusoids. If two of the roots are $\lambda_1 = r + i\omega$ and $\lambda_2 = r - i\omega$, the solution will have components of the form $Ae^{rt}\sin\omega t + Be^{rt}\cos\omega t$. In order that there be complex roots it is necessary that the order n be at least two.

Example 20.1 (Pendulum). Consider a pendulum of length L and mass of its bob M. Let the angle away from center be $\theta(t)$. The gravitational force pulling it back to the center[1] is $-Mg\theta(t)$, where g is the gravitational constant. This force must, according to Newton's law, be mass times acceleration, $ML\,d^2\theta(t)/dt^2$. Hence, after dividing by ML,

$$\frac{d^2}{dt^2}\theta(t) + \frac{g}{L}\theta(t) = 0.$$

The characteristic equation is $\lambda^2 + g/L = 0$, with solutions $\lambda = \pm i\sqrt{g/L}$. This means that the motion will be of the form $A\sin(gt/L) + B\cos(gt/L)$.

---

[1] Actually it is $Mg\sin\theta(t)$, but for small angles $\sin\theta(t) \approx \theta(t)$.

Nonhomogeneous Solutions

Generally, a system not only responds to its initial conditions, but also to input signals or forces. The relevant equation then has the form

$$\frac{d^n x(t)}{dt^n} + a_{n-1}\frac{d^n x(t)}{dt^n} + \cdots + a_0 x(t) = s(t). \tag{20.3}$$

This is termed a **nonhomogeneous equation**. The equation may be difficult to solve directly—except when $s(t)$ is exponential.

Assume that $s(t) = e^{qt}$ where q may be a complex number. Then trying a solution of the form $x(t) = Ae^{qt}$ one finds that all terms in (20.3) have this same factor. Hence canceling that common term gives

$$A(q^n + a_{n-1}q^{n-1} + \cdots + a_0) = 1.$$

The solution therefore has

$$A = \frac{1}{q^n + a_{n-1}q^{n-1} + \cdots + a_0}. \tag{20.4}$$

In general, there would be added to this solution some solution to the homogeneous equation as determined by the initial conditions. Notice that if q is close to a root λ of the characteristic equation, then A will be large, since the denominator is zero at a root.

This analysis reflects the importance of sinusoids in analysis of linear systems. If an input signal is a sinusoid, the response will be sinusoids of the same frequency, but perhaps with a different magnitude and phase, depending on the coefficients A and B of the sine and cosine terms. If there are nonzero initial conditions, or if the signal is suddenly applied at, say, time zero, there will also be sinusoids that are characteristic of the homogeneous system. The general rule is that a system responds to a sinusoid with the same frequency and with frequencies characteristic of the homogeneous system.

Example 20.2 (Lord Kelvin's system). Consider the simplified model of a telegraph transmission line of figure 19.2. The equation for the voltage v_1 is

$$\frac{dv_1}{dt} = -\frac{1}{RC}(v_1 - v_0).$$

The characteristic equation is $\lambda + 1/(RC) = 0$. Hence the homogeneous solution is of the form $v(t) = e^{-t/(RC)}$. The input signal, defining the nonhomogeneous part, is $v_0/(RC)$, which is a sinusoid of zero frequency. The part of the solution corresponding to this will be a constant. It will be of magnitude $v_0/(RC) \times A$, where A is given by equation (20.4), as $A = RC$. Hence, $v_1 = v_0 + ce^{-t/RC}$. The constant c is chosen to make $v_1(0) = 0$. Thus $v_1(t) = (1 - e^{-t/RC})v_0$.

Example 20.3 (Tuned circuit). Consider the classic RLC circuit of figure 20.1.

The equations of the RLC circuit can be derived by using the fact that the current must be equal at every point in the circuit and the total voltage drop around the circuit

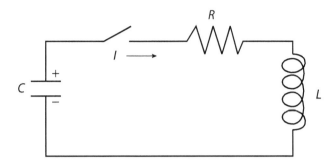

FIGURE 20.1 A classic RLC circuit. For small values of resistance R, a circuit with capacitance C and inductance L will oscillate.

must be zero. The three voltage relations (measured clockwise around the circuit) are (with V's for **volts** and I for current in **amperes**)

1. $V_R = I(t)R$ (Ohm's law, with R in **ohms**)

2. $V_L = L\dfrac{\mathrm{d}}{\mathrm{d}t}I(t)$ (changing current produces voltage, with L in **henrys**)

3. $V_C = \dfrac{1}{C}\displaystyle\int_0^t I(\tau)\mathrm{d}\tau - V_0$ (changing voltage produces current, with C in **farads**). The initial voltage V_0 across C is regarded as $-V_0$ when moving clockwise around the circuit.

It is always true that

$$V_R + V_L + V_C = 0, \tag{20.5}$$

or upon differentiation, $\dfrac{\mathrm{d}}{\mathrm{d}t}[V_R + V_L + V_C] = 0$. This yields the differential equation

$$\frac{\mathrm{d}^2 I}{\mathrm{d}t^2} + \frac{R}{L}\frac{\mathrm{d}I}{\mathrm{d}t} + \frac{1}{LC}I = 0,$$

for the current $I(t)$. The characteristic equation is therefore

$$\lambda^2 + \frac{R}{L}\lambda + \frac{1}{LC} = 0.$$

Using the techniques discussed above leads to the solution

$$I(t) = \frac{V_0}{\omega_1 L}e^{-Rt/2L}\sin\omega_1 t, \tag{20.6}$$

where

$$\omega_1 = \sqrt{\frac{1}{LC} - \left(\frac{R}{2L}\right)^2}.$$

If R is zero, the circuit will oscillate like a pendulum. The addition of positive resistance has two main effects. It lowers the oscillation frequency, and it damps the oscillations so that they die out exponentially.

20.2 Resonance

A tuning fork stimulated by a sharp tap responds by vibrating at its characteristic frequency. Likewise, an LC circuit with low resistance stimulated by a sudden voltage will "ring" at its characteristic frequency. It is also true that if a continuous sound tone is played near a musical string, the string will vibrate in sympathy to the impressed tone, especially if the pitch of the impressed tone is near the characteristic pitch of the string. An RLC circuit behaves in much the same way. If a sinusoidal voltage is impressed upon it, it will respond at the frequency of the impressed sinusoidal voltage. However, an RLC circuit with small R will, like a tuning fork, respond vigorously when the impressed frequency is close to the characteristic frequency of the circuit. This can be seen from equation (20.4), which shows that the magnitude of the response will be large if q is near the characteristic frequency. This phenomenon is termed **resonance**, and an RLC circuit with low resistance is often called a **resonant circuit**.

20.3 The Birth of Radio

Hertz must be credited with construction of the first radio, although he did not have that in mind. His experimental apparatus generated electromagnetic waves and he was able to detect them. His experimental setup is shown schematically in figure 20.2.

The inductance of the circuit was supplied by a coil. The capacitance was obtained from a small gap in the antenna and large (30 cm) zinc disks at the ends of the antenna. The overall capacitance was low, so that the characteristic frequency would be high.

Across his lecture hall he set up wall-sized plates of zinc, capable of reflecting electromagnetic waves (if they existed). When the capacitor was discharged through the antenna and spark gap, the oscillator drove current up and down the antenna.

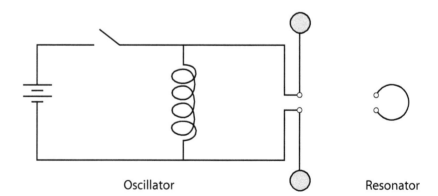

Oscillator   Resonator

FIGURE 20.2 The Hertz oscillator and resonator. The oscillator circuit was the transmitter. Inductance was supplied by a coil, capacitance was supplied by two large plates attached to the ends of the antenna, and resistance was present in the spark gap and the wiring. The spark gap was at the center of a dipole antenna.

This fluctuating current produced a magnetic field and generated the radiating electromagnetic waves predicted by Maxwell's theory. These waves were reflected by the zinc, and standing waves were established. Hertz measured the amplitude of the waves with a small circle of wire with a gap—a device he termed a **resonator**. If the magnetic field were sufficiently strong, a spark would form in the gap of this resonator.

According to Maxwell's theory all electromagnetic waves travel at the speed of light. If λ denotes the wavelength, and f the corresponding frequency in cycles per second (hertz), then $\lambda f = c$, where c is the speed of light. Since the speed of light is $c = 3 \times 10^8$ meters per second, the radiation of frequency 100 MHz (100 megahertz), a frequency Hertz used, has a wavelength of $\lambda = 3 \times 10^8 / 10^8 = 3$ meters.

Greatest radiation efficiency of an antenna like Hertz's is achieved when the rods have a length equal to one-quarter of a wavelength. At 100 MHz, the wavelength is 3 meters and hence each side of the dipole should be 3/4 meters. If he had used low frequencies of, say, 1 MHz, characteristic of the now familiar AM band, the best length for each side of an antenna would be 75 meters, or about 245 feet. Today it is common to use a grounded monopole antenna where one side of the dipole is ground. The single arm must be about 245 feet, and this explains why radio station towers you may see near highways are about 245 feet tall.

After Hertz's successful experiment his students asked what use there might be for these waves. Hertz responded, "It's of no use whatsoever. This is just an experiment that proves Maestro Maxwell was right—we just have these mysterious electromagnetic waves that we cannot see with the naked eye, but they are there."

"So what is next?" asked one of this students.

Hertz shrugged. "Nothing I guess."

But of course, there was a great deal more. Hertz's experiment ushered in the modern age of electromagnetic waves.

20.4 Marconi's Radio

Guglielmo Marconi was only 22 years old in 1896 when he presented his version of a radio to William Preece, chief engineer of the British General Post Office. Preece, himself, had experimented extensively with wireless communication, and in fact had achieved it with underwater conductors communicating for short distances. Preece's system was based on inductive waves as opposed to the radiated waves of Hertz's experiment. Inductive waves do not radiate energy and their effect falls off as the square of distance instead of in direct proportion to distance, as is the case for radiated waves. Preece apparently did not understand the difference, although physicists had attempted to explain it to him. Nevertheless, for whatever reasons, Preece received Marconi enthusiastically, and arranged to have his system tested.

Marconi had experimented with radio since he was 19, using Hertz's basic experimental design, which he read about while vacationing in the Alps. He had moderate success at first, transmitting several yards, but his father advised him that nothing would come of it unless he could achieve communication over much greater distances. In response, Marconi changed his antenna design. Instead of the two equal rods of a dipole that Hertz had used, Marconi used a single vertical rod attached to one side of the spark gap, and a metal plate on the ground attached to the other side of the gap. He found that the waves emitted from this antenna bent with the earth and traveled farther than those emitted from a Hertz dipole. He achieved distances of about two miles, and these results were confirmed by the tests under the auspices of Preece. With the support of the British General Post Office, Marconi was able to complete development of a practical radio system.

Sending the result of a single spark is not of much use, for the amplitude of its associated wave dies out quickly. Even if there is little actual resistance in the LC circuit, the oscillations are damped the same as if there were large resistance because the transmitting circuit loses energy by producing sound and light at the gap, and by radiating electromagnetic energy from the antenna. Indeed, the whole objective is to radiate energy, and that dissipates energy from the oscillator circuit with a net effect that is identical to loss due to resistance. Hence the wave pattern generated by a single spark is short—on the order of a dozen or so cycles. To obtain usable signals, therefore, early radios were designed to produce a rapid sequence of sparks (keyed by electro-mechanically opening and closing the circuit switch with a vibrator). These rapid pulses (repeated at audible frequencies of about 500 Hz) were heard at the receiver as a continuous pitch, and thus dots and dashes could be sent at this pitch. This then is the basic **spark radio transmitter** diagrammed in figure 20.3. The figure shows the vibrator and shows also that the switching circuit is coupled to the antenna circuit with a coupling transformer. The capacitance is associated with the path between the antenna and the earth.

Because Marconi's antenna was vertical, he could extend its length and thereby reduce the basic frequencies of transmission from the MHz (megahertz) region that Hertz worked with to the hundreds of kHz (kilohertz) that is now familiar in AM radio.

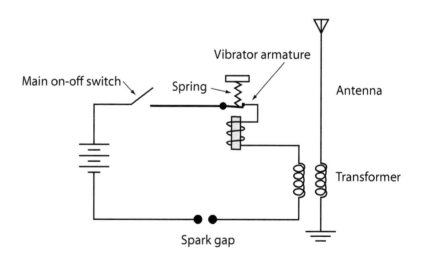

FIGURE 20.3 A simple spark radio. Closing the main switch attaches the battery to the circuit, causing a spark to jump across the gap and generating a wave pulse that is transmitted through the transformer to the antenna. A short time later the electromagnet pulls down the armature of the vibrator, breaking the circuit. A spring then pulls the armature back up, closing the circuit and enabling another pulse to be transmitted. The electromagnet vibrator has a frequency of about 500 cycles per second.

It was found that these waves bend with the earth, while higher-frequency waves tend to follow straight lines (like light, which is after all a very high-frequency electromagnetic wave). In Marconi's first transatlantic transmission, one end of his transmitter was attached to a high-flying kite, and the other was attached to a plate in the sea. The low frequencies bent around the earth and were received well below the horizon seen at the transmitter. (The existence of the ionosphere, which is responsible for the bending, was not known at the time, but its existence was conjectured by Heaviside and Kennelly.)

Sea Rescues

By 1909 it was fashionable for seagoing luxury liners to carry "wireless," capable of sending Morse code from sea to land, up to about 200 miles using a Marconi spark radio. Passengers could send and receive greetings, follow the stock market, and get important news. In January of 1909 the *Republic*, a ship of the White Star line heading for Europe off of the Nantucket sound, collided in the fog with the Italian ship *Florida*. The young wireless operator on the *Republic*, Jack Binns, working with a damaged wireless unit, managed to contact the operator at the Nantucket station, and eventually the not too distant *Baltic*, which came to the rescue. Although the *Republic* sank, everyone was saved, except the few who died as a direct result of the collision. Jack Binns became a national hero, and wireless took on new respect as an indispensable technology.

Three years later, in 1912, Jack Binns was offered the position of wireless operator on another luxury liner—the *Titanic*—but he turned down the offer because he was going to be married. After the iceberg collision, the *Titanic* wireless operator tried in vain to get immediate help, but tragically, the operator on the *Californian*, just 20 miles away, had gone to bed. Fortunately, the *Carpathia*, 58 miles distant, heard the distress call and arrived three and a half hours later, in time to save the passengers in lifeboats. Shortly thereafter Congress passed legislation requiring wireless units on all ships.

20.5 The Spark Bandwidth

A spark radio generates frequencies across an extremely broad band even if its underlying characteristic frequency is stable. This makes it essentially impossible for different stations to broadcast simultaneously since even if they use different base frequencies their outputs will interfere with each other. This "frequency pollution" greatly limited radio's utility. Indeed, in its early years radio was largely confined to the transmission of telegraphic messages to and from ships at sea, where interference was avoided because of the great geographic distance between different parties.

A wave generated by a single spark is a heavily damped sinusoid having the general form shown in figure 20.4. Such a signal with unit initial amplitude can be represented mathematically as

$$x(t) = e^{-\alpha t} \sin 2\pi f_0 t, \ t \geq 0.$$

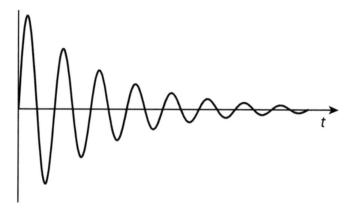

FIGURE 20.4 The wave generated by a spark circuit. If the spark circuit contains capacitance and inductance, a sinusoidal wave will be generated; however, circuit losses and radiation severely damp the oscillations.

The energy spectral density of this spark wave can be calculated by evaluating its Fourier transform. Specifically,

$$X(f) = \int_0^\infty e^{-\alpha t} \sin 2\pi f_0 t \, e^{-i2\pi ft} \, dt.$$

This can be evaluated from integral tables or by using the identity $\sin \omega_0 t = \frac{1}{2} i[e^{-i\omega_0 t} - e^{i\omega_0 t}]$, in which case every term in the integral is of the form e^{bt}, with different values of b. The integral of such a term is e^{bt}/b.

The overall result is

$$X(f) = \frac{i}{2} \left\{ \frac{1}{\alpha + i2\pi(f - f_0)} - \frac{1}{\alpha + i2\pi(f + f_0)} \right\}. \tag{20.7}$$

Notice that if α is close to zero, corresponding to light damping, the Fourier transform has a sharp peak at $f = f_0$, the frequency of the underlying sine wave. Indeed, if $\alpha = 0$, the value of $X(f_0)$ is infinite. The transform is nonzero at other frequencies because even if $\alpha = 0$, the signal is not a pure sine wave because it is zero for $t < 0$. The sudden start introduces other frequencies.

The energy spectral density, $|X(f)|^2$, can be approximated near f_0 as

$$|X(f)|^2 \approx \frac{1/4}{\alpha^2 + 4\pi^2(f - f_0)^2}. \tag{20.8}$$

For spark radiation the value of α is rather high, and hence the amplitude of the wave dies down considerably after only a few cycles. The period of a cycle is $T = 1/f_0$. If a signal of unit magnitude is reduced to $.05 \approx e^{-3}$ in seven cycles the corresponding α satisfies $e^{-7\alpha T} = e^{-3}$. Hence $\alpha = 3/(7T) \approx .43 f_0$. The AM radio band, at which spark radios operated, extends from roughly 500 kHz to 1500 kHz. Consider a spark radio operating at the middle of the band, at 1000 kHz. A plot of the corresponding spark energy spectral density when $\alpha = .43 f_0$ is shown in figure 20.5. Eighty percent of the energy is spread over the range between 800 kHz and 120 kHz—a band 400 kHz wide; still 20 percent of the energy is outside this broad band. Today AM stations

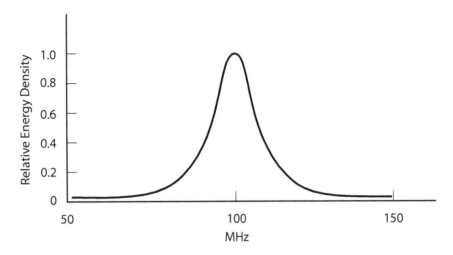

FIGURE 20.5 Normalized energy spectral density of a spark. The damped wave produces a very broad spectrum, so that it is almost impossible to distinguish different radio transmissions even if their base frequencies are widely separated.

operate completely within a band of only 10 kHz on each side of the base frequency. The spark transmitter with the stated α has an energy spectral density 40 times as broad. That is, the spark radio would be heard in 40 contiguous stations of today's radio.

Fortunately, a better method of transmission was developed, and spark radios were outlawed in 1923.

20.6 The Problems

Several problems plagued radio at the end of the 19th century. One was the inherent properties of the spark transmitter. Its broad spectrum polluted the frequency band and made it difficult for more than a few stations to transmit at a time. Another problem was that radio was essentially limited to the transmission of Morse code dots and dashes. For these reasons alone, radio was at that time considered to be a version of the telegraph, restricted to transmissions from one point to another, between a sender and one intended receiver. Indeed, radio was primarily used at sea as a substitute for telegraph. The fact that other stations might also receive an outgoing message was regarded as a disadvantage, which often required encryption for security.

A final problem lay with the receiver. In order to receive a message it was necessary to detect the presence of waves. It was impractical to use the simple loop with spark gap that Hertz had used in his experiments. Instead a **coherer** invented by Oliver Lodge, professor of experimental physics in Liverpool, was used even in early Marconi radios. This was a small glass tube filled with iron filings. The presence of electromagnetic waves caused the filings to cohere, thus decreasing the resistance through the tube. To restore the original state, the tube was mechanically tapped, and of course this took time. Accordingly, message transmission rates were very slow.

These limitations were surmounted over a period of less than 20 years, but the lives of some of those who achieved it were filled with litigation and tragedy, as well as inspiration.

20.7 Continuous Wave Generation

There were two early approaches to the generation of continuous waves. One was to build a very high-frequency alternator, which was essentially a high-speed rotating power generator. Alternators were being produced by companies such as General Electric, but they operated at much lower frequencies. Amazingly, however, suitable high-frequency alternators were built. A GE transmitter designed by Fessenden (who had previously worked for Edison) was tested by the navy in 1906. On Christmas and New Year's Eve that year, music and speech were transmitted to several navy ships in the North Atlantic.

A second method for generating continuous waves was to use an arc, familiar as the source of strong light in streetlamps and searchlights. As we know, the reason that the LC circuit of a spark transmitter cannot sustain its oscillations is that the circuit loses power due to circuit resistance and radiation. Those losses can be canceled out if some form of negative resistance is inserted in the circuit. At a certain region of its

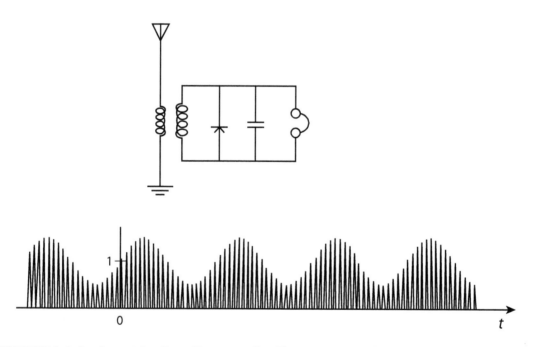

FIGURE 20.6 A simple crystal radio and its output. *Top*: The antenna is coupled with a transformer to the receiver. The crystal diode (whose symbol is arrow-like) shorts out the negative half of the signals, leaving positive pulses that charge the capacitor and are then discharged through the headphones. *Bottom*: The signal with the negative portion chopped off.

operating parameters, an arc exhibits negative resistance, and this can be used in an LC circuit to generate continuous waves. By 1912 this concept had been perfected to the point where arc radio transmitters were practical. From 1912 to 1917 they were, in fact, considered the best radios.

Improvement of wave detection progressed simultaneously with development of wave generation. The best detectors acted as diodes, letting current flow in one direction only. It was found that Carborundum and other crystals have this diode property, and based on this, the crystal detector was invented by Dunwoody of the De Forest Wireless Company in 1906. To operate a simple crystal radio, one moves a **cat's whisker** (a thin wire probe) on the surface of the crystal until a spot with diode properties is found, and radio speech or music is heard in the earphones.

With a diode in the circuit, only the positive part of a wave is transferred to a capacitor, which smoothes the wave and passes it to the headphones at the spark frequency. Or in the case of a continuous wave, the diode-capacitor combination fully recovers the amplitude of the AM signal. (See section 20.9.) Such a receiver is shown in figure 20.6.

20.8 The Triode Vacuum Tube

While working with his early electric lightbulbs, Edison noticed that the bulbs' interiors blackened with a thin layer of carbon dust. Investigating this, he constructed a bulb with an additional wire inserted a short distance from the bulb's filament. He found that if he connected this wire in the circuit so that it was positive relative to the filament, a small current flowed between the wire and the filament. If, on the other hand, the voltage of the wire was negative relative to the filament, no current flowed. He could not explain this result, nor put it to use, but nevertheless he patented it in 1884. The electron had not yet been discovered, so it was impossible for him to conceive the explanation, which is that the hot filament boils off electrons (with some carbon attached) and these are attracted to a positive element in the bulb and repelled from a negative element. Regardless of the explanation, it was clear that the device served as a one-way electric valve—a diode. One can only guess why Edison dismissed it as relatively unimportant, considering that his first inventions, the duplex and quadruplex telegraph systems, relied on the one-way devices of polarized electromagnets. But this was only 1884. Radio was young, and Edison had many other things to attend to.

Edison's experiment was repeated by others—one of these being Sir William Preece, chief engineer of the British General Post Office. It was Preece who called the phenomenon the **Edison effect**. The design of the bulb as a working diode was made by Fleming and patented in 1904; however, it found little use as a detector since it was less effective than crystal detectors that by then were available. The Edison effect, nevertheless, would later form the basis of modern electronics and open the possibility of much fuller exploitation of the gifts of frequency.

Lee de Forest was born in 1873, eight months earlier than Marconi. He possessed a high level of internal drive, and was determined to invent something important. Borrowing funds for subsistence, he put together a radio system and formed a small three-person company. Noting that Marconi had achieved great fame in 1899 for providing coverage of a yacht race in New York, de Forest arranged to provide coverage of the America's Cup races in 1901. However, Marconi also provided coverage of the race. The result was a disaster—due to the fundamental problem of spark radio. The signals from the two competitors interfered with each other, and consequently neither was able to successfully transmit news of the races.

In 1906, while working to design a better receiver detector by improving the diode characteristics of an Edison effect bulb, de Forest had the inspiration to add an additional element between the cathode (heated by the filament) and the anode (the positive electrode). This new element, shaped by zigzagging a wire back and forth a few times, he termed the **grid**. A small negative voltage applied to the grid shielded the electrons from the anode and hence drastically reduced the current between cathode and anode. Small variations in grid voltage controlled similar but stronger current variations in the anode-to-cathode circuit. Therefore, if a receiving antenna were attached to the grid, the tiny voltage could indirectly drive a large current in the headphones circuit. He called his device an **audion**.

De Forest used the audion only as a superior detector, which it certainly was. But he did not immediately seriously pursue its development.

Edwin Armstrong was the true four-time genius of radio. He was full of energy; he studied hard, rode a red motorcycle, and enjoyed climbing high towers. In 1912 while a student at Columbia University, he undertook a serious investigation of the audion, meticulously measuring currents and voltages. It occurred to him one day that if the audion produced an amplification of signal, then perhaps he could take the resulting signal and send it back through the audion again for greater amplification. Perhaps the signal could be sent around and around hundreds or thousands of times. During the summer he accomplished this by placing an output circuit coil in close proximity to the antenna coil so that by induction the output would reenter the circuit at the antenna. He had invented **feedback**, although he called it **regeneration**. The results were astounding. His regenerative receiver was able to bring in signals from Ireland and Hawaii, and he was able to shed the earphones in favor of a loudspeaker. This was a great leap forward in radio, and was Armstrong's first stroke of genius.

Armstrong noticed that if he increased the regeneration, the circuit oscillated at radio wave frequency. Using this knowledge, he modified his regenerative circuit to serve as a transmitter, generating continuous waves. Now the massive alternators or spark-gap machines could be replaced by a small circuit containing an audion. This revolutionized radio transmission, and was Armstrong's second stroke of genius. There were more yet to come.

Armstrong faced nasty patent disputes, especially with Lee de Forest. It was clear that de Forest did not understand the workings of the audion, believing incorrectly that its action depended on the presence of special gases in the tube. He also did not seem to understand its significance as an amplifying device, at least initially. But later, after Armstrong's results were established, de Forest claimed to have discovered feedback and oscillation years earlier. The two of them battled in court repeatedly.

De Forest developed another innovation that is now part of daily life. In New York in 1907 he acquired an arc generator and sent music over the airwaves to "distribute sweet melody broad-cast over the city & sea." He may have been the first to use the term *broadcast* and, more importantly, the first to see the potential of radio as more than a wireless telegraph or telephone designed for point-to-point communication between two parties. He saw that radio could reach everyone. He extended his broadcasts to news events and even to live opera.

The first real radio broadcast station was set up in 1909 by Charles Herrold and Frank Schmidt in San Jose, California. The first major news broadcast (by another station) was on election eve in 1916. At 11:00 PM Eastern Standard Time it was (incorrectly) declared that Charles Evans Hughes won the presidential election! Radio, as conceived by de Forest and made possible by his invention, introduced instantaneous mass communication.

Although de Forest perhaps did not understand the physics of his triode or the full significance of the feedback that it made possible, it was the single most important technological invention associated with the mastery of frequency, leading to continuous-wave transmission, modulation, heterodyning, efficient receivers, oscillators, telephone repeaters, audio equipment, and of course the future developments of television, computers, and all other miraculous devices dependent on amplification of electric signals.

The electromagnetic relay used in Morse's telegraph, the liquid microphone used in Bell's revised telephone, and the triode audion developed by de Forest—all three of these were amplification devices that enabled a small signal to control a larger one. Morse, the least technical of the three inventors, seems to have grasped this immediately for the electromagnet. Bell gave the microphone only secondary importance in his patent application, and it was not embodied in his original telephone. De Forest did not at first appreciate the amplification feature of his audion, for he used it only as an improved diode. So here we have three of the greatest inventions in electrical communication all based, at root, on the common principle of amplification, yet even when the principle was at hand, its significance was poorly understood by two out of three of the very individuals whose inventions depended on it.

20.9 Modulation Mathematics

Continuous waves are useful only if information can be efficiently attached to them. This is accomplished by **modulating** the waves. The most obvious method is **amplitude modulation** (AM) in which the amplitude of the continuous (carrier) wave is varied in proportion to the communicated signal of speech, music, video, or a train of pulses. Using the principle that any signal can be decomposed into sinusoids and then reconstituted, AM is analyzed by supposing that the signal is itself a sinusoid, but of much lower frequency than the radio carrier frequency. For example, the carrier may have a frequency $f_c = 1,000$ kHz (which is the about the middle of the standard AM band), while the signal may have a frequency somewhere between 100 and 10,000 hertz, the frequency range of speech and low-fidelity music.

If a radio-frequency carrier $\cos 2\pi f_c t$ at frequency f_c is amplitude modulated by a signal $s(t)$, the result is the product wave

$$v(t) = [1 + as(t)]\cos 2\pi f_c t,$$

where a is a modulation coefficient. Generally a is chosen so that $1 + as(t)$ is always positive. For the signal $s(t) = \cos 2\pi f_s t$, the modulated wave is

$$m(t) = [1 + a\cos 2\pi f_s t]\cos 2\pi f_c t, \qquad (20.9)$$

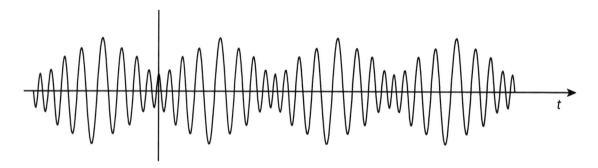

FIGURE 20.7 The result of amplitude modulation. The amplitude of the carrier frequency is proportional to the signal. In this figure the signal is itself a low-frequency sinusoid.

and a is taken with $0 < a < 1$. See figure 20.7. Using the identity

$$\cos A \, \cos B = \frac{1}{2}[\cos(A + B) + \cos(A - B)],$$

we see that the modulation term introduces sum and difference frequencies, often called **beat frequencies**. Explicitly, for the modulated wave (20.9)

$$m(t) = \cos 2\pi f_c t + \frac{a}{2}\cos 2\pi (f_c + f_s)t + \frac{a}{2}\cos 2\pi (f_c - f_s)t.$$

Hence, overall, power components are present at the carrier frequency f_c and at the sum and difference frequencies $f_c + f_s$ and $f_c - f_s$.

For a general signal made up of many sinusoids as expressed by its Fourier transform, each frequency component contributes two beat frequencies to the final composite. In this final composite the original Fourier transform will be copied twice, as illustrated in figure 20.8.

The frequency spectrum (that is, Fourier transform) will in general be complex valued, so the figure shows only magnitudes. In essence, the original spectrum is shifted upward to the carrier frequency f_c in two copies, termed **sidebands**. The sideband above the carrier corresponds to the sum frequencies, and the lower sideband is reversed in shape corresponding to the difference frequencies. There is also a signal at the original carrier frequency f_c, corresponding to the background level of the carrier.[2] If the frequency width (the **bandwidth**) of the original signal is W, then the bandwidth of the modulated wave, including its two sidebands, is $2W$.

Figure 20.8 actually shows only one-half of the spectrum. Recall that there is a mirrored copy of it in the negative part of the frequency axis. This is a complex conjugate of the spectrum in the positive half. Typically, only the positive axis need be shown.

Although the frequency spectrum of an AM wave is complicated, it is relatively easy to recover the original signal with the simple radio receiver shown earlier in

---

[2] It is assumed in the figure that the carrier frequency does not overlap with the signal spectrum, and this assumption is certainly satisfied by standard AM radio.

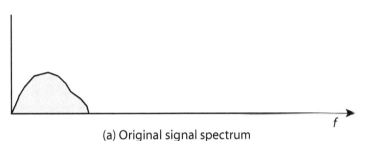

(a) Original signal spectrum

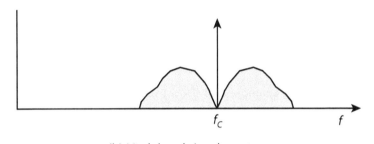

(b) Modulated signal spectrum

FIGURE 20.8 The frequency spectrum of AM. The spectrum of the modulated wave consists of the carrier and two sidebands. The upper sideband is a copy of the original spectrum, translated up by the carrier frequency. The lower sideband is a reflected complex conjugate of the upper sideband.

figure 20.7. The diode shorts out the negative part of the signal, leaving a chopped-off version of the modulated signal that charges up the capacitor. The high frequencies are suppressed by the RC circuit, consisting of the capacitor and the headphones, in much the same way that telegraph signals are smoothed as they traverse undersea cables. The result is that audio frequencies pass through to the headphones in nearly original form.

20.10 Heterodyne Principle

Edwin Armstrong joined the army in 1917 and was stationed in France, where he assisted with military radio operations as an expert. He was, however, perplexed by two related issues. The Germans were communicating at frequencies of 500,000 to 3 million hertz, and the Allies' equipment could not effectively tune to frequencies that high. The other issue was that Armstrong wondered whether the high-frequency (10 million Hz or more) radiation generated by enemy aircraft engines could be tracked and thus used to guide antiaircraft guns. He addressed both of these questions by using the concept of heterodyning invented by Fessenden in 1901.[3]

---

[3] Heterodyne: from the Greek meaning "other forces."

The **heterodyne principle** is actually a specialized version of AM modulation. If a high-frequency wave is mixed (that is, multiplied) by a wave of lower frequency, one result is a sideband at the difference frequency. In this way, a high-frequency wave can be shifted down to a lower frequency.

Armstrong used this principle to not only solve the immediate issues, but also as the basis of an entirely new receiver for standard AM radio: the **superheterodyne** receiver. It works by shifting the incoming modulated signal to a standard intermediate frequency. The remainder of the receiver design can then be optimized to amplify and detect at that one frequency. Virtually all commercial AM radios made since that time are based on Armstrong's design. The superheterodyne was Armstrong's third stroke of genius.

Example 20.4 (How to raise your voice). Imagine a baritone who sings in a narrow range below the middle C tone. Suppose a tape recording of one of his songs is made and is played back at double speed. You know what will happen. The pitch of his music will be raised an octave and the whole song will be twice as fast as before. Suppose, alternatively, that the recording is multiplied by a pure middle C note.[4] If this new recording were played, an unpleasant mixture of high and low notes would be heard. Since the new tape represents pure amplitude modulation, its signal will contain two copies of the original spectrum: one above middle C and a reversed one below (because a signal at f is transformed to $C + f$ and $C - f$). If the result is passed through a filter that greatly attenuates the frequencies below middle C (as by playing through a tiny tweeter loudspeaker), the song will be heard, sung by a baritone with his voice raised by approximately an octave,[5] and the song will progress at normal speed.

FIGURE 20.9 A baritone raises his voice. By application of the heterodyne principle, a baritone can raise his voice without changing the speed.

---

[4] The A above middle C is the concert tuning note of 440 Hz. Middle C is nine notes below that, so its frequency is $440 \times 2^{-9/12} = 261.625565$ Hz.

[5] The notes other than C will be distorted somewhat because the transformation adds 261.625565 Hz to each note rather than multiplying their frequencies by two.

20.11 Frequency Modulation

Frequency modulation (FM) was proposed in the early days of radio as a possible way to reduce reception static, but the method was supposedly proved to be worthless by several theoretical investigations. Armstrong believed, perhaps intuitively, that it could work. He embarked on a multiyear intensive exploration of the concept, even in the face of continued skepticism by established experts. The original idea was that FM could reduce the required bandwidth and thereby limit the noise power. Finally, after years of work, Armstrong decided instead that the bandwidth should be widened! He went on to develop a system that was nearly impervious to noise and had greater fidelity than normal AM radio. This was Armstrong's fourth stroke of genius.[6]

In FM, the frequency of the carrier is varied in proportion to the amplitude of the low-frequency audio signal. The explicit form of the transmitted wave is, accordingly,

$$v(t) = \cos\left(\omega_c + as(t)\right)t,$$

where for convenience we use $\omega_c = 2\pi f_c$, and where again a is a design factor that determines the degree of modulation. The effect of FM modulation on a simple sinusoidal signal is shown in figure 20.10. Note that the amplitude of the transmitted wave is constant, and only the frequency varies. Most noise is an amplitude effect, and hence by clipping a received FM signal so that the amplitude is constant, the pure noise-free signal will be recovered.

The Fourier series associated with a carrier frequency modulated by a sinusoid can be calculated, but it is quite complicated. Consider a carrier modulated by a single sine wave producing the composite signal

$$v(t) = \sin\left(\omega_c + a\sin\omega_s t\right)t, \tag{20.10}$$

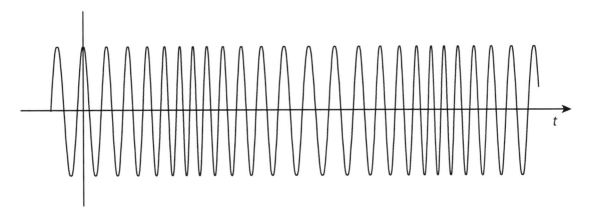

FIGURE 20.10 Frequency modulation. In FM the instantaneous frequency of the carrier is shifted in proportion to the amplitude of the signal.

---

[6]There was yet another earlier stroke of genius, the **superregenerative** receiver that, by squelching possible oscillations, amplified signals by a factor of 100,000 in one stage. But it was not commercially successful.

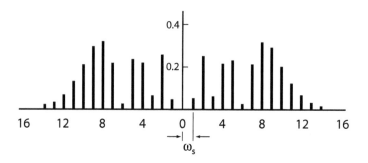

FIGURE 20.11 A spectrum of FM. Shown are the magnitudes of the Fourier coefficients when the carrier is modulated by a single sinusoid. The frequencies are relative to the carrier ω_c. The pattern is the result of a particular design choice ($a = 10$). The individual spectral lines are separated by the frequency of the modulating sinusoid.

where ω_c is the carrier frequency and ω_s is the information frequency. The single modulating sinusoid $\sin \omega_s t$ produces an infinite set of frequencies in the FM wave. These frequencies are located symmetrically above and below the carrier frequency ω_c, so just as in the case of AM, there are two sidebands, although the structure is more complicated in FM than in AM.

One possible pattern, due to a single modulation frequency, is shown in figure 20.11, where the magnitudes of the Fourier coefficients are shown. In commercial FM transmission, most power is in the range below 7.5 kHz, and regulations allow a frequency band of 200 kHz.

The Belated Success of FM

FM was fully developed by Armstrong in 1933, a time when the huge company RCA was making good profit with the sale of superheterodyne AM radios and was about to launch television. FM would be used for the sound portion of TV, but RCA and other companies did everything they could to thwart FM radio and bypass Armstrong's FM patents. Armstrong continued to deploy FM and is thus responsible for both its inception and commercialization. But it was not a happy existence. Armstrong constantly fought patent battles. In a fit of despondency, he took his own life in 1954 by jumping from his apartment in New York. His wife continued his legal battles with up to 18 companies and eventually won or settled them all. Today a majority of all radio sets sold include FM, and television, microwave links, and space communications all use FM.

More recently, the complexity of the FM spectrum suggested to Professor John Chowning of Stanford University that the sounds produced directly by FM waves might be musically interesting, since they contain a rich pattern of overtones. He designed a music synthesizer based on that idea, and patented the concept. Today, most sophisticated music synthesizers use that system.

20.12 EXERCISES

1. (Homogeneous) Find the solutions to the equations below with the given initial conditions (at $t = 0$).

   (a)
   $$\frac{d^2 x(t)}{dt^2} + \frac{dx(t)}{dt} - 2x(t) = 0, \ x(0) = 1, \ \frac{dx(0)}{dt} = 4.$$

   (b)
   $$\frac{d^2 x(t)}{dt^2} + x(t) = 0, \ x(0) = 1, \ \frac{dx(0)}{dt} = 0.$$

2. (Nonzero R^*) Derive equation (20.6).

3. (Spark limit) What is the energy spectral density of a (hypothetical) spark radio pulse with zero damping? What is the energy spectral density magnitude at $\omega = 0$? How much energy is radiated by such a pulse? Hint: consider it over time, not frequency.

4. (Antenna height) About how tall is a quarter-wave monopole antenna for FM radio?

5. (Circuit Q) A measure of the quality of an oscillatory system is its Q, defined as

   $$Q = 2\pi \frac{\text{maximum energy stored}}{\text{energy dissipated per cycle}}$$

   at the resonant (assuming no dissipation) frequency. For a pendulum, the stored energy is transferred back and forth between potential energy (the height of the pendulum bob) and kinetic energy (the speed of the bob). The energy dissipated is due to friction. In a series RLC circuit, the stored energy is transferred back and forth between the capacitor and inductor and the energy dissipated is the loss due to the resistor. The Q of many practical resonant circuits is several hundred or more. The energy stored in an inductor when the current through it is I is $\frac{1}{2}LI^2$. The energy stored in a capacitor when the voltage across it is V is $\frac{1}{2}CV^2$. The rate of energy dissipation is RI^2. The energy dissipated is approximated by assuming that the current is an undamped sine wave over a resonant cycle. Show that Q for the series RLC circuit of figure 20.1 is $Q = \omega_0 L / R$. Hint: To calculate the maximum energy stored, consider the point in the cycle where the current is maximum and the voltage across the capacitor is zero. Use $\int_0^{2\pi} \sin^2 t \, dt = \pi$ in the calculation of dissipation.

6. (Resonance calculation) Consider the circuit of figure 20.12, which is a series RCL circuit with sinusoidal input voltage applied. Suppose this voltage is represented in complex form as $e^{i\omega t}$.

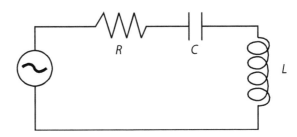

FIGURE 20.12 A stimulated resonant circuit.

(a) Derive the second-order differential equation for current I

$$L\frac{d^2 I(t)}{dt^2} + R\frac{dI(t)}{dt} + \frac{I(t)}{C} = \frac{de^{i\omega t}}{dt},$$

and find a solution of the form $A(\omega)e^{i\omega t}$.

(b) With ω_0 equal to the ideal resonant frequency, show that

$$\frac{A(\omega)}{A(\omega_0)} = \frac{1}{1 + iQ\left(\frac{\omega}{\omega_0} - \frac{\omega_0}{\omega}\right)},$$

where Q is defined as in exercise 5.

(c) Notice that at ω_1 and ω_2 satisfying

$$\frac{\omega}{\omega_0} - \frac{\omega_0}{\omega} = \pm\frac{1}{Q}$$

the power (proportional to the square of the magnitude of current) is one-half of what it is at ω_0. The two frequencies ω_1 and ω_2 with $\omega_1 < \omega_0 < \omega_2$ are called the half-power points. The range $\omega_1 \leq \omega \leq \omega_2$ is called the **bandwidth** of the circuit. (See figure 20.13.) Calculate the bandwidth in terms of Q and ω_0; that is, find $\omega_2 - \omega_1$.

7. (*Q* of a radio) The AM frequency band is centered at 1 MHz, and the bandwidth of each station is 10 kHz. What order of magnitude do you estimate as the Q of many of the resonant circuits in AM radios?

8. (AM wave) Consider the ideal AM modulated wave $v(t) = s(t)\cos 2\pi f_c t$. Using the Fourier transform $S(f)$ of the applied signal, show explicitly that the Fourier transform of the AM wave is

$$V(f) = \frac{1}{2}[S(f_c + f) + S^*(f_c - f)].$$

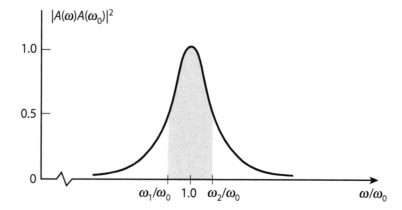

FIGURE 20.13 Power ratio. *Q* determines the bandwidth of a resonant circuit. The higher the *Q*, the narrower the bandwidth.

9. (Speech scrambler) A speech scrambler is a security device that attaches to the mouthpiece of a telephone. It scrambles the speech so that casual eavesdroppers cannot understand the message. The voice message is assumed to have a spectrum falling between −5 kHz and 5 kHz. The magnitude of the spectrum is depicted in figure 20.14 as a triangular shape. The actual shape may be different but, since the message is real, the shape is always symmetric about the vertical axis.

The scrambler works by multiplying the signal by a continuous wave of 20 kHz and then passing the result through a perfect high-pass filter that only accepts frequencies of $|f| \geq$ 20 kHz. The result is then multiplied by a signal of 25 kHz and then passed through a perfect low-pass filter that only accepts frequencies of $|f| \leq$ 20 kHz. The result $y(t)$ is the scrambled voice message.

(a) Draw a diagram of the magnitude of the spectrum of the signal, say $x(t)$, at the point right after the high-pass filter.

(b) Draw a diagram of the magnitude of the spectrum of $y(t)$.

(c) How can the scrambled signal be unscrambled?

10. (Parallel circuit) Assume that the parallel circuit of figure 20.15 generates a current of $e^{i\omega t}$ upward through the generator shown at the left side of the circuit. Using the fact that the sum of the currents through the three circuit elements must equal this generated current, find the voltage of the form $B(\omega)e^{i\omega t}$ across the circuit.

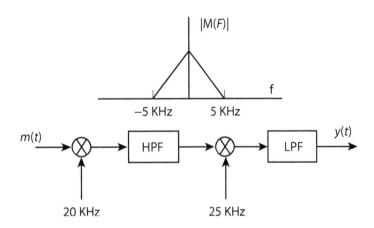

FIGURE 20.14 Speech scrambler.

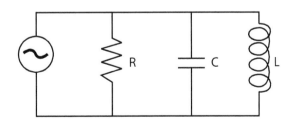

FIGURE 20.15 A parallel RLC circuit.

20.13 Bibliography

The quoted conversation of Hertz upon the success of his experiment is from [1]. For history of more general advances in electricity and electromagnetic theory, see [2], [3], and [4]. A wonderfully fascinating account of early radio is [5]. A pleasant and accessible introduction to the theory of radio is [6], from which exercise 9 is adapted. The story about Jack Binns and the *Republic* is presented dramatically and informatively in [7]. See [8] for a popular introductory text that, among many other things, explains Q.

The history of the early days of radio after continuous waves were produced is beautifully presented in [9]. Other good historical accounts are [10], [11], [12], and [13]. A lively account of the relations between Edwin Armstrong, Lee de Forest, and David Sarnoff (of RCA), recounting their contributions and differences, is [14]. See also [15]. The details on the FM frequency spectrum are adapted from [16].

References

[1] "Heinrich Hertz." http://www.webstationone.com/fecha/hertz.htm.

[2] Hunt, Bruce J. *The Maxwellians*. Ithaca: Cornell University Press, 1991.

[3] Appleyard, Rollo. *Pioneers of Electrical Communication*. Freeport, N.Y.: Books for Libraries Press, 1968.

[4] Skilling, H. H. *Exploring the Electrical Age*. New York: Ronald Press, 1948.

[5] Aitken, Hugh G. J. *Syntony and Spark—the Origins of Radio*. New York: Wiley, 1976.

[6] Nahin, P. J. *The Science of Radio*. 2nd ed. Woodbury, N.Y.: American Institute of Physics Press, 2001.

[7] *Rescue at Sea*. PBS Home Video, WGBH Educational Foundation, 1999.

[8] Smith, Ralph J. *Circuits, Devices, and Systems*. 4th ed. New York: Wiley, 1984.

[9] Aitken, Hugh G. J. *The Continuous Wave: Technology and American Radio, 1900–1932*. Princeton: Princeton University Press, 1985.

[10] MacLaurin, W. J. *Invention & Innovation in the Radio Industry*. New York: Macmillan, 1949.

[11] Hijiya, J. A. *Lee de Forest and the Fatherhood of Radio*. Bethlehem, Pa.: Lehigh University Press, 1992.

[12] Wedlake, G.E.C. *SOS: The Story of Radio-Communication*. New York: Crane, Russak, 1973.

[13] Bray, John. *The Communications Miracle: The Telecommunication Pioneers from Morse to the Information Superhighway*. New York: Plenum, 1995.

[14] Lewis, Tom. *Empire of the Air*. New York: Edward Burlingame Books, Harper Collins, 1991.

[15] Burns, Ken, dir. *Empire of the Air*. PBS Home Video, Florentine Films, 1991.

[16] Seely, S. *Radio Electronics*. New York: McGraw-Hill, 1956.

21

SAMPLING AND CAPACITY

This chapter brings us back to the beginning—back to Shannon's theory. Shannon's simple definition of entropy for a source with a discrete channel produced the notion of channel capacity, the framework for data compression, a theory of classical encryption, and an understanding of the limits to information organization and retrieval. In this chapter, Shannon's work once again is central. His formula for the capacity of a continuous channel in the presence of additive white Gaussian noise is the basis for analysis of many communication technologies. As with his theory for discrete channels, his theory of continuous channels captures the essence of communication issues. Today it can be regarded as a crucial step toward the mastery of frequency.

The chapter begins with the extension of entropy to sources having a continuum of possible values, rather than a finite number. Such sources occur whenever the possible message is any real number—for instance, a source reporting outside temperature, rather than simply whether it is sunny or cloudy; or signals derived from speech, pictures, or other continuous values.

Shannon went further and derived a formula for the capacity of a channel that operates continuously in time, within a frequency bandwidth. This is considered by many to be Shannon's most important result.

A particular conclusion of Shannon's capacity theorem is that capacity increases (but not linearly) with bandwidth. This led to the development of spread-spectrum technology, which as we shall see was first proposed by the beautiful movie star Hedy Lamarr.

21.1 Entropy

The definition of entropy for continuous channels is straightforward, but subtle.

Let $p(x)$ be a probability density on the real numbers. This is a function $p(x)$ for $-\infty < x < \infty$ that is never negative and integrates to 1. Suppose that this probability density is associated with a possible signal x that will be sent.

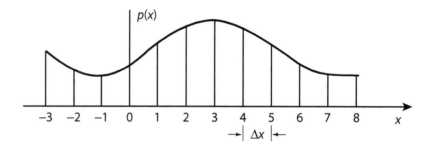

FIGURE 21.1 Discrete approximation. A continuous variable x is approximated by points Δx apart.

As an approximation to this continuous signal, the axis of real numbers is discretized into equally spaced steps of width Δx, and accordingly, the possible signals are restricted to the magnitudes: ... $-2\Delta x$, $-\Delta x$, 0, Δx, $2\Delta x$ .... The corresponding values of x are denoted $x_i = i\Delta x$, for $-\infty < i < \infty$. The probability of level x_i is defined to be $p(x_i)\Delta x$. See figure 21.1.

The entropy H_D of a signal discretized this way is easily calculated (using the basic definition of section 2.3) as

$$H_D(p) = -\sum_{-\infty}^{\infty} p(x_i)\Delta x \, \log\left[\, p(x_i)\Delta x\,\right]$$

$$= -\sum_{i=-\infty}^{\infty} p(x_i)\log\left[\, p(x_i)\,\right]\Delta x - \sum_{i=-\infty}^{\infty} p(x_i)\log\left[\Delta x\,\right]\Delta x$$

$$= -\left[\sum_{i=-\infty}^{\infty} p(x_i)\log\left[\, p(x_i)\,\right]\Delta x\right] - \log\left[\Delta x\,\right] \tag{21.1}$$

because it can be assumed that $\sum_{i=1}^{n} p(x_i)\Delta x = 1$. Notice that the first term in the last expression can be approximated by an integral as $\Delta x \to 0$. However, the second term goes to infinity because $\log\left[\Delta x\,\right]$ goes to minus infinity as Δx goes to zero. Hence the entropy H_D goes to infinity. This reflects the fact that with an infinite number of possibilities, an infinite amount of information is conveyed by transmission of any particular point.

The entropy of the continuous distribution is, however, defined to be the finite part, corresponding to the integral, since the other part does not depend on the probabilities.

It may seem odd to throw out the infinite part and keep only the finite part, but this is why the definition is subtle. The reason this is a useful definition is that calculations of channel capacity involve differences of entropy—the extra term, common to all entropies in the capacity expression, cancels out and hence can be ignored. The formal definition is given in what follows.

Entropy of a continuous random variable. The entropy of a random variable x with probability density $p(x)$ is

$$H(p) = -\int_{-\infty}^{\infty} p(x)\log\left[\, p(x)\,\right]\mathrm{d}x.$$

If one wishes to keep the discretized version in mind, the discretized version can be approximated as

$$H_D(p) = -\int_{-\infty}^{\infty} p(x)\log\,[p(x)]\mathrm{d}x - E,$$

where E is the extra term $\log[\Delta x]$, which is common to all entropies with that same discretized Δx.

Example 21.1 (Uniform density). Suppose that

$$p(x) = \begin{cases} 1/a & 0 \le x \le a \\ 0 & \text{otherwise.} \end{cases}$$

Then

$$H(p) = -\int_0^a \frac{1}{a}\log\left(\frac{1}{a}\right)\mathrm{d}x = \log a.$$

This is in accord with the entropy of an analogous finite source. A source with n equally likely symbols is $\log n$. For a uniform density, the width of the density a is analogous to n.

Example 21.2 (Gaussian density). A most important continuous variable density is the Gaussian (or normal) density. With zero mean value and standard deviation σ it is

$$p(x) = \frac{1}{\sqrt{2\pi}\,\sigma}e^{-x^2/(2\sigma^2)}. \tag{21.2}$$

The corresponding continuous entropy can be computed by a series of simple steps (remembering that $\log x \equiv \log(e)\ln x$, $\ln e = 1$, and that $E(x^2) = \int_{-\infty}^{\infty} x^2 p(x)\mathrm{d}x = \sigma^2$). Thus

$$\begin{aligned}
H(p) &= -\int_{-\infty}^{\infty} p(x)\log\left[\frac{1}{\sqrt{2\pi}\,\sigma}e^{-x^2/(2\sigma^2)}\right]\mathrm{d}x \\
&= -\int_{-\infty}^{\infty} p(x)\log(e)\ln\left[\frac{1}{\sqrt{2\pi}\,\sigma}e^{-x^2/(2\sigma^2)}\right]\mathrm{d}x \\
&= -\log(e)\int_{-\infty}^{\infty} p(x)\left[\frac{-x^2}{2\sigma^2} - \ln\sqrt{2\pi\sigma^2}\right]\mathrm{d}x \\
&= \log(e)\left[\frac{1}{2} + \ln\sqrt{2\pi\sigma^2}\right] = \log(e)\left[\frac{1}{2} + \frac{1}{2}\ln 2\pi\sigma^2\right] \\
&= \log(e)\left[\frac{1}{2}\ln e + \frac{1}{2}\ln 2\pi\sigma^2\right] = \frac{1}{2}\log(e)\ln 2\pi e\sigma^2 \\
&= \frac{1}{2}\log 2\pi e\sigma^2 \ \text{bits.} \tag{21.3}
\end{aligned}$$

This final result is surprisingly simple.

The Maximum Property of a Gaussian Density

The Gaussian density has a maximum property that is important because it is the density that one must strive for to attain a maximum rate of information transmission.

Specifically, the Gaussian density maximizes entropy under a constraint on the variance that can be considered the level of average energy, or power. In mathematical terms the maximization problem is

$$\text{maximize} - \int_{-\infty}^{\infty} p(x) \log [p(x)] \, dx \tag{21.4}$$

$$\text{subject to} \int_{-\infty}^{\infty} p(x) \, dx = 1$$

$$\int_{-\infty}^{\infty} x^2 p(x) \, dx \le P.$$

The first line is the formula for the value of continuous entropy in terms of the density function $p(x)$. The first constraint states that the area under the density function must be 1, so that $p(x)$ is a legitimate density function. The second constraint is the average energy or power constraint. Energy is proportional to the square of the signal, so the constraint is on expected energy. Power is energy per unit time; hence, by suitable normalization the constraint can be considered to be on average power.

This problem is not difficult to solve (see exercise 1). The result is the Gaussian density (21.2) with $\sigma^2 = P$.

Recall that for a finite set of symbols, entropy is maximized when the underlying probabilities are all equal. A uniform density does not make sense if the symbols consist of all real numbers $-\infty < x < \infty$. In practical situations, an energy or power constraint is often a necessity, inherent in the technology. When there is such a constraint, entropy is maximized by a Gaussian density.

21.2 Capacity of the Gaussian Channel

Let us turn now to one of the most important results in Shannon's information theory—the formula for the capacity of a channel subjected to additive Gaussian noise. The derivation is remarkably simple and has a nice interpretation.

Consider a general channel that can transmit continuous-valued signals, but that is subject to the requirement that the average energy per signal, the power, must be less than P. Signals are corrupted by additive noise with average power N. The noise is statistically independent of the signal.

Let X represent the transmitted signal, Y the received message, and Z the noise. Then $Y = X + Z$, and X and Z are independent.

The capacity is the maximum value of[1]

$$I(X; Y) = H(Y) - H(Y|X).$$

[1] Note that this is the difference of two entropies, so the extra term in equation (21.1) cancels.

We have

$$
\begin{aligned}
I(X;Y) &= H(Y) - H(Y|X) \\
&= H(Y) - H(X+Z|X) \\
&= H(Y) - H(Z|X) \\
&= H(Y) - H(Z) \\
&= H(\text{signal} + \text{noise}) - H(\text{noise}). \qquad (21.5)
\end{aligned}
$$

The first step is a definition, and the second step should be clear. The third follows because the information about X, given that X is known, is zero. Hence, $H(X+Z|X) = H(Z|X)$. The fourth step follows from the independence of Z and X.

Let us look at some special cases. Suppose, for example, that the noise power is very small. Then the information transmitted through the channel is nearly equal to $H(\text{signal})$ and the entropy of the noise is very small. (Remember that these entropy values omit the discritization terms that go to infinity.) Hence, $I(X;Y) \approx H(\text{signal})$. Conversely, suppose that the noise power is very large and the signal power is small. Then $I(X;Y) \approx H(\text{noise}) - H(\text{noise}) = 0$, and essentially no information is transmitted. Note that this also explains why both the signal and the noise are in the first term. If only the signal were in that term, large noise would produce negative entropy, which is impossible. The expression $I(X;Y) = H(\text{signal} + \text{noise}) - H(\text{noise})$ is therefore a highly intuitive result for channels with independent additive noise.

Formula (21.5) can be used to find the capacity of a channel with additive Gaussian noise under a power constraint. The capacity is found by maximizing $I(X;Y)$ subject to the constraint that the signal have average power of S. The noise is given as Gaussian with average power N. Its entropy is therefore, from equation (21.3), $H(\text{noise}) = \frac{1}{2}\log 2\pi eN$.

The power of the signal plus the noise is $S + N$. From the maximum property of the Gaussian density under a power constraint, the maximum of $H(\text{signal} + \text{noise})$ is achieved by making the signal Gaussian with average power S, for this will render the resulting signal plus noise Gaussian with average power $S + N$ (since the sum of two Gausian random variables is itself Gaussian). Therefore the capacity is

$$
C = \frac{1}{2}\log\left[2\pi e(S+N)\right] - \frac{1}{2}\log\left[2\pi e(S+N)\right] \qquad (21.6)
$$

$$
= \frac{1}{2}\log\left(S+N\right)/N) \qquad (21.7)
$$

$$
= \frac{1}{2}\log\left(1 + \frac{S}{N}\right). \qquad (21.8)
$$

This result is formalized by the following statement.

Capacity of a channel with Gaussian noise, subject to a power constraint.
The capacity of a continuous channel with signal power constrained to S, but subject to independent additive Gaussian noise of power N, is

$$
C = \frac{1}{2}\log\left(1 + \frac{S}{N}\right). \qquad (21.9)
$$

Example 21.3 (S/N = 1). If the signal and noise have equal average power, then $C = \frac{1}{2}\log 2 = \frac{1}{2}$ bit per symbol.

Example 21.4 (California temperature). The autumn temperature T in California is likely to be somewhere in the range of 65°–80° Fahrenheit. Suppose a friend measures the temperature as 72° and reports that value to you. How much information has been transmitted? Suppose that your friend's crude thermometer is accurate to about 1°. In that case the standard deviations of the temperature T and the measurement noise N are about $80° - 65° = 15°$ and 1°, respectively. The variances are the squares of these. Hence, the mutual information, the information transmitted, is from (21.5) equal to $I(T; T + N) = \frac{1}{2}\log(1 + 15^2) \approx 3.5$ bits.

As a comparison, you could assume that the 16 temperatures $65°, 66°, \ldots, 80°$ were equally probable and that the measurement accurately described the correct integer value. The corresponding discrete-value entropy would then be $\log 16 = 4$ bits.

21.3 Sampling Theorem

A continuously varying signal can be converted to a series of separate values by periodically sampling the continuous signal as illustrated in figure 21.2.

It seems clear that, in general, the samples provide only an approximate representation of the original continuous message. The **Nyquist–Shannon sampling theorem** states that, in fact, a band-limited signal can be accurately reconstructed from its samples provided that the sampling rate is high enough. Specifically, if the signal bandwidth is W cycles per second, the sampling rate must be at least $2W$ per second.

Theorem 21.1 (Sampling theorem). Suppose a function $x(t)$ has a frequency spectrum (Fourier transform) that is band-limited to frequencies less than W cycles per second. Then the function can be completely reconstructed from samples taken at the uniform rate of $2W$ samples per second. (That is, the samples are $1/(2W)$ seconds apart.)

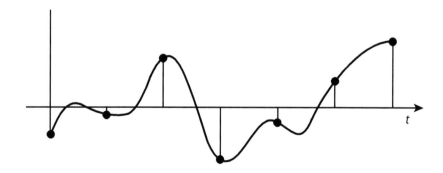

FIGURE 21.2 Sampling a continuous signal. Sampling takes place at regular intervals of time.

Proof: Let $X(f)$ be the Fourier transform of $x(t)$. Using the inverse Fourier transform, the function x can be written as in equation (19.3)

$$x(t) = \int_{-\infty}^{\infty} X(f)e^{i2\pi ft}\mathrm{d}f \tag{21.10}$$

$$= \int_{-W}^{W} X(f)e^{i2\pi ft}\mathrm{d}f \tag{21.11}$$

because $X(f)$ is zero outside the band $-W \le f \le W$.

The samples can be expressed as

$$x\left(\frac{n}{2W}\right) = \int_{-W}^{W} X(f)e^{i2\pi \frac{n}{2W}f}\mathrm{d}f.$$

The theory of Fourier series can now be used in a kind of backward manner, by considering a Fourier series of a function of the variable f rather than of t. The right-hand side of the above equation defines the coefficients of a Fourier expansion of a periodic repetition of the function $X(f)$ (with period $1/(2W)$), with the interval $-W \le f \le W$ as the basic interval. (The sign of n must be changed to get an exact correspondence with the Fourier series.) See figure 21.3. The samples $x(\frac{n}{2W})$ determine the Fourier coefficients of the periodic version of $X(f)$, and hence they determine $X(f)$. From $X(f)$, the original function $x(t)$ can be recovered. This proves the theorem. ∎

Using the sampling theorem the function $x(t)$ can be expressed in terms of its samples. Consider the function

$$\mathrm{sinc}(2\pi Wt) \equiv \frac{\sin(2\pi Wt)}{2\pi Wt}.$$

This function, shown in figure 21.4, has value 1 at $t = 0$ and value 0 for all other sampling points $t = n/2W$ (with $n \ne 0$). The Fourier transform of this function is constant with value $1/(2W)$ in the band $-W \le f \le W$ and zero outside that band. (See example 19.1 in chapter 19 but where the roles of t and f are reversed compared to here.) Hence the sinc function is band-limited. If, when the original function $x(t)$ is sampled, all the samples are zero except the one at $t = 0$, which is 1, it would follow that $x(t) = \mathrm{sinc}(t)$ since that is appropriately band-limited and agrees with

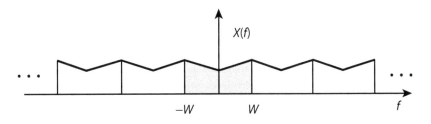

FIGURE 21.3 Fourier series of transform. Samples of the function x provide the coefficients for a Fourier series of the Fourier transform $X(f)$. Here the Fourier transform is the function indicated in the shaded region. Its copies make up the periodic extension of that transform.

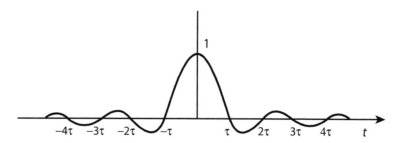

FIGURE 21.4 The function sinc $(2\pi Wt)$. This function is zero at every sampling point $t = n\tau, n \neq 0$, where $\tau = 1/(2W)$. The Fourier transform of the function is constant over the frequency interval $-W \leq f \leq W$.

the obtained samples. If the only nonzero sample were at a different point, at say $\tau \equiv 1/(2W)$, then likewise it would follow that $x(t) = \text{sinc}(t - \tau)$, since this shifted version of the sinc function is also band-limited to between $-W$ and W. In general, therefore, any nonzero sample corresponds to a sinc function shifted so that it is centered at the sample point. Therefore the original function $x(t)$ can be expressed as below.

Recovery from samples. Suppose a function x is band-limited to frequencies between $-W$ and W cycles per second. Suppose also that this function is sampled at every $t = n\tau, n = \cdots -2, -1, 0, 1, 2 \ldots$ where $\tau = 1/2W$. Then x can be recovered from the samples by the expression

$$x(t) = \sum_{n=-\infty}^{\infty} x(n\tau)\,\text{sinc}(t - n\tau). \tag{21.12}$$

21.4 Generalized Sampling Theorem*

A generalization of the standard sampling theorem is closely related to the theory of amplitude modulation, and this viewpoint enhances the understanding of both sampling and modulation.

Consider a periodic pulse train $p(t)$ with period T as shown in figure 21.5(a). If this pulse train is multiplied by the message signal $s(t)$, whose Fourier transform is band-limited to $-W \leq f \leq W$, the result is the new signal $v(t) = s(t)p(t)$.

Since $p(t)$ is periodic, it can be expressed as a Fourier series. Thus, denoting $f_p = 1/T$, we use the complex coefficient version of the Fourier series as given in secion 19.3 to write

$$p(t) = \sum_{n=-\infty}^{\infty} c_n e^{in2\pi f_p t}.$$

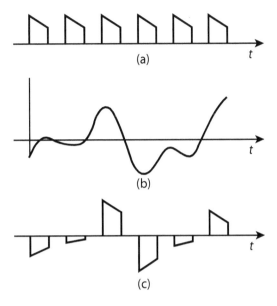

FIGURE 21.5 Sampling pulse train. The pulse train of (a) is multiplied by the signal (b) to produce the sampled signal (c).

Hence the Fourier transform of $v(t)$ is

$$V(f) = \int_{-\infty}^{\infty} \left[s(t) \sum_{n=-\infty}^{\infty} c_n e^{in2\pi f_p t} \right] e^{-i2\pi f t} dt. \qquad (21.13)$$

Or, interchanging the order of integration and summation,

$$V(f) = \sum_{n=-\infty}^{\infty} c_n \int_{-\infty}^{\infty} s(t) e^{-i2\pi(f-nf_p)t} dt. \qquad (21.14)$$

The integral in the last expression is recognized as $S(f - nf_p)$, where $S(f)$ is the Fourier transform of $s(t)$. Hence (21.14) becomes

$$V(f) = \sum_{n=-\infty}^{\infty} c_n S(f - nf_p).$$

This result says that the Fourier transform of $v(t)$ is simply several copies of the Fourier transform of $s(t)$ spread along the f axis as shown in figure 21.6. Each copy is separated by a distance f_p. In the figure it is assumed that f_p is greater than $2W$ so that the Fourier transform copies do not overlap. This condition, $f_p \geq 2W$, corresponds to $T \leq 1/(2W)$, which is the condition that the pulses repeat at least as fast as the Nyquist–Shannon sampling rate for a signal band-limited between $-W$ and W cycles per second.

The specific values of the c_n's depend on the pulse shape, but they do not matter for most purposes. It is clear that as long as one of the c_n's is nonzero and the copies

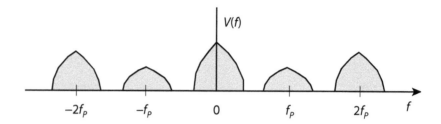

FIGURE 21.6 Fourier transform of a signal multiplied by a pulse train of frequency f_p. Copies of the Fourier transform of the original signal are obtained.

do not overlap, the Fourier transform of the original signal $s(t)$ is available in $V(f)$, and hence $s(t)$ can be recovered from $v(t)$. This conclusion holds no matter what the shape of the pulse train $p(t)$ as long as it is periodic with a period $T \le 1/(2W)$. The classic sampling theorem corresponds to the limiting case of vanishing small pulse width.

Relation to AM

The result stated above can be related to amplitude modulation by letting the pulse train be a (carrier) sinusoid of frequency f_c. The coefficients c_n of the Fourier series of the carrier are all zero except for c_{-1} and c_1. Hence in this case there are only two copies of the Fourier transform of $s(t)$, one centered at $-f_c$ and the other at f_c.

Sampling provides a simple way to shift the Fourier transform of a signal. If the signal is sampled at least as fast as the Nyquist–Shannon rate and uses any pulse shape that has all its Fourier coefficients nonzero, the resulting Fourier transform will have copies of the signal transform spread f_c apart. Using a selective bandpass filter, all copies can be eliminated except the one in a particular region. The result is a shifted version of the original Fourier transform. This surprising and powerful result was not known to early researchers.

Example 21.5 (How not to raise your voice). Consider again the baritone who sings below the middle C tone, as in example 20.4. Suppose a tape recording of one of his songs is multiplied by a pulse train with frequency one octave higher than middle C. If this recording were played, a horrible whistle at two octaves above middle C would be heard.

However, if the result is passed through a filter that greatly attenuates the frequencies above middle C (as by playing through a woofer loudspeaker), the song will be heard, sung by a baritone, his voice apparently normal.

When a pure tone is used as the multiplying signal, it is only necessary to raise the pitch one octave, as in example 20.4. Figure 21.7 makes this clear.

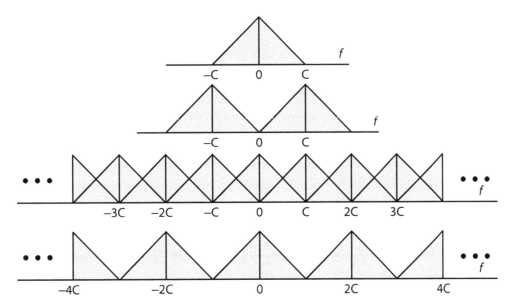

FIGURE 21.7 Not raising your voice. *Top*: The original spectrum. *Second*: The result of multiplying by a middle C tone. Since the pure tone has components only at C and -C, the sifted copies do not overlap. *Third*: The result of multiplying by an arbitrary pulse train of frequency of middle C. There are copies at every multiple of C, and hence there is overlap, which means that the original shape cannot be recovered by filtering. *Bottom*: The result of multiplying by an arbitrary pulse train of frequency two times middle C. Now there is no overlap and the original spectrum can be recovered.

21.5 Thermal Noise

Noise is present in every form of electronic transmission, and it causes errors. There are numerous sources of noise: interfering transmissions (as in spark radios), lightening, crosstalk through adjacent telephone wires, multipath disturbances, thermal disturbances at the electronic level, and approximations in a sampling process. One of the most ubiquitous forms is thermal noise, discussed in this section.

Thermal noise is due to the inherent random movements of electrons, which intensify as temperature is increased. This noise is present in every form of electronic media, and it is, for practical purposes, completely random. The Fourier apparatus can be extended to treat random waveforms. The principal tool for this purpose is the **power spectral density**, which as the name implies is the density of power (energy per unit time) at various frequencies.

Thermal noise is usually modeled as **white noise**, defined as a process with a power spectral density that is constant across all frequencies. This means that theoretically the total average power is infinite, but in actuality the power spectral density can be observed only over a finite range.

The level of white thermal noise power in bandwidth W is $N_0 W$, where N_0 is given by the formula

$$N_0 = kT.$$

N_0 is in watts per Hz, k is Boltzman's constant 1.3803×10^{-23} joules per degree kelvin, and T is the temperature in degrees kelvin. The power spectral density of white thermal noise is constant at level $N_0/2$ since a bandwidth of W defines the frequency range $-W \leq f \leq W$ of twice the width of W itself, and hence the total noise in the band $|f| \leq W$ is $N_0 2W/2 = N_0 W$.

Example 21.6. Suppose a transistor radio sitting in your room operates at a temperature of 20°C. That corresponds to $20° + 273° = 293°$ kelvin. The thermal noise power is therefore $N_0 = 293 \times 1.3803 \times 10^{-23} = 4.05 \times 10^{-21}$ watts per Hz. If this receiver is tuned to receive a bandwidth of 20 kilocycles per second, the total thermal noise power in that bandwidth is $20 \times 10^3 \times 4.05 \times 10^{-21} = 8.1 \times 10^{-17}$ watts. That may not seem like much, but it can be a significant fraction of the total power in that band received by the radio's antenna.

21.6 Capacity of a Band-Limited Channel

It is now possible to put everything together and present Shannon's celebrated formula for the capacity of a band-limited channel.

Suppose the channel is capable of transmitting continuous-time signals with a bandwidth of W, and there is (independent) additive white Gaussian noise. From section 21.2, the capacity of a Gaussian channel consisting of a single sample is

$$C = \frac{1}{2} \log \left(1 + \frac{S}{N} \right),$$

where S and N are the average powers of the signal and noise, respectively. Since the signal can be completely generated by samples defined at a rate of $1/(2W)$ samples per second, the channel can send up to $2W$ independent samples per second. Thus the capacity of this channel is

$$C = W \log \left(1 + \frac{S}{N} \right)$$

bits per second, where now S and N are the average powers of signal and noise, respectively.[2] This leads to the famous capacity formula stated below.

Capacity of band-limited channel. The capacity of a continuous channel, band-limited to W hertz and subject to additive white Gaussian noise of average power spectral density $N_0/2$ (and hence total power over $-W \leq f \leq W$ of N_0) is

$$C = W \log \left(1 + \frac{S}{N_0 W} \right), \tag{21.15}$$

where S is the average power of the signal.

Shannon also explained how this capacity can, in principle, be achieved. The transmitted signals must approximate white Gaussian noise within the given bandwidth, and when those signals are perturbed by additive white Gaussian noise, the received signals will also be approximately white noise in that band.

---

[2] Since both S and N have the same units, the renormalization in going from energy per sample to energy per second cancels out. Hence it can be assumed that they are in standard units of average powers of signal and noise, respectively.

Example 21.7 (Modem capacity). The V.34 telephone modem operates in a bandwidth of approximately 3,400 Hz. The signal-to-noise level is usually about 3,000. Hence the theoretical capacity is $3,400 \log 3,001 = 39,274$ bits/second. This modem actually achieves about 33,400 bits/second.

21.7 Spread Spectrum

Hedy Lamarr, the glamorous Hollywood star often called "the most beautiful girl in the world" during World War II, invented a concept of spread spectrum that is the basis for much of modern communication.

She was born Hedwig Maria Eva Kiesler, in Vienna, Austria. In 1933 she married Fritz Mandl, who was one of Europe's leading armaments manufactures. Mandl began selling arms to Hitler and displayed Lamarr as a showpiece in business meetings but kept her as a virtual prisoner. One issue that Mandl was working on was the remote control of weapons such as torpedoes by use of radio signals. It was an attractive alternative to wire control, but had the severe disadvantage that an adversary could also tune into the transmitting frequency and jam the signal.

In 1937 Lamarr escaped from her husband by drugging her watchful maid, and made it to London. There she met Louis B. Mayer of Metro Goldwyn Mayer (MGM); he brought her to the United States and gave her the stage name Hedy Lamarr.

In Hollywood she met George Antheil, a charming musician known for technical innovations in music, such as use of airplane propellers and player pianos in concerts. As a result of personal experience in Europe, he was firmly anti-Hitler.

Hedy mentioned the radio control problem to Antheil and explained her solution. The transmitter should randomly and rapidly change its transmitting frequency, selecting various frequencies over a wide band. The enemy would not be able to jam all frequencies at once, so 99 percent of the signal would get through. She explained also, however, that the scheme would require that the frequency changes in the transmitter and receiver be perfectly synchronized, and she did not know how to achieve that.

Antheil proposed that identical piano roll mechanisms be installed in both the transmitter and receiver. Together Lamarr and Antheil worked out a system, and obtained a patent in 1942. Rather than exploit the patent commercially, they gave it to the U.S. government as a contribution to the war effort. However, because additional technical problems would have to be worked out, the navy never used the idea.

The patent expired in 1959, but soon after that engineers at Sylvania Electronic Systems Division developed a digital version of synchronization and a corresponding system that supplied secure communications during the Cuban missile crisis of 1962.

Lamarr and Antheil's concept of randomly changing the frequency of transmission is today called **frequency hopping** and is one version of a class of techniques called **spread spectrum** that employ a broad spectrum in transmission. Frequency hopping continues to be used, especially by the military, as an effective method of secure wireless transmission. Since reception of a message that is transmitted by frequency hopping is based on a prearranged random hopping schedule, the method was one of the first effective encryption techniques for radio. It is also, as we shall see, an efficient method of communication from an information theory viewpoint.

Capacity Gain

The advantage of wide spectrum communication is revealed by Shannon's formula for the capacity of a band-limited channel:

$$C = W \log \left(1 + \frac{S}{N_0 W} \right). \tag{21.16}$$

Suppose that the available average signal power S is constrained, which is reasonable since it is derived from transmitter power resources. Suppose though, that by appropriate system design, the system bandwidth can be varied. As bandwidth W is increased, the signal-to-noise ratio worsens because the signal power is fixed but the noise power increases. However, the channel capacity actually increases because the leading W term in the capacity formula outweighs the decrease inside the log term. Figure 21.8 shows the capacity as a function of W for a ratio of $S/N_0 = 10^7$, meaning that the signal-to-noise ratio is 10 at a bandwidth of $W = 1$ MHz (megahertz). There is then a limiting capacity of $C = S/[N_0 \log_2(e)]$, achieved as $W \to \infty$. For the example of figure 21.8 this limiting capacity is $C = 14.4$ megabits/sec, which is 14.4 times what it is at $W = 1$ megahertz.

Early spark radio was troublesome precisely because each transmission occupied a broad spectral band. This led to inefficient use of power and exceedingly vexing problems of transmitter interference. The solution, made possible by a continuous carrier and amplitude modulation, was to slice the available spectrum into separate narrow bands that could be assigned to different stations. This restriction of transmission bandwidth, as we now see, sets perhaps a low limit on the capacity that can be achieved by any one station.

Recall that when Edwin Armstrong was designing FM, conventional wisdom said that he should reduce the broadcast bandwidth in order to reduce the total noise power received. Instead, as another indication of Armstrong's genius, he decided that it was better to increase bandwidth, in accord with what is now clear from Shannon's result.

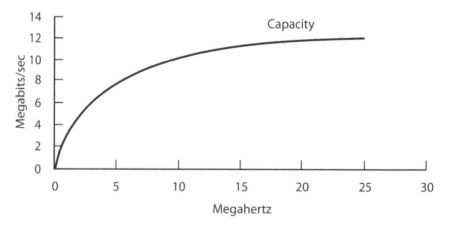

FIGURE 21.8 Spread spectrum capacity. The capacity increases as a function of the bandwidth when average signal power is fixed and noise power is proportional to bandwidth.

21.8 Spreading Technique

A wide spectrum is desirable, but how is it to be obtained? One way is by frequency hopping, jumping randomly every split second among narrow bands within a larger one. Another method is by **direct spreading**, which spreads the spectrum by multiplying the signal by a random signal of broad bandwidth.

To see how direct spreading works, consider figure 21.9. The first panel shows a low-bandwidth signal. The second shows a high-bandwidth random signal (like pure static). The third panel shows the result of multiplying the two together. This resulting waveform is also essentially purely random and hence has a broad and nearly flat power spectral density similar to that of static. The signal has been spread across the wide bandwidth. The signal can be recovered if the static part of the signal is known, for the static part can be divided out of the combined signal.

In modern practice, this spreading technique is implemented digitally with pulses. The original signal (such as a voice waveform) is sampled and the samples are coded with one of the binary codes discussed in chapters 3 and 6. This binary code is then converted to a series of 1's and -1's (rather than 0's and 1's). If sent directly through an electrical communication system, these binary symbols would be pulses of magnitude 1 or -1. Each pulse would have a duration roughly equal to $1/(2W_s)$, where W_s is the bandwidth of the signal. The resulting pulse train will also have a bandwidth of roughly W_s.

Next, the pulse train is multiplied by a **chipping code**, which is generated as pseudorandom noise (alternatively termed **pseudonoise** or PN for short). The chipping code pulses have duration much shorter than the signal pulses. See figure 21.10. The resulting high-frequency pulse train inherits the pseudorandom character of the chipping PN, and occupies the same broad frequency band. Of course, the signal can be recovered by dividing out the PN. For binary -1's and $+1$'s, division is equivalent to multiplication, so the spread signal can be multiplied by the PN series to recover the original signal.

The PN sequences can be generated by shift registers of the kind discussed in chapters 6, and 11. There are two commonly used sequences: a long code and a short code. The long codes are 2^{42} bits long. If an original digital bit stream runs at 9,600 bits/second and a long code spreading stream runs at 1.2288 megabits/second, it takes about 41 days for the code to complete a single cycle. By contrast the short codes are of length 2^{15}, and in the same situation they cycle every 26.7 milliseconds.

FIGURE 21.9 Spreading a signal. The first graph shows a low-bandwidth signal. The second graph shows a broad bandwidth random waveform. The third graph shows the result of multiplying the two together. The result is again essentially random with a broad bandwidth. The original signal can be recovered by dividing out the random waveform.

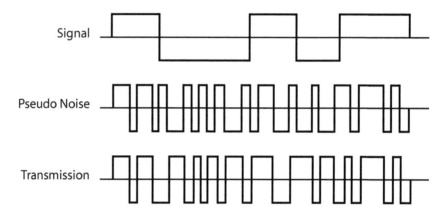

FIGURE 21.10 Spreading of binary pulses. The original binary coded message (of 1's and −1's) is multiplied by a pseudonoise chipping code. The resulting signal inherits the random nature of the PN, and hence has a wide power density spectrum. In practice, the pulse width of the spreading PN is orders of magnitude shorter than that of the signal pulses.

21.9 Multiple Access Systems

Many modern communication networks are designed to be shared by several users, all operating simultaneously. Cell phones are a principal example. There are four basic methods for sharing bandwidth.

1. **Time Division Multiple Access (TDMA).** In this method a basic time duration, a fraction of a second, is itself divided into a number K of short segments, and these are assigned to K different users. Each user therefore has command of the entire bandwidth for that short period. This method is commonly used in data and digital voice transmission.

2. **Frequency Division Multiple Access (FDMA).** Here the available bandwidth W is divided into K segments that are assigned to K users. This is the method used to allocate frequency bands among radio and television stations and in some digital communication systems.

3. **Code Division Multiple Access (CDMA).** In this system all users operate over the entire bandwidth W and at all times. Users spread their signals to fully occupy W by using a chipping code. Interference is minimized by the incorporation of orthogonal chipping codes, as discussed below. This system is used in some cellular telephone systems.

4. **Nonspread Random Access.** In this method users access the common channel randomly. Some messages collide and must be retransmitted, as discussed in chapter 22.

Orthogonal Codes

CDMA relies on orthogonal chipping codes so that a given receiver can screen out signals from all users except the one desired.

Here is the general idea. A PN chipping code is of the form $c = (b_1, b_2, b_3, \ldots, b_n)$, where each b_i is 1 or -1. Two codes c_u and c_v are **orthogonal** if

$$c_u \cdot c_v \equiv \sum_{i=1}^{n} b_{ui} b_{vi} = 0.$$

For example $c_u = (1, 1, -1, -1)$ and $c_v = (1, -1, 1, -1)$ are orthogonal.

A message signal s is also a code, although in the chipping time frame the message signal has long blocks of constant values corresponding to the long duration message pulses. A signal s_u of a sender u is spread by multiplying it by the sender's chipping code c_u, and the result is transmitted as $t_u = s_u \cdot c_u$. The message is recovered by multiplying by the code c_u giving $t_u \cdot c_u = s_u \cdot c_u \cdot c_u = s_u n$ (where the factor of n is due to $c_u \cdot c_u = n$).

When there are several users u, v, w with orthogonal chipping codes, the overall signal received by all is $t = s_u \cdot c_u + s_v \cdot c_v + s_w \cdot c_w$. The message from u can be recovered by multiplying by c_u as before, since

$$t \cdot c_u = (s_u \cdot c_u \cdot c_u) + (s_v \cdot c_v \cdot c_u) + (s_w \cdot c_w \cdot c_u) = s_u n$$

as before because of the orthogonality of the chipping codes.

A basic component of such a system is a large set of orthogonal PN chipping codes. It is standard practice to use **Walsh codes**, which are available in sets with lengths that are powers of two, and they are easily constructed. (See exercise 4.) For example, the IS-95 standard for CDMA specifies a set of 64 orthogonal Walsh codes of length 64.

In practice, things are not as perfect as in the simplified theory. Various delay times, multipath transmission, and other factors degrade the orthogonality relation. A practical system must incorporate many compensating features in its design. But the basic principle is elegant and has been made very effective.

Capacity Comparisons

Shannon's formula for the capacity of a channel subject to additive white Gaussian noise (AWGN) can be used to analyze the capacity of multiple access systems. We consider them in turn.

Suppose there is available a bandwidth of W and there are K users each having average power S. There is white Gaussian noise of power spectral density $N_0/2$, but with a total power of $N_0 W$.

First consider TDMA. Assume that each user transmits a fraction $1/K$ of the time, with the average power during that time being KS (so that the average overall time is S). The capacity of user k is therefore

$$C_k = \frac{W}{K} \log \left[1 + \frac{KS}{N_0 W} \right],$$

since the user only gets $1/K$ of the capacity. The total capacity for K users is accordingly

$$C = W \log \left[1 + \frac{KS}{N_0 W} \right],$$

which is the same as the capacity that would accrue to a single user with average power KS. It is clear that the capacity increases with K (because the total signal power used is proportional to K).

Next consider FDMA. Assume that the bandwidth is divided into K equal segments, each allocated to a different user. The capacity enjoyed by user k is

$$C_k = \frac{W}{K} \log\left(1 + \frac{S}{N_0(W/K)}\right).$$

The total capacity is therefore

$$C = W \log\left(1 + \frac{KS}{N_0 W}\right),$$

which is exactly the capacity that would accrue to a single user with average power equal to the sum of the average powers of the K users, and is the same as the capacity of TDMA.

Finally, consider CDMA. Each user uses the entire bandwidth W and transmits a PN signal (a message multiplied by a chipping code) of average power S. A receiver is subject to the ambient Gaussian noise and interference from $K - 1$ users' PN signals. Hence the capacity for user k is

$$C_k = W \log\left[1 + \frac{S}{N_0 W + (K - 1)S}\right].$$

Hence the total capacity for K users is

$$C = KW \log\left[1 + \frac{S}{N_0 W + (K - 1)S}\right].$$

For large K, $\log(1 + 1/K) \approx (\log_2 e)/K$, and hence

$$C \rightarrow W \log e.$$

Hence, unlike TDMA or FDMA, the capacity does not increase to infinity as K is increased.

CDMA has other features that make it attractive. It is easy to add or subtract users (since no new time or frequency divisions are required). In addition, when all users are not active (which is the normal case for cell phones), the capacity of CDMA can be superior to that of other forms of multiple access.

The design of multiple access systems requires knowledge of Fourier transforms, power spectral density, random processes, noise characteristics, effective coding of message sources, orthogonal chipping codes, the sampling theorem, and Shannon's theory of capacity. All of these represent a great mastery of frequency beyond the pioneering efforts of Fourier, Lord Kelvin, Hertz, Bell, and others who contributed to the progress of the communication field.

21.10 EXERCISES

1. (Maximum entropy) Solve problem (21.4) by the following steps:
   (a) Introduce Lagrange multipliers for each of the two constraints, and differentiate the resulting Lagrangian integral with respect to p at each point x. Set this derivative to zero.
   (b) Show that $\ln p(x) = c - dx^2$ for some constants c and d.
   (c) Evaluate the constants to satisfy the two constraints.

2. (Two models of information transfer) Consider a random variable whose value is transmitted.
   (a) Suppose that the variable has standard deviation n and that the true value is transmitted with noise of standard deviation 1. What is the amount of information transmitted?
   (b) Alternatively, suppose that the variable has n equally probable values, and the actual value is transmitted exactly. How much information is transmitted?
   (c) As $n \to \infty$, how do these values compare?

3. (Noise loss) A delicate electronic instrument normally operates at $20°$ C with a signal-to-noise ratio of 100. If it is taken to the desert and operated at $40°$ C, by what percentage will its capacity be reduced?

4. (Walsh codes) Walsh codes are generated from Hadamard matrices consisting of 0's and 1's. Hadamard matrices of order 2^k can be found by a simple recursion. Given a Hadamard matrix H_k of order 2^k, the matrix H_{k+1} is formed as

$$H_{k+1} = \begin{bmatrix} H_k & H_k \\ H_k & \overline{H}_k \end{bmatrix},$$

   where $\overline{H}_k$ denotes the matrix complementary to H_k, with 1's and 0's interchanged.
   (a) Starting with $H_0 = [0]$, find H_1, H_2, and H_3.
   (b) A Hadamard matrix defines a set of codes from the rows of the matrix. One way is to convert 0's to -1's and use ordinary multiplication. Show that the codes resulting from H_1, H_2, and H_3 each define orthogonal sets.
   (c) Argue that orthogonality applies to codes derived from any Hadamard matrix.

5. (CDMA and FDMA) Suppose that there are 100 users but on average only 10 of them use the system at any one time. Each has average power S when active, and the background noise has power spectral density of $N_0/2$.
   (a) Suppose that in FDMA the available bandwidth is divided into 10 equal segments and 10 people are assigned to each of them. Any active user can expect that one other user will be active on the same sub-band. Write an expression for the capacity enjoyed by a user.
   (b) In CDMA an active user can expect to see pseudorandom noise from 10 other users. Write an expression for the capacity enjoyed by a user.
   (c) At low signal-to-noise levels, which system gives users the greater capacity?
   (d) *Does one of these systems always have more capacity than the other?

21.11 Bibliography

The theorems on capacity for continuous channels were first presented in Shannon's classic monograph [1]. An excellent general text on continuous information theory is [2]. See [6] for a nice presentation of the generalized sampling theorem. See [4] and [5] for more about Hedy Lamarr's invention. Two excellent general texts on modern communication are the basic text [6] and the more advanced [7] (from which the discussion of capacity for multiple access systems was adapted). Overviews of CMDA, including examples, background theory, and practical details, are found in [8] and [9] which has a good discussion on long codes and gives examples.

References

[1] Shannon, Clande E. *The Mathematical Theory of Communication*. Urbana: University of Illinois Press, 1949.
[2] Cover, T. M., and J. A. Thomas. *Elements of Information Theory*. New York: John Wiley and Sons, 1991.
[3] "Female Inventors—Hedy Lamarr." http//www.inventions.org/culture/female/lamarr.html.
[4] Hughes, David. *Nomination of Hedy Lamarr and George Antheil for Achieves of Science Award*, 1997.
[5] Nahin, P. J. *The Science of Radio*. 2nd ed. Woodbury, N.Y,: American Institute of Physics, 2001.
[6] Stallings, W. *Data and Computer Communications*. 6th ed. Upper Saddle River N.J.: Prentice-Hall, 2000.
[7] Proakis, John G. *Digital Communications*. 4th ed. New York: McGraw-Hill, 2002.
[8] Garg, V. K., K. Smolik, and J. E. Wilkes. *Applications of CDMA in Wireless/Personal Communications*. Upper Saddle River, N.J.: Prentice-Hall, 1997.
[9] Lee, Steve. *Spread Spectrum CDMA*. New York: McGraw-Hill, 2002.

NETWORKS

22

A great portion of the organized information we receive is transmitted through networks that can interconnect various combinations of users. The telephone network, wide area computer networks such as ARPAnet, local area networks such as ethernets, and the World Wide Web are examples.

The earliest networks sliced up the available channel capacity and allocated the pieces to various users, the actual information flow being guided by circuit switches or frequency allocation. Today, many large networks allocate shares of capacity by packet switching, whereby a message is itself divided into small individual packets that are sent separately through the network. This has the advantage that capacity is allocated dynamically according to demand. The Internet is the prime example of a network based on packet switching.

Today's Internet grew out of the ARPAnet. Although several individuals envisioned something like it, the most influential vision of distributed computing through a network was proposed in 1962 by Joseph C. R. Licklider of MIT in the form of a "Galactic Network." The first paper on packet switching theory was published in 1961 by Leonard Kleinrock based on his MIT Ph.D. dissertation. His colleague, Lawrence Roberts, soon went to ARPA (Advanced Research Projects Agency) to implement Licklider's network concept using Kleinrock's packet technology.

The basic concept of packet switching is simple—but development of a truly effective system is far from trivial. How large should the packets be? How often will packets collide? How long will the queues be when packet congestion is high? What route should the packets take? The quantitative analysis of these questions makes up much of the modern theory of communication networks.

Queueing theory provides the basis for a large portion of the required analysis. And fortunately for our purposes, some important aspects of packet technology can be explored with only the simplest concept from queueing theory; namely, a model describing the probabilistic character of packet origination.

22.1 Poisson Processes

The generation of packets by network users is frequently modeled as a **Poisson process**. Such a process describes the random times at which events (such as packet origination) occur. The process is defined by a single parameter $\lambda > 0$ that is the average rate of event occurrence. Specifically, in a Poisson process events are assumed to occur independently in time, with the probability of occurrence within any small interval Δt being[1]

$$\text{Prob\{occur\}} = \lambda \Delta t + o(\Delta t).$$

The probability of two or more occurrences in an interval Δt is assumed to be on the order of $o(\Delta t)$.

From this basic definition, the probability of any number of events occurring within a finite interval τ can be computed. For our purposes, however, it is only necessary to know the probability $P_0(\tau)$ of no occurrence within a finite time segment of length $\tau > 0$.

To find $P_0(\tau)$, the time segment $[0, \tau]$ is divided into m small intervals of length τ/m. The probability of no occurrence in all of these together is the product of the probabilities that no event occurs in each of the small segments, which is $(1 - \lambda\tau/m + o(1/m))^m$. Letting $m \to \infty$ and using the general fact that for any x, $\lim_{m\to\infty}(1 + x/m)^m = e^x$, it follows that

$$P_0(\tau) = e^{-\lambda\tau}. \tag{22.1}$$

Although $P_0(\tau)$ is the only probability that is needed in the following section, the probability $P_n(\tau)$ of n occurrences in an interval of length τ also can be found, as shown in exercise 1, to be

$$P_n(\tau) = \frac{\lambda^n t^n}{n!} e^{-\lambda\tau}.$$

The average time between event occurrences can also be easily calculated. Suppose there is an occurrence at time 0, and let t be the time of the next occurrence. Clearly Prob $(t \geq \tau) = P_0(\tau)$. Hence, Prob$(t \leq \tau) = 1 - e^{-\lambda\tau}$. The probability density of t is by definition

$$\frac{d}{d\tau}\text{Prob}(t \leq \tau) = \lambda e^{-\lambda t}.$$

Hence the average value of t is

$$\bar{t} = \int_0^\infty t\lambda e^{-\lambda t} dt = 1/\lambda. \tag{22.2}$$

Therefore, the average time between events is $1/\lambda$ and, correspondingly,[2] events are generated at an average rate of λ per unit of time.

---

[1] As before, $o(\Delta t)$ represents terms that go to zero faster than Δt.
[2] See exercise 2.

22.2 Frames

In packet switching each message is broken into a number of packets of fixed length. Each packet is treated by the network as a separate unit and passed along from one network node to another until it reaches its intended destination. In some cases the packets of a given message are sent on different routes and may in fact arrive at the destination out of order, and thus must be reassembled to produce the original message.

To aid in the handling of packets, there is appended to each packet additional information such as beginning and end indicators, and a header that includes source and destination addresses as well as the sequence number of the packet. The specific requirements and configuration depend on the system, but in general the overall large package consisting of message packet and additional information, as illustrated in figure 22.1, is termed a **frame**.

It is assumed that frames are generated randomly by a large number of users as shown in figure 22.2. If the frames of m users are generated independently, each according to a Poisson process with parameter μ, then the events corresponding to initiation of any frame are themselves a Poisson process with parameter $\lambda = m\mu$.

It is convenient to express time in frame units; that is, the duration of a frame is taken to be one unit. Suppose that the collection of all users generates frames at a rate of λ per unit time (frame time). Clearly if $\lambda > 1$, frames are generated faster on

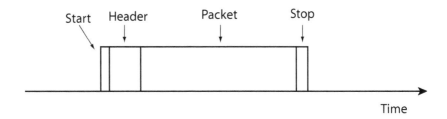

FIGURE 22.1 A frame. A frame consists of a packet of message information as well as additional information to assist with proper identification and transmission of the packet.

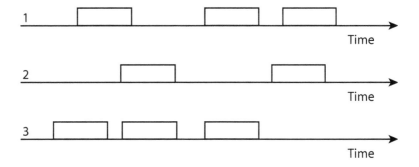

FIGURE 22.2 Frames generated by three users. If the users each generate packets according to independent Poisson process, the resulting overall process is also Poisson.

average than they can be transmitted even if they are sequenced one after the other. So, $\lambda < 1$ is necessary for stable operation of the network. Nevertheless, some frames will collide. These frames must be retransmitted, which increases the effective rate of frames entering the network. This rate, which includes original frame generation and the rate of retransmission, is denoted G. Clearly $G \geq \lambda$.

22.3 The ALOHA System

The ALOHA system was one of the first networks, after the ARPAnet, to use packet switching. It was built under the leadership of Professor Norman Abramson as a network to allow people at the several campuses of the University of Hawaii on various islands to communicate with the main computer using local terminals. It was not practical to use undersea cables to connect the islands, so radio links were used. This meant that, unlike the ARPAnet, which used telephone lines, everyone shared a common channel, defined by a radio frequency. Hence there was a strong possibility of frame collision. If there was a collision, the two frames were sent again, after a random delay. (In that system, a packet corresponded to one line of text, and it was sent to the network when the carriage return key was pressed.)

The first step in the analysis of the ALOHA system is to determine the probability that any frame introduced for transmission (or retransmission) is successfully transmitted without collision. In a **pure ALOHA** system, frames are initiated at any time, without restriction. It is like drivers entering a freeway while blinded, not being able to detect existing traffic. Certainly there is a high chance of collision unless the traffic level is low.

The situation for data frames is illustrated in figure 22.3. Imagine that frame A is introduced into the system as shown. If another frame, frame B, is introduced anytime during the period of frame A and one entire period before frame A begins, it will collide with frame A and both frames will be sent back for retransmission. The vulnerability period is therefore two frame periods in duration.

The throughput S of the system is the rate at which frames are successfully transmitted through the channel, measured in frames per unit frame time. (According to this convention, if frames were sent one right after the other, continuously, the throughput would be $S = 100$ percent.) In the case analyzed above, the throughput is equal to the rate that frames are presented for transmission times the probability

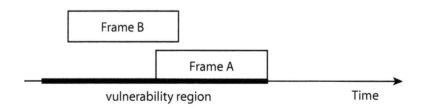

FIGURE 22.3 Vulnerability region. The vulnerability region is twice the duration of a single frame.

that a frame is in fact successfully transmitted. Thus, using the formula for $P_0(2)$, the probability that no additional packet is generated within two frame periods, we find

$$S = G \times P_0(2) = Ge^{-2G}.$$

The point of maximum throughput can be found by setting the derivative to zero, as

$$\frac{dS}{dG} = \frac{dGe^{-2G}}{dG} = e^{-2G} - 2Ge^{-2G} = 0,$$

which gives $G = 1/2$. The corresponding maximum throughput is $S = 1/(2e) \approx 0.184$. In other words, the system has a maximum efficiency of only 18.4 percent. The value of S versus G for pure ALOHA is shown as the lower curve in figure 22.4.

An alternate version of ALOHA, termed **slotted ALOHA**, has greater throughput. In this system, time is divided into a series of discrete intervals each of which can accommodate one frame. Each sender can initiate a frame only at the beginning of an interval (a slot). Hence if a packet is ready for transmission, it must wait until the current slot is completed before actually entering the network. A collision occurs only if two or more senders present a packet during a given slot. If a collision should occur, access to a later slot is purposely delayed according to a random draw, so that the colliding frames will not immediately collide again.

Since a slot interval is one frame in length, the vulnerability period is only one frame length instead of two as with pure ALOHA. Hence the throughput of the slotted system is

$$S = Ge^{-G}. \tag{22.3}$$

This achieves a maximum at $G = 1$ with a corresponding value of $1/e \approx .368$, which is twice the throughput of pure ALOHA. The throughput as a function of G is shown in figure 22.4.

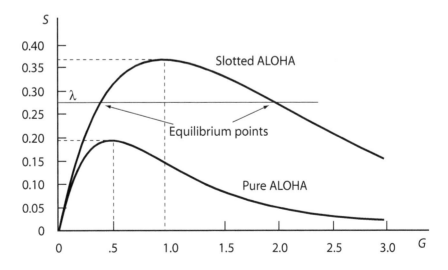

FIGURE 22.4 Throughput for ALOHA systems.

The preceding analyses leads to the concept of equilibrium for an ALOHA system. In equilibrium, the throughput rate S must equal the rate λ at which frames are originated. The equilibria for slotted ALOHA with a certain parameter λ are shown in figure 22.4. There are two equilibria, corresponding to the two places where S crosses the horizontal line at height λ. The one with the lowest value of G is the most desirable, since the one with a high value of G implies that there is, on average, a large backlog of frames waiting to be retransmitted, and hence long average delays.

However, the simple analyses and the corresponding curves based on averages ignore important dynamic effects, and in reality these equilibria are not stable. After a heavy run of collisions, the backlog may increase to the point where more transmissions are attempted than can be served, which means that practically all transmissions will result in collision.

Several schemes have been proposed to stabilize slotted ALOHA. Most are based on the idea of varying the characteristics of the random delay purposely introduced before retransmission, increasing the average delay when the system is congested and decreasing it when the system is idle.

22.4 Carrier Sensing

The maximum throughput of the slotted ALOHA system is $1/e$, which is rather disappointing, but it is perhaps not surprising, since there is no real opportunity for coordination by the various sources. Each source simply transmits when it has a packet, independent of what others are doing; as a natural consequence, collisions are highly likely. In some situations, such as local area networks, it is possible for senders to detect the carrier signal corresponding to a busy channel. Each user can then wait until the channel is clear before attempting transmission. If all parties behave this way, higher throughput is possible.

The situation is analogous to a driver who is able to see directly in front of a freeway ramp when planning to enter traffic. This is better than being completely blinded (analogous to ALOHA); nevertheless, the entrance may look clear, but when moving forward, the car may be hit by an oncoming vehicle.

Data network systems of this type are termed **Carrier Sense Multiple Access** (CSMA) systems, and there are several versions. To analyze these systems, we assume that the frame duration is the same for each packet, and that the one-way propagation of signal between any source–destination pair is a, measured in frame units. Thus if a transmission is initiated by a source, other users detect its presence in the system only after a delay of a. Typically, a is small.

If a source detects that the channel is clear, it might attempt to transmit. However, the attempt will fail if another source began to transmit a frame a short time (less than a) earlier.

Various versions of CSMA follow different strategies for transmission and retransmission. In **persistent CSMA**, a sender whose frame collides retransmits the frame at the next sensed idle period. In **nonpersistent CSMA**, a sender whose frame collides schedules a future attempt according to a specified random delay function, repeating this process until successful.

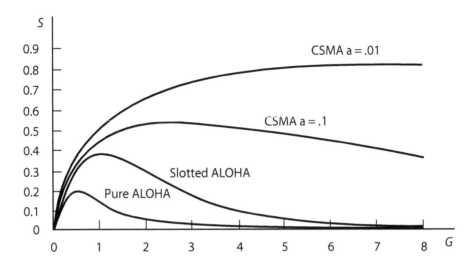

FIGURE 22.5 Comparison of throughput for various packet-switched systems.

The throughput of nonpersistent CSMA can be shown to be

$$S = \frac{Ge^{-aG}}{G(1 + 2a) + e^{-aG}}.$$

(22.4)

This value assumes that attempted transmission or retransmission occurs according to a Poisson process with parameter G. The value of G is determined by the underlying rate λ of frame generation and the retransmission delay strategy.

Depending on the delay time a, the throughput S for nonpersistent CSMA can be much greater than for ALOHA systems. For example, at $a = .01$, which is perhaps typical, the throughput can approach 83 percent. Figure 22.5 compares the throughput of various systems.

22.5 Routing Algorithms

Good network management attempts to minimize the transit time of packets while maximizing the throughput of the system and allocating capacity to different parties in a fair manner. Such management requires a variety of tools. It is much like the management of a road system: it is useful to have signals, carpool lanes, speed limits, and one-way streets rather than allowing a free-for-all that leads to congestion and collisions. And like a road system, there is no simple plan for communication networks that accommodates all of the sometimes conflicting goals.

The overall issue is typically approached by studying the different aspects independently. This alone produces a good first step, provides valuable insight, and is the foundation for further work. The remainder of this chapter discusses the routing problem: determining which path to use for each packet.

There are two major distinctions for how a path is selected. In the **virtual circuits** approach, a route is chosen for an entire message; that is, all packets that make up that message follow the same path. In the **datagram** approach, packets are considered to be separate entities and are individually routed, different packets perhaps following different paths. Many of the principles used for determining paths are common to both approaches.

Flooding

One of the simplest routing methods is **flooding**. In this method every packet arriving at a node is forwarded along every outgoing link except the one by which the packet arrived. It is clear that every packet is sure to eventually get to every node provided only that the network is fully connected so that it is possible to reach every node from any node. However, it is also clear that a great number of duplicate packets are likely to be sent. This duplication can be ameliorated by including in the frame header a counter that is decremented at each node. This counter is initialized at a number known to be greater than the minimum number of links separating the sender and receiver; then later as frames are sent scurrying through the network, those frames for which the counter is zero are discarded. Although flooding is inefficient, its simplicity is an advantage and may be useful in networks with relatively low levels of traffic. Flooding is also useful for **broadcasting** a common message to all sites.

22.6 The Bellman–Ford Algorithm

Optimal routing methods are typically based on some form of generalized path length. This may be the actual length, or more often, a measure of transit time, including delays due to buffering at congested nodes. Once a measure of length is defined, the objective of the routing procedure is, ideally, to minimize the total (generalized) distance traveled by each packet. There are a number of algorithms designed for this purpose, but one of special interest is the Bellman–Ford algorithm described in this section.

The network is assumed to consist of a number of nodes, numbered 1 through n. These nodes are connected by links. The distance of a link from node i to node j is denoted d_{ij}. It is assumed that all $d_{ij} \geq 0$. If there is no direct link from i to j, then $d_{ij} = +\infty$. It is allowed that $d_{ij} \neq d_{ji}$; but if there is a link from i to j (that is, $d_{ij} < \infty$), then there is a reverse link ($d_{ji} < \infty$). It is also assumed that $d_{ii} = 0$ for all i.

The total length of a path is assumed to be the sum of the lengths of the links traversed.

Consider a certain node, say node 1. Suppose we wish to find the shortest path length to node 1 from every other node. Denote the shortest path length from node i to node 1 by D_i. The Bellman–Ford algorithm finds the D_i's by first considering only paths traveling over at most one link. The corresponding minimum distance is denoted D_i^1. Clearly $D_i^1 = d_{i1}$.

Next, given D_i^k (the minimum distance from i to 1 using at most k links), the distance D_i^{k+1} is defined as

$$D_i^{k+1} = \min_j (d_{ij} + D_j^k). \tag{22.5}$$

This says that the minimum path length using at most $k + 1$ links is equal to the minimum path length from a node j using at most k links plus the distance from i to j; and this is minimized with respect to all intermediate nodes j. The iteration (22.5) converges to the actual minimum path length D_i from i to 1 in at most n steps, since a minimum path will have no more than n links.

The algorithm can be carried out separately for each of the destination nodes in order to calculate the shortest path lengths to every node.

Note that in the minimization step $D_i^{k+1} = \min_j (d_{ij} + D_j^k)$, the values of j can be restricted to those j's that are directly linked to i, for otherwise $d_{ij} = +\infty$. It is this property that allows the algorithm to be implemented in a somewhat decentralized form. Node i only needs to know (1) the value of the d_{ij}'s for its neighboring j's, and (2) the values D_j^k which can be reported from the neighboring j's. Thus a node need only communicate with its neighbors to compute the minimum path length.

22.7 Distance Vector Routing

The Bellman–Ford algorithm was the basis of the routing method originally used in ARPAnet, and it has been used in the Internet under the name RIP. It is understood that traffic conditions in the network constantly change, which means that the effective link lengths (defined usually as transit time) also change, and this in turn means that the shortest paths change.

The actual routing algorithm based on the Bellman–Ford method, termed **distance vector routing**, periodically repeats one step of the iteration (22.5) so that each node can determine an estimate of the minimum path lengths. The values determined in practice are always estimates because the algorithm never actually converges, owing to the fact that network conditions change between iteration steps. Nevertheless, the estimates obtained are useful and the method produces good, if not ideal, routing.

The calculation employed in the Bellman–Ford algorithm and its dynamic version of distance vector routing can be supplemented by a bookkeeping operation that keeps track of the first link in a shortest path—that is, it keeps track of the minimizing j in (22.5). This is all that is needed to be known at a node, for that node can send a packet along that first link and let the receiving node decide where to send it next.

Example 22.1 (Small network). An example of the operations carried out at a node is shown in figure 22.6. The left side shows the network. We look at the situation from the point of view of node D. D has three neighbors: A, C, G. D receives from each of these their estimates of the shortest path lengths to every node in the network. For example, node A reports that the length from A to A is 0 (by definition), the length from A to B is 30, and so forth.

Node D uses this information to compute its own best estimate of the minimum path lengths by using one step of the Bellman–Ford algorithm. D first obtains direct estimates of the lengths to its three neighbors. In the case where length is transit time, estimates might be obtained by sending a test packet that should be immediately returned by the neighbor, and the round-trip time can be divided in half to estimate the one-way transit time. In the example, it is assumed that these direct estimates are $d_{DA} = 18, d_{DC} = 14, d_{AG} = 10$, indicated in the shaded boxes. D finds its updated

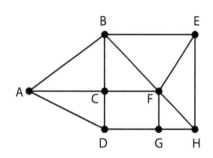

To	**From**			Estimated Length	Start
	A	C	G		
A	0	24	30	18	A
B	30	15	35	19	C
C	26	0	20	14	C
D	20	16	10	0	-
E	42	22	17	27	G
F	38	12	8	18	G
G	30	18	0	10	G
H	45	33	15	15	G

FIGURE 22.6 Distance vector routing. Node D receives shortest length estimates from its three neighbors, as shown in the first table. D also obtains direct estimates of the lengths to its neighbors. All these estimates are combined to produce a new table of estimates for D.

estimates of shortest paths from D to other nodes by using the Bellman–Ford update step. For example, D calculates the shortest path from D to B as the minimum of $d_{DA} + 30$, $d_{DC} + 15$, $D_{DG} + 35$; in other words, the minimum of 48, 19, 45, which is 19. This minimum is achieved by going first to C, and hence A will route to C all packets destined for B.

At the next update cycle D will send the result of its calculations (the first column in the last table) to the three neighbors A, C, and G, and they will use these estimates as part of their own next updating procedure.

An important issue associated with this method is that in pure form it is subject to instability. If a certain link is not congested and has a short length, all nodes may send traffic to that link, causing a high degree of congestion. When this condition is reflected in the next stages of the algorithm, traffic will not be sent that way. Hence, a link's traffic level can oscillate wildly, and the paths used will be far from optimal. Such oscillation can be smoothed by damping the updating algorithm by lessening the impact of new estimates. The full procedure when modified by damping has proved to be reasonably effective in major networks.

22.8 Dijkstra's Algorithm

Dijkstra's algorithm is another shortest-path algorithm and has the advantage that, in theory, it requires less computation than the Bellman–Ford algorithm.

Suppose one wishes to find the shortest path from node 1 to all other nodes. The idea of the algorithm is to find the shortest paths from 1 to nearby nodes and then expand outward.

In outline, the general procedure is this: Nodes are successively labeled with the current best estimate of the minimum path length from node 1. Some of these labeled nodes are marked **permanent** since their estimates are actually optimal. The estimation process progresses until all nodes are labeled permanent.

Example 22.2 (A Dijkstra solution). A network and successive stages of the algorithm are shown in figure 22.7. The objective is to find the shortest path from A to G.

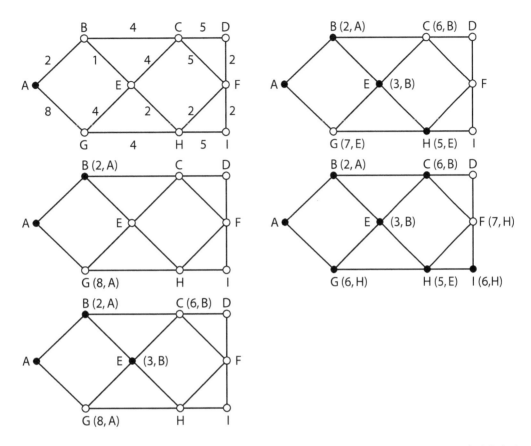

FIGURE 22.7 Dijkstra algorithm to find the shortest path from G to A. Nodes are successively labeled and successively become permanent.

In this example, initially the shortest path to A is from A, and so that node is labeled permanent with a solid circle. Next, the neighbors of A are considered and labeled with their distances from A, and the previous node in the corresponding path is also noted, as shown in the diagram below the first one. Of these labeled nodes, B has the shortest path, so it is labeled permanent. Next the neighbors C and E of this newly labeled permanent node B are labeled with current estimates. E is the shortest of all labeled but nonpermanent nodes, so it is marked permanent. Notice that on the next step, the estimate at G and its noted predecessor node are revised, but G is still not permanent. Finally, at the next step G is revised again and becomes permanent. The path to G can be found by going backward through the labels: G ← H, H ← E, E ← B, B ← A. If the algorithm is continued for another two steps, all nodes will be permanent and hence the shortest paths from A to all nodes will be found.

In 1979 the routing algorithm of ARPAnet was changed to what is termed **link state routing** in which each node individually computes the shortest path from it to every other node. Nodes gather information about the transit times to its neighbors by direct experimentation. Then on a periodic basis (roughly once a minute), these estimates are broadcast to every other node by flooding. Each node then uses a variant of Dijkstra's algorithm to compute the minimum path to every other node, and uses that to route the packets that it receives. Like distance vector routing, this algorithm also must be supplemented with stabilization measures.

22.9 Other Issues

This chapter provides only a brief introduction to some of the issues associated with packet-switched networks. In addition, the queues that form at nodes due to backlogs must be accounted for in design, algorithms for allocating capacity to various users must be implemented, stabilization procedures must be incorporated, protocols must be standardized, coding and encryption techniques must be added, and of course the physical characteristics of a link (such as bandwidth and noise) must be related to capacity requirements.

The principal lesson from this chapter, however, is that packet switching represents yet another way to use frequency for communication. Packets are not merely isolated little bundles of information; at root they are signals that use frequency spectrum and a share of channel capacity.

Most importantly, the study of packet switching illustrates that an advanced mastery of frequency involves electrical engineering, economics, optimization algorithms, queueing theory, and many other disciplines from engineering and science. It illustrates that depth in any one aspect of information science is likely to require broad understanding of all five E's.

22.10 EXERCISES

1. (Poisson probabilities) Let $P_n(t)$ be the probability that a Poisson process with parameter λ has n events in an interval of length t. Let h denote a very small interval of time (like Δt).

   (a) Argue that

   $$P_n(t + h) = P_n(t)P_0(h) + P_{n-1}(t)P_1(h) + o(h).$$

   (b) Substitute expressions for $P_1(h)$ and $P_0(h)$.
   (c) Let $h \to 0$ and derive a differential equation involving $P'_n(t), P_n(t)$, and $P_{n-1}(t)$.
   (d) Try a solution of the form $P_n(t) = c_n t^n e^{-\lambda t}$ and evaluate c_n.

2. (Poisson rate*) Given a Poisson process with parameter λ, use the probabilities $P_n(t)$ of exercise 1 to find the expected number of events in unit time.

3. (Finite users) Suppose there are a finite number M of users of a slotted ALOHA system. User m generates a frame (original or retransmitted) in any given slot with probability G_m. The users' generations are independent of one another and independent frame by frame. Let S_m be the probability that a packet from user m is successfully transmitted in a slot.

   (a) Find S_m in terms of the G_i's.
   (b) Assume that the users are statistically identical, and each $G_m = G/M$. Find the overall throughput S in terms of G and M.
   (c) Let $M \to \infty$ and show that the limiting throughput is the same as that of equation (22.3).

4. (Limiting throughput) What is the throughput (equation 22.4) of CSMA in the limiting cases $a \to 0$ and $G \to \infty$?

5. (Bellman–Ford) For the network of figure 22.8 find the shortest path from node A to node C using the Bellman–Ford algorithm.

6. (Dijkstra example) For the network of figure 22.8, find the shortest path from node A to every other node using Dijkstra's algorithm.

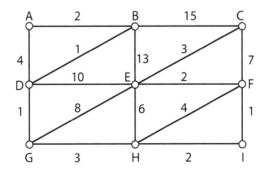

FIGURE 22.8 Find the shortest path.

22.11 Bibliography

The classic reference on network communication is Kleinrock's pioneering [1]. An outstanding and comprehensive text is [2]. A quite accessible and thorough introduction is the popular text [3]. For information about ALOHA see [4]. For the basic theory of CSMA see [5]. For a history of the Internet, see [6] and many sites on the Internet.

References

[1] Kleinrock, Leonard. *Queueing Systems.* Vol. 2, *Computer Applications.* New York: Wiley, 1976.

[2] Bertsekas, Dimitri, and Robert Gallager. *Data Networks.* 2nd ed. Upper Saddle River, N.J.: Prentice-Hall, 1992.

[3] Tanenbaum, Andrew S. *Computer Networks.* 4th ed. Upper Saddle River, N.J.: Prentice-Hall, 2003.

[4] Abramson, Norman. "Development of the ALOHANET." *IEEE Transactions on Information Theory* 31 (1985): 119–23.

[5] Kleinrock, Leonard, and Fouad Tobagi. "Random Access Techniques for Data Transmission over Packet-Switched Radio Channels." *Proceedings of the National Computer Conference* (1975): 187–201.

[6] Leiner, Barry, et al. "A Brief History of the Internet." *Communications of the ACM* 40 (1997): 102–8.

SUMMARY OF PART V

Shannon's theory of communication includes a theory for continuous random variables and for continuously time-varying signals. This theory contains a formula relating the **capacity** of a channel to available **bandwidth** and **signal-to-noise** level.

Now, in hindsight, the relations are seen to have played a role even during the development of the telegraph, especially in undersea cables. The blurring of dots and dashes can be traced to the loss of bandwidth due to the **capacitance** of the line. William Thompson was able to analyze these lines using the then relatively new tool of **Fourier series**, which is the mathematical device for transforming analysis to the frequency domain.

Alexander Graham Bell struggled tirelessly to understand the role of frequency in human voice and how it could be transmitted electrically. He was not an academic, but he consulted the works of the great professor Helmholtz, who had carried out extensive experiments that showed that indeed the human voice could be considered to be composed of various frequencies. This led Bell to conceive of a **harp telephone** that would decompose a human voice message into a set of frequencies, transmit them separately, and then recombine them at the receiver. Bell's actual first telephone used the duality of the electromagnetic effect, whereby movement of a coil through a magnetic field produces current, and dually, a current produces movement. However, the phone system he built on this principle had low signal power and hence low communication capacity. Amplification of the signal at the sending end was what finally produced a workable telephone.

Edison's phonograph taught everyone that music and the human voice can be represented as a single wavy line. The very name phonograph suggests this visualization.

For centuries people have experienced continuous single tones through plucked strings, flutes, and so forth, but although radio waves were predicted by Maxwell's theory, no one had concrete evidence that such waves existed—until Hertz. Hertz's demonstration of radio waves ushered in the field of modern communication. The first practical radio system was that of Marconi, who was able to send a radio signal across the Atlantic. Early radios like Marconi's were **spark radios** that radiated a series of short radio frequency pulses generated by the discharge of a large capacitor across a spark gap. The **Fourier transform** of such a pulse shows that the pulse occupies a broad bandwidth. At the time, this was inimical to wide use of radio, for pulses from a number of different radios, even if tuned differently, competed and polluted the airwaves, causing chaos. Today, it is understood that a broad bandwidth has greater capacity than a narrow one, but it takes techniques more sophisticated than those available to Marconi to efficiently share this bandwidth.

Order was instilled in the field of radio by the development of generators and later by circuits that could produce continuous radio waves of a single constant frequency. This basic **carrier** wave could be modulated in amplitude with the resulting wave occupying only a relatively narrow bandwidth. This innovation allowed several stations to broadcast simultaneously on nonoverlapping channels, and for signals to be received as continuous waveforms rather than bursts. Both the generation and the reception of radio waves were substantially improved by de Forest's invention of the **triode** vacuum tube and the discovery that it could be used in a **feedback** circuit to produce oscillations as well as to greatly amplify the signal.

Edwin Armstrong's several inventions transformed the young field of radio into a sophisticated, practical, and ubiquitous part of life. His **superheterodyne** receiver design remains even today the standard for AM receivers. He learned to control frequency: modulating it, shifting it with the heterodyne principle, and developing feedback oscillation circuits. His greatest technical achievement was the development of FM modulation, which he believed, contrary to the theoretical analyses of others, would be more resistant to noise than AM. He also believed that a broad bandwidth would be better than a narrow one, which contradicted conventional wisdom that claimed that the greater total noise in a broad bandwidth would be detrimental. Armstrong's belief was born out in practice and is now substantiated by Shannon's theory.

Shannon developed information theory for continuous signals by logically extending his theory for finite sources: he approximated continuous variables by discretized versions and took the limit. He used the **sampling theorem** to represent a continuous wave by a series of discrete samples. Putting it all together, he developed the capacity theorem for a channel subject to additive Gaussian white noise that is the basis for much of modern communication theory. One innovation that surely would have been inconceivable to early pioneers is multiaccess systems that purposely add pseudonoise to spread the spectrum of a signal.

It is perhaps also an interesting twist of history that modern network communication is based on **packets**. Early telegraph and telephone systems provided a degree of multiplexing, but most pairwise connections were established by circuit switching, providing, at least temporarily, a dedicated connection between the pair. Now, almost like dots and dashes, information is sent in discrete packets. Packets from different messages may collide, requiring them to be sent again. The performance of these systems is largely based on **queuing theory**. Large communication networks, such as the Internet, also use a minimum-distance routing algorithm to send each packet efficiently to its intended receiver. Two such algorithms are based on the **Bellman–Ford** and the **Dijkstra** shortest path algorithms.

Years ago, who could have imagined that modern communication systems would exploit light frequencies, communicate to remote planets, have telephones without wires, send pictures electronically, purposely add pseudonoise to signals to spread the spectrum, and send competing packets through a network? We enjoy the fruits of those before us who had the brilliance and perseverance to improve our mastery of the gift of frequency.

AUTHOR INDEX

Index contains names found in end-of-chapter references.

SUBJECT INDEX